The Palgrave Atlas of Byzantine History

The Palgrave Atlas of Byzantine History

John Haldon

First published in hardback 2005 and in paperback 2010 by
PALGRAVE MACMILLAN

Palgrave Macmillan in the UK is an imprint of Macmillan Publishers Limited,
registered in England, company number 785998, of Houndmills, Basingstoke,
Hampshire RG21 6XS.

Palgrave Macmillan in the US is a division of St Martin's Press LLC,
175 Fifth Avenue, New York, NY 10010.

Palgrave Macmillan is the global academic imprint of the above companies
and has companies and representatives throughout the world.

ISBN 978-1-4039-1772-8 hardback
ISBN 978-0-230-24364-4 paperback

This book is printed on paper suitable for recycling and made from fully
managed and sustained forest sources. Logging, pulping and manufacturing
processes are expected to conform to the environmental regulations of the
country of origin.

A catalogue record for this book is available from the British Library.

A catalog record for this book is available from the Library of Congress.

10 9 8 7 6 5 4 3 2 1
19 18 17 16 15 14 13 12 11 10

Printed and bound in Malta by Gutenberg Press

Contents

List of Maps, Figures and Tables

Maps

Figures

Tables

Preface

This Historical Atlas is an attempt to represent graphically some of the major developments in the history and evolution of the medieval eastern Roman or Byzantine empire. It may be seen as both an introduction to the history of the Byzantine empire in its own right and as an accompaniment to general histories of the empire. It cannot, of course, illustrate all facets of the empire's development, and in particular it can say very little, without gross over-simplification, about the culture, beliefs and social or economic relationships and structures of the empire. Nevertheless history books are all too rarely accompanied by useful and detailed maps, and I hope that this short volume of maps with parallel explanatory texts will at least put Byzantium more clearly in its geopolitical context and show how its internal history is interlinked with and influenced by developments among the peoples and political formations which surrounded it.

A word of caution is in order, however. The breadth of coverage of the Atlas inevitably means that the maps are drawn to a relatively small scale. Absolute exactitude in respect of the relationship between physical features and historical or cultural features such as frontiers is not, in consequence, attainable. This is especially true given the lack of precise information for, or the ambiguity pertaining to, many such features. It is also the case that historians disagree among themselves about such features, while the line of a particular treaty frontier, for example, or the lines of provincial and state boundaries or frontiers must be guessed from often very general information. Users should be aware of these limitations from the beginning, and while I have tried to base all the maps on the results of the most recent research, there will inevitably be disagreement about the exact location of many features.

I have appended a brief time-line or chronology, a glossary of Byzantine technical terms and a short bibliography, the last including the works from which the information contained in the different maps is drawn and representing also appropriate further reading.

I owe thanks in particular to my colleagues in the Centre for Byzantine, Ottoman & Modern Greek Studies at the University of Birmingham, as well as to Henry Buglass for his excellent cartography and to Graham Norrie for much valuable help with technical matters, both of the Institute of Archaeology & Antiquity at Birmingham. I am particularly indebted to my friend Meaghan McEvoy, who found the time to act as a generous and invaluable commentator on the texts, to Ruth Macrides and Dimiter Angelov, who also commented on sections of the text, and in particular to Rosemary Morris, who went through maps and texts and saved me from many a blunder. All of their views helped me fashion the whole into a more useful form than it might otherwise have been. Needless to say, any shortcomings are mine alone.

Finally, thanks are also due to the editorial team at Palgrave for their patience and co-operation in producing this volume.

A Note on Placenames

In rendering placenames appropriately across time and across a cultural milieu in which several languages were used, the historian is confronted by a number of difficulties. I have chosen to adopt in this atlas the simple expedient of using common English versions of the best-known places – thus Constantinople, Thessalonica, Rhodes, rather than Konstantinoupolis, Theassalonike/ Thessaloniki, Rhodos – for the whole period, and otherwise to transliterate the names according to the common usage of the dominant culture of the area in question. Chronologically this means that up to the seventh century most names within the Roman world are given in their Latin form; thereafter in their Greek form. There will undoubtedly be some inconsistencies, but I hope this will at least allow a clear identification of the places in question.

1 General Maps

Physical Geography and Climate

The late Roman world from the sixth century was dominated initially by four land-masses (Asia Minor or Anatolia, very roughly modern Turkey; the Levant or Middle Eastern regions down to and including Egypt; North Africa, from Egypt westwards to the Atlantic; and the Balkans). The Mediterranean and Black Seas united these very different regions, and after the loss of much of Italy and all of North Africa during the seventh and eighth centuries, acted as a connecting corridor between east and west. The climate of these very different regions determined the patterns of agricultural and pastoral exploitation within the empire's borders and the nature of the state's surplus-extracting activities.

Asia Minor can be divided into three zones: central plateau, coastal plains, and the mountain ranges which separate them. The plateau rises from about 1,000 metres in the west to over 1,800 metres feet in the east and is typified by extremes of hot and cold temperatures in summer and winter (altitude and the effect of the northern Pontic range of mountains promotes in effect a continental, steppe-type climatic system). Four climatic sectors are usually identified: the Pontic (Black Sea coastal) region has warm summers, mild winters, and a regular rainfall across the year – temperatures range from 23° C in midsummer to some 14° C in the winter; the south and west coastal regions have a Mediterranean climate, with mild, wet winters and hot dry summers – temperatures range from 12° C in winter to 20° C in summer; the semi-arid plateau and interior have cold, wet winters and hot, dry summers, with temperatures ranging from freezing and below in mid-winter to 23° C in the summer. Finally, the north-eastern plateaux have warm summers but severe winters, with winter temperatures reaching –12° C to 18° C in summer. This pattern reflects the physical geography, for the relief of the whole peninsula is dominated by ranges in the north and south of over 3,000 metres that encircle the central plateau. To the north the Pontic Alps follow the line of the southern shore of the Black Sea; to the south the Taurus and Anti-Taurus ranges extend along the Mediterranean coast and across northern Syria curving north-eastwards into the Caucasus region. All the mountain zones, but particularly the southern and eastern regions, are characterised by smaller plateaux dissected by crater lakes, lava flows and depressions, producing a highly fragmented landscape. The central plateau itself is divided into several large basins and salt lakes, with extensive eroded areas

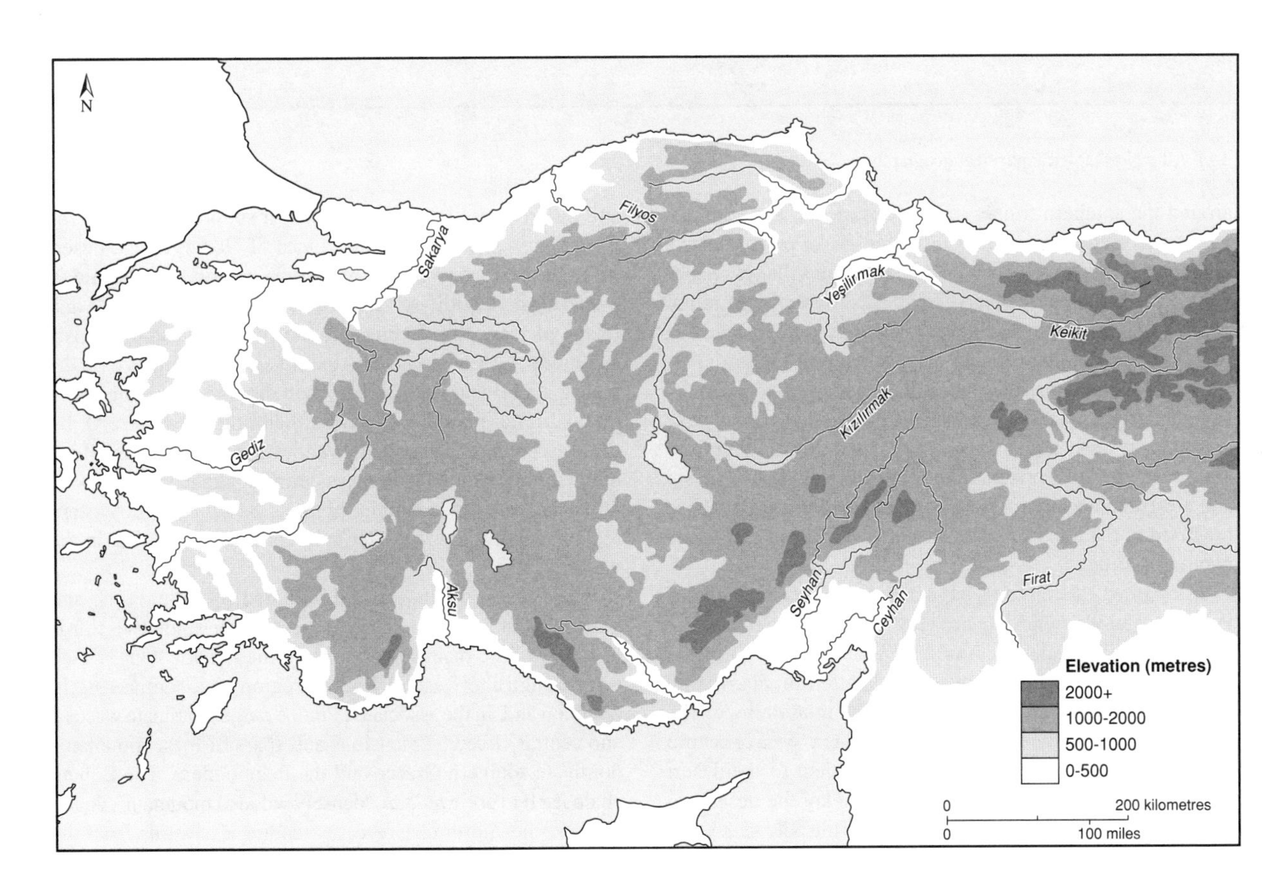

Map 1.1 Asia Minor: physical geography.

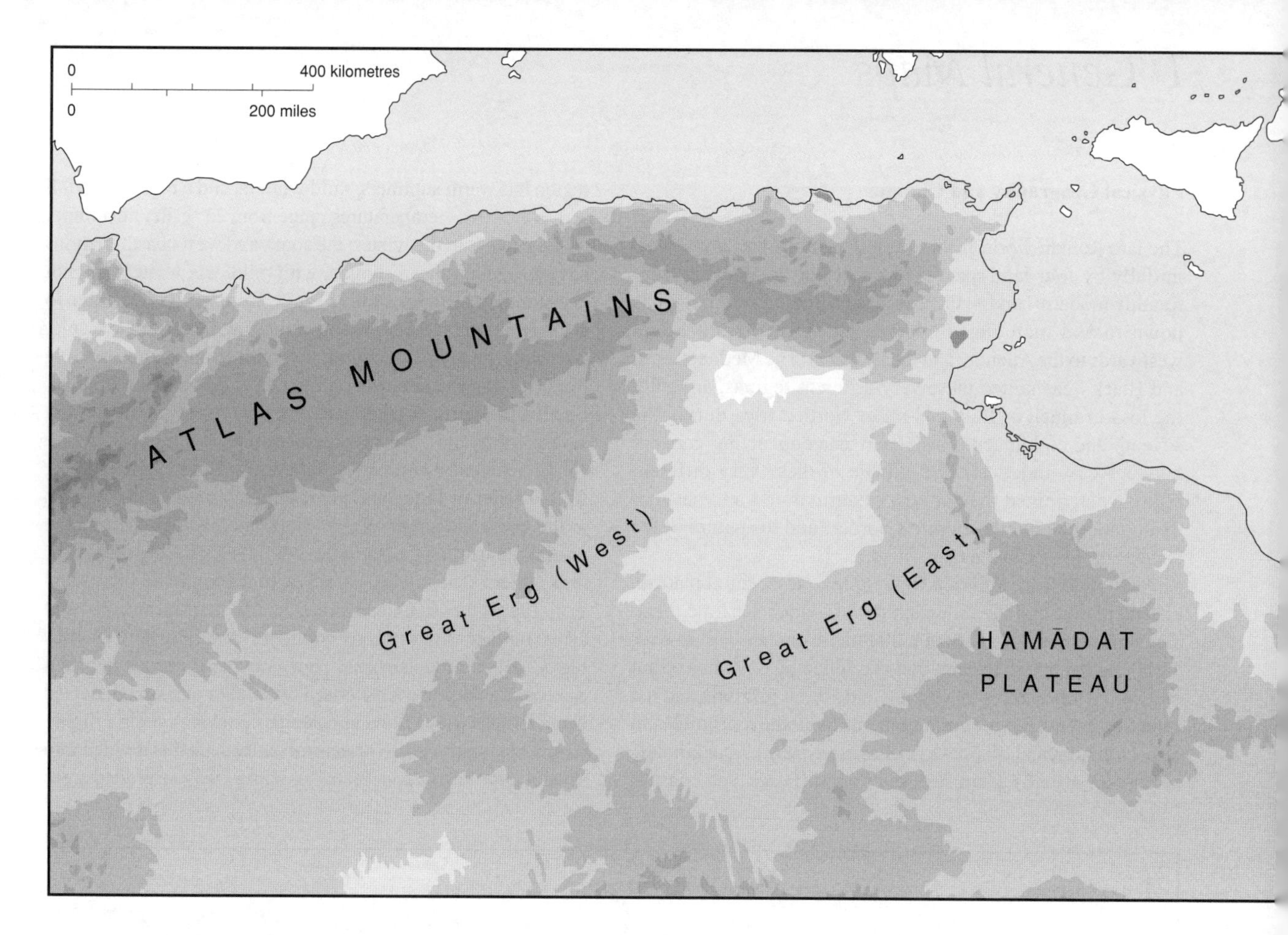

Map 1.2 North Africa: physical geography.

around the southern fringes, as in Cappadocia, for example, where the eroded limestone formations have permitted the creation of cave dwellings and subterranean villages. Land-use is determined very clearly by these differences in relief. Agricultural production is limited to the coastal regions – often quite extensive in the Cilician plain or western lowlands, for example, and to the fertile river valleys which cut through the central plateau or coastal ranges. The uplands and plateau have traditionally been exploited by pastoral activity, ranging from sheep and goats to horses and in some areas cattle. In ancient and pre-Islamic medieval times extensive pig rearing was also practised in the transitional zones between plateau and fertile agrarian districts.

In contrast, the limited but fertile agricultural lands of Palestine and western Syria are very much wealthier. Greater Syria, including Palestine and the Lebanon, incorporates a number of very different landscapes, the terrain alternating from rugged highlands (for example the mountains of the Lebanon), through the fertile plains of northern Syria or central Palestine, the hilly uplands around Jerusalem to the desert steppe of central Syria. south of Palestine lay the deserts of the Sinai peninsula, leading then into the fertile Nile valley and delta regions – an area of fundamentally different character, heavily dependent on the annual flooding of the great river and the irrigation agriculture which it supported. Westwards from Egypt stretched the provinces of North Africa, desert through the eastern sector of Cyrenaica and Tripolitania in modern Libya with very limited fertile coastal stretches and inland plateaux, graduating into the coastal plains of Tunisia and modern Algeria. This was in turn clearly delineated by the plateaux and sandy desert regions in the south-east, including the al-Jifarah plain (and beyond them, the great desert), by the Aures range in the centre, and the Saharan Atlas. Mean temperatures along the northern coastline range from a low of 16° C in winter to a summer high of 38–40° C in the eastern region (slightly lower winter temperatures of 8–12° C in the western sector).

The Balkan peninsula is dominated by mountains, and although not particularly high, these cover some two thirds of its area. The main formations are the Dinaric Alps, which run through the western Balkan region in a south-easterly direction and, in the associated Pindos range, dominate western and central Greece. Extensions and spurs of these mountains dominate southern Greece and the Peloponnese. The Balkan chain itself (Turkic *balqan*, 'densely wooded mountain'; Greek *Haimos*) lies north of Greece, extending eastwards from the Morava river for about 550 kilometres as far as the Black Sea coast, with the Rhodope range forming an arc extending

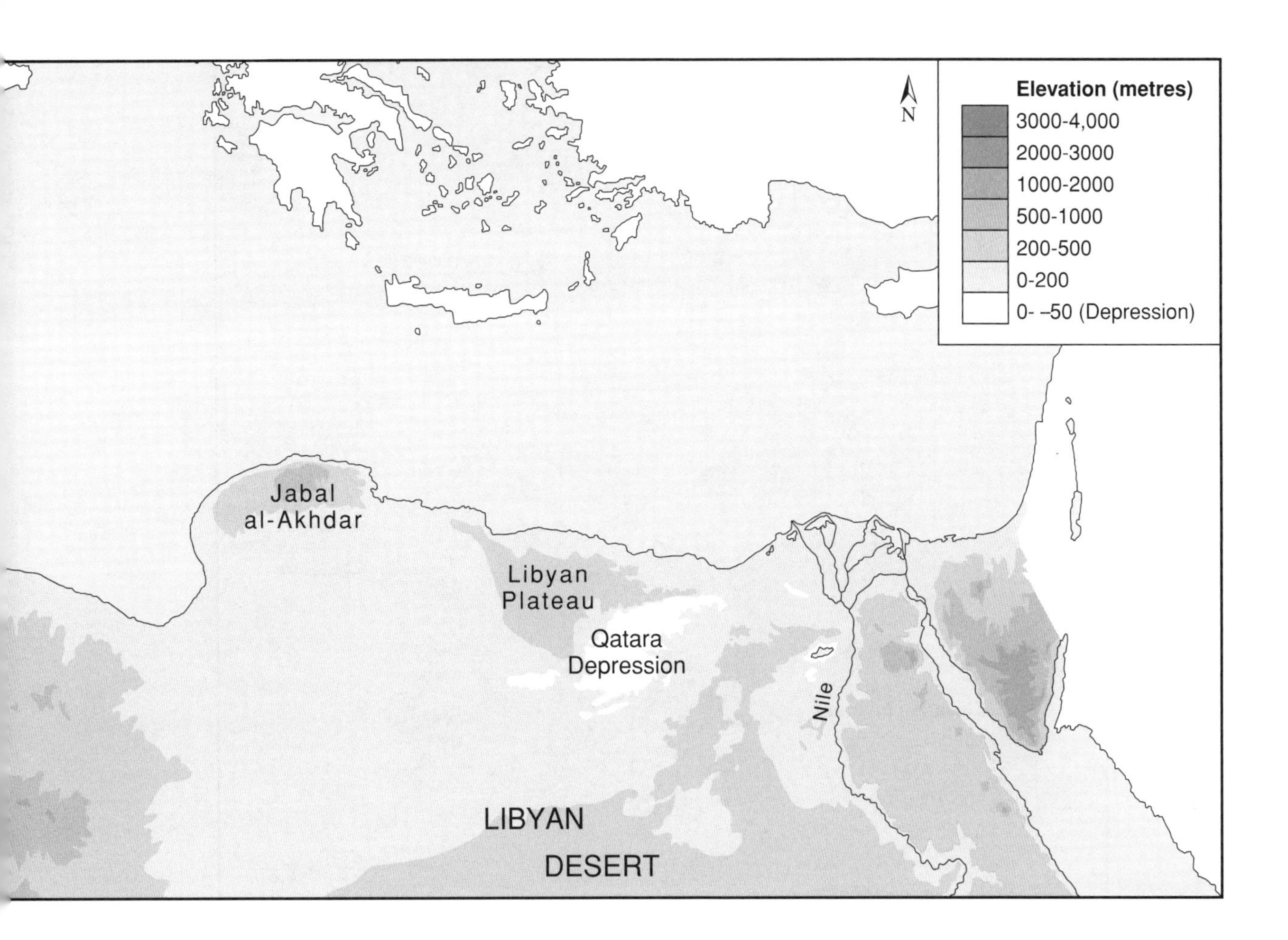

southwards from this range through Macedonia towards the plain of Thrace. River and coastal plains are relatively limited in extent. There are thus very distinct climatic variations between the coastal, Mediterranean-type conditions and the continental-type conditions of the inland and highland regions. Mean temperatures in the Peloponnese and in the coastal regions of southern and north-western Greece range from 5–10° C in winter to 25–30° in the summer, contrasting with northern and central upland temperatures of from 10 to –5° C in winter and 10–15° C in the summer. Rainfall patterns are similarly accentuated, although with a stronger differentiation between those areas west and south of the Dinaric and Rhodope ranges and those to the east – means of about 100 centimetres per annum in the former and of as little as half that much in some parts of the latter have been recorded in modern times. This has in turn generated a very accentuated settlement-pattern consisting in a series of fragmented geopolitical entities, separated by ridges of highlands, fanning out along river-valleys towards the coastal areas.

The highland regions are dominated by forest and woodland; the lower foothills by woodland, scrub and rough pasturage. Only the plains of Thessaly and Macedonia offered the possibility of extensive arable exploitation; the river plains, and the coastal strips associated with them (such as the region about the gulfs of Argos and Corinth, much more limited in extent), present a similar but more restricted potential. Here were to be found in ancient and medieval times orchards, as well as vine and olive cultivation. The relationship between this landscape of mountains, valleys and coastal plains and the sea is fundamental to the political, military and cultural history of the region, in particular in the southern zone. Surrounded by the sea, for example, except along its northern boundary, the extended coastline, with its gulfs and deep inlets serves as a means of communication with surrounding areas and for the dissemination of common cultural elements even to the interior districts of the Balkans. But equally, easy sea-borne access from the west, the south or from the north-east via the Black Sea made the southern Balkan peninsula – in particular Greece and the Peloponnese – vulnerable to invasion and dislocation.

Climate has remained, within certain margins, relatively constant across the late ancient and medieval periods, yet there are a number of fluctuations that need to be borne in mind and which, in conjunction with natural events such as earthquakes, man-made phenomena such as warfare, and catastrophes such as pandemic disease, could have dramatic short- to medium-term effects on the human populations of the region, and thus patterns of settlement, land-use, the extraction, distribution and consumption of resources, and political systems. The climate

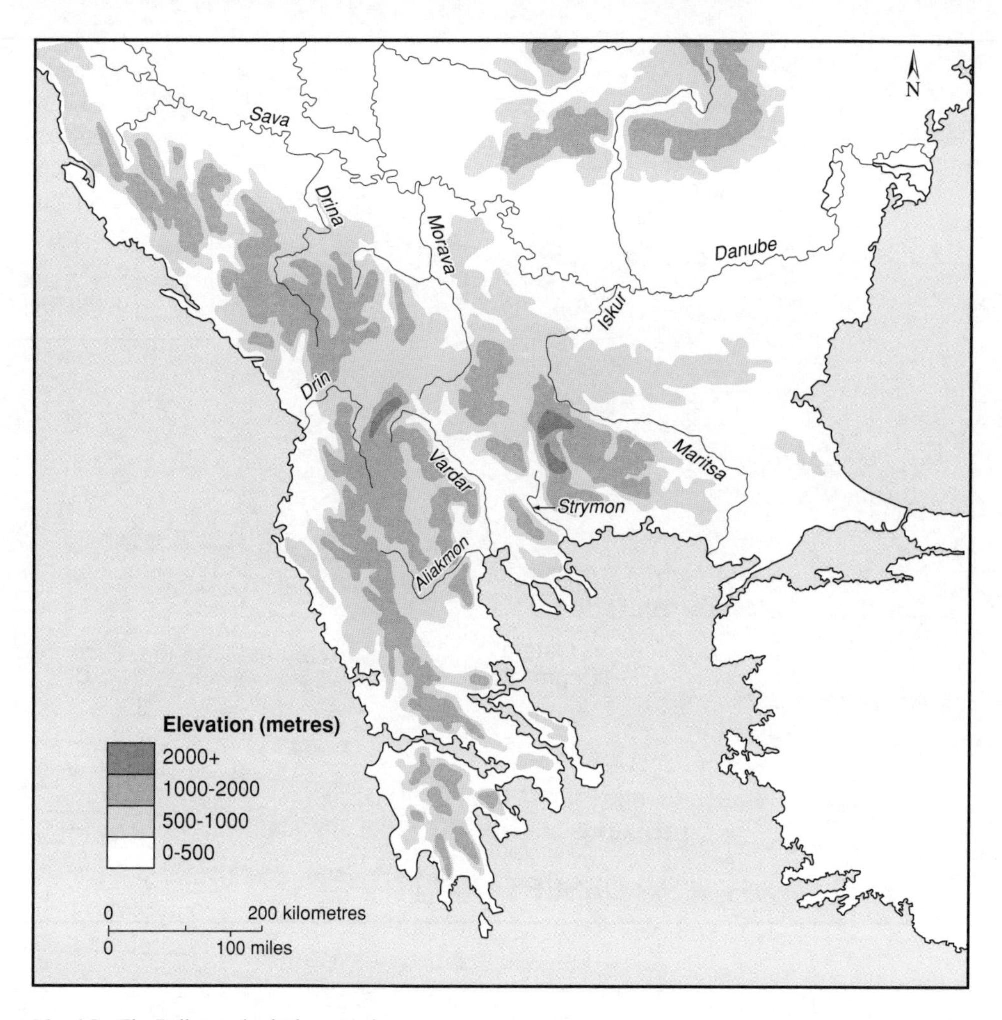

Map 1.3 The Balkans: physical geography.

throughout much of the late Hellenistic and Roman imperial period was relatively warmer and milder than in the period which preceded it, and constituted a 'climatic optimum' which favoured the expansion of agriculture. By about 500 CE this situation was changing, with colder conditions persisting up to the mid-ninth century. The human environment of the later fifth to seventh centuries thus became both more challenging and the economy of existence more fragile. Combined with the great plague of the middle of the sixth century this may have affected the human population in a number of ways, although these remain unclear and the subject of continuing debate. Some marginal lands were abandoned, soil erosion increased where agriculture receded, the colder, wetter climate generated increasing water volume in rivers and watercourses, contributing to alluviation and lowland flooding in many more exposed areas. It remains difficult to disentangle the effects of climatic and human factors on the changing landscape. During the ninth century this trend was reversed – and is paralleled by an extension of agriculture and of human exploitation of woodland and scrubland, strong demographic growth and an increasing density of settlement and rate of exploitation of agrarian resources. But from the fourteenth century once more this tendency was halted, and with lower temperatures, increased glaciation in high alpine zones (in particular the European Alps), a growth in the rate of afforestation, a reduction in agricultural exploitation, and a demographic decline, the fragile conditions of existence of the human populations of the region were once more thrown into disequilibrium, with phenomena such as the fourteenth-century Black Death one of the most obvious accompanying developments. All these phenomena thus form the background to the 'little ice age' of the sixteenth and seventeenth centuries. It is against this background that we must understand and interpret the social, economic and political history of the late Roman and Byzantine worlds.

Land-use and Resources

Land-use and the exploitation of natural resources are closely determined by the geophysical and climatic framework described above. Four basic types of productive exploitation occur – arable farming, pastoral farming, the exploitation of woodland and scrubland, and the extraction and working of mineral resources. The extent of agricultural activity, of the

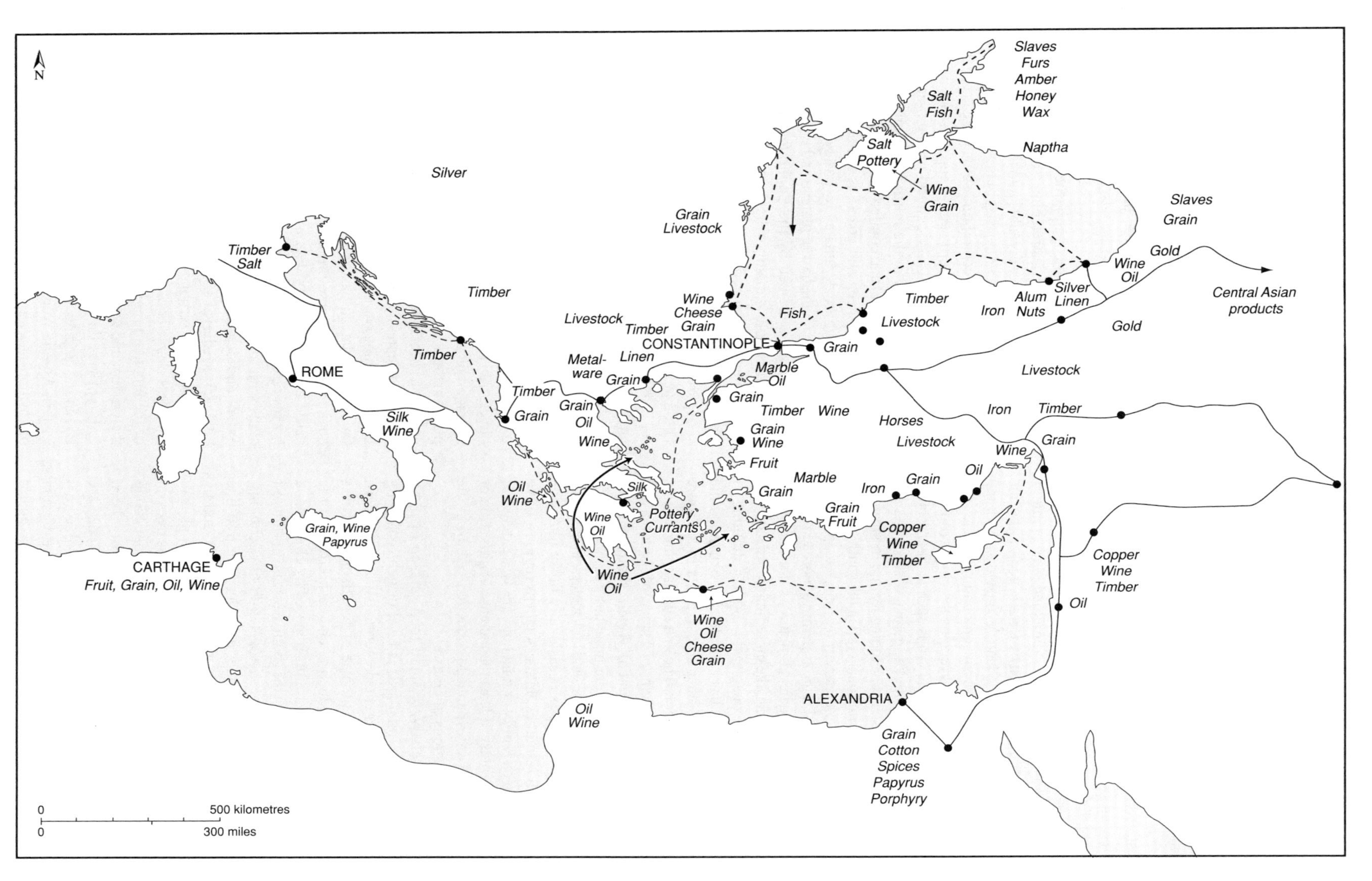

Map 1.4 Land-use and resources.

exploitation of natural resources such as woodlands, and of particular crops such as cereals or grapes, is reflected also in the climatic fluctuations and shifts which took place across the period in question.

The modern Balkan regions have changed very dramatically since the Second World War, the result of both mechanisation and intensification of production, on the one hand, and of political reform and change on the other. In Bulgaria and Romania and some of the western Balkan countries, for example, collectivisation encouraged a considerable improvement in output and efficiency, although the longer-term social and economic results were less fortunate. In spite of these changes in the organisation of production, however, the patterns of land-use themselves remained very stable, a reflection of the constraints imposed by terrain, geography and climate – approximately 30% of the land devoted to agricultural production, with pasture and meadow amounting to one fifth of the total. In the western coastal regions the chief crops are grains (wheat and corn), industrial crops such as beet, cotton and tobacco, and, usually on a market-garden basis, fruits and vegetables. Vineyards are also a developing feature in the west, although they have been a traditional crop in the south. Similar ratios prevail in the central and eastern regions, except where the broader plains and alluvial regions permit a more extensive cultivation of cereals. The medieval picture is not dissimilar, except for the absence of cotton, the more limited surface area devoted to agriculture (for example, the modern draining of the Danube delta marshlands has considerably expanded the land available for cereal and other crops), and a much more fragmented pattern of production. The rich alluvial plains along the southern Danube, and the plains of Thrace, Macedonia and Thessaly, offered the main potential. Again, sheltered river valleys and depressions within the mountain regions permit settlement and agrarian production, and archaeological evidence for settlement density suggests occasionally fairly intensive exploitation of such resources. In the southern regions, olive and vine production on family or joint holdings was extensive; and from the tenth century at least the increased cultivation of the mulberry allowed an expanded production of silk in the central and southern regions of Greece.

Asia Minor has a relatively small total surface of plain – in fact, only 9% of the total area is level or gently sloping land. Modern Turkey has benefited enormously from modern mechanised techniques and the use of fertilisers, and this has helped expand cereal production and cash crops on the central plateau beyond the constraints imposed by climate and geography. Considerable areas in the south, west and north-west are dominated by a Mediterranean vegetation of deciduous, coniferous or mixed forest at higher altitudes (the tree-line is between 1,800 and 2,100 metres above sea level), and by scrub and brush in the lowlands. Whereas the central plateau is a region of steppe, with forest of oak and coniferous trees on the higher parts, the damper and warmer northern zone along the Black Sea coastline is densely wooded and has always been a source of timber. The main products in this region today are tea (in the eastern districts), hazelnuts and tobacco, with corn – maize – dominating as the main cereal crop. The degree of grain production increases markedly towards the west, with a greater proportion of wheat to maize. In the Marmara region, which is also the most heavily urbanised, a very mixed agriculture has developed – wheat, rice, tobacco, sunflower, maize, olives and vines, and silk. The Aegean zone, stretching down as far as the island of Rhodes to the south, produces a large number of cash crops – cotton, tobacco, vines, olives, figs in the coastal regions, with cereal and livestock (and a controlled opium crop) predominating in the hill country inland. The plateau, with its steppe climate and limited rainfall, is dominated today by pastoral production (a third of the sheep and three quarters of the Angora goat population are raised in this region) and cereals – some 40% of the country's wheat is based here, occupying 90% of the arable. To the south, the Mediterranean region is dominated by the Taurus, stretching from Rhodes to the border with Syria, and is further divisible into three sectors – the fertile and intensively cultivated coastal plains (citrus fruits, sesame, vegetables, cotton) the central limestone plateaux in the centre (pastoral), and the western semi-steppe district of the lakes, where cereals dominate agricultural production. The eastern highlands, dominated in the north by mountain pastureland (beef and dairy cattle) and coniferous forest and in the south by wooded steppe (sheep and goats), is sparsely populated, with a limited agriculture dominated by barley and summer wheat. To the south again the barren plateau at the foot of the southern Taurus range is drained by the Euphrates and Tigris rivers, where agriculture – mainly wheat, vegetables, rice and vines – is limited to sheltered or irrigated valleys and depressions. The population is largely semi-nomadic or nomadic.

Apart from the introduction of different crops from the Ottoman period onwards (cotton and flax in the west and north, for example), the basic pattern of agricultural production from late Roman times through the Byzantine period was much the same, with the key difference that lack of modern technology meant that levels of production were very much lower, and the possibilities for cereal production on the central plateau were also very much more limited. But it is clear that the production of cereals – wheat and barley – on the one hand, and vines, olives, fruit (especially in the south-west) and vegetables played an important part in the economy of the river valleys and coastal plains in the north, west and south-west, while inland the cereal and fruit/vegetable producing areas were limited to sheltered zones and depressions on the plateau (such as the district around Konya/Ikonion) or along river valleys. In the uplands and on the central plateau pastoral economies had dominated since ancient times – horse breeding in Cappadocia, for example, cattle and pigs in Paphlagonia and Cappadocia, sheep and horses elsewhere, and long before the arrival of the Türkmen clans with their central Asian pastoral tradition (although the extent and degree of pastoralism before the Turks remains unclear). Medieval sources – Greek, Latin and others – all stress the arid or scrubland nature of much of the plateau and the waterless character of considerable stretches, the inhospitability of the mountain regions, and the productivity and fertility of the western and southern plains and coastal districts.

Egypt was the bread-basket of the late Roman and early Byzantine empire, although the coastal regions of Tunisia and eastern Algeria were the source of very considerable cereal production also, along with vegetables, fruit, olives and grapes.

After these regions were lost to Islam during the course of the seventh century, the eastern empire turned to Asia Minor in particular, and to the southern plain of Thrace for its staples, especially wheat.

The exploitation of woodland and scrubland has only recently attracted the attention of historians and archaeologists, and it is clear that in the middle Byzantine period certainly, and probably from late Roman times and before, there was a well-structured pattern of extracting resources in timber and other products from the western Anatolian, southern Balkan and Pontic regions under imperial control. Mineral resources were also extracted either through state-controlled operations (sometimes quite extensive), especially in the late Roman period, or through smaller, more fragmented private enterprise and state contracting in the Byzantine period. Iron was a key resource, and deposits in Palestine, the Pontic region, the Taurus/Anti-Taurus and the Caucasus, the eastern Danube, the Crimea, Macedonia, and the north-western Balkans were exploited in the late Roman period. Copper was extracted from Cyprus, the Caucasus and the Pontic mountains; gold was obtained either directly or by trade from the Caucasus (Armenia), by trade from west Africa, and directly from deposits in the Rhodope mountains and Thrace in the southern Balkan region. Silver likewise came from Armenian sources and from Cyprus, but there is some evidence that the silver deposits in Attica continued to be exploited, while in the later period Serbian and Caucasian silver was also obtained. It is possibly indicative of the proportions of precious to non-precious ores available to the empire that there are many more place-names with the element 'iron' or 'copper' in them than there are with that for 'gold' or 'silver'.

Population and Settlement

Estimating pre-modern population numbers and densities is notoriously difficult and fraught with dangers, methodological and factual, so while the distribution of settlement and settlement densities represented in Maps 1.5–1.8 give a reasonably accurate picture of the proportions between different areas of the empire, the numbers suggested below for mean population levels must be taken with a considerable degree of caution, however credible they may appear to be. On the whole, I have erred on the cautious, but even here exactitude is impossible.

The climatic and geographical features which determined land-use likewise determined where populations were concentrated and how many people the land could support. The degree of continuity from medieval to modern times is, in this respect, considerable. But there were within our period very considerable fluctuations, both in respect of the relationship between the populations of urban and rural regions, on the one hand, and in terms of their density. Broadly speaking, there appears to have been a long downward curve in population during the late Roman period, reaching a nadir in the later seventh and eighth centuries, followed by a slow recovery into the later ninth and tenth centuries, with a fairly dramatic rise in the twelfth century. It has been estimated that the population of Roman Europe (including Britain and the Balkan provinces) was in the order of approximately 67–70 million at the end of the second century CE, falling to around 27–30 million by the early eighth century, rising again by 1300 to some 73 million, with a particularly noticeable rise about 1200 CE. All the evidence suggests a similar curve in the near eastern and – in the later centuries – Islamic world, and these accord with the minor climatic changes described above. The catastrophic slump of the mid-fourteenth century, which saw the population of Europe drop to somewhere in the region of 45 million, was made good within a century. While these figures are necessarily crude approximations, in view of the nature of the available sources and the problems of their interpretation, and while one can point to a number of exceptions, quite apart from a differential rate of change from east to west, and including important regional and local variations, they seem now generally agreed, at least in their broad outlines. The most recent estimates for the late Roman and Byzantine areas propose a population for the empire's eastern provinces, of some 19–20 million just before the middle of the sixth century (before the plague of the 540s), with a further 7 million in the west; of 17 million in the early seventh century, with a reduction to about 7 million by the middle of the eighth century, and a gradual rise to about 10 million in the mid-ninth century, 12 million by the time of Basil II, falling again to about 10 million (after the loss of central Anatolia to the Turks) in the mid-twelfth century, 9 million in the early thirteenth century, 5 million by about 1280 and a consistent downward trend thereafter as the empire's territorial extent was reduced. Slightly higher figures for the tenth to twelfth centuries have also been proposed, with a population of some 18 million in the 1020s, for example. All can be challenged on various grounds, but they provide some very crude totals in respect of the amount of agrarian produce consumed and available for, for example, the support of armies or similar transient population groups.

Given the geographical constraints described already, it is apparent that the pattern of settlement, and in particular the density of settlement, will reflect this environment very closely, and this is indeed the case both in modern times as well as in the pre-modern and pre-industrial world. A comparison of the areas of settlement density as reflected in the presence of cities (as defined in the Roman legal context) in the late Roman and early Byzantine world with one showing modern demographic patterns demonstrates a remarkable continuity in both the Balkans and Anatolia. Such a map can tell us little about absolute numbers, of course, nor about the fluctuations across time in density and extent of settlement; but it does point to the relationship between human populations and the ability of the land to support them. A glance at the demographic situation in Turkey before the Second World War (representing the mid-1930s) shows this relationship quite clearly (Map 1.8). A map showing the density of Roman cities and Byzantine Episcopal sees highlights the fact that it is more or less the same areas which could maintain substantial populations in ancient and medieval times, which saw the densest concentration of urban centres, and which may thus be taken to have remained the most productive and heavily-settled regions of the Byzantine period after the transformation of the late ancient city network after the seventh century. A similar pattern emerges from a comparison of Roman and medieval population centres with

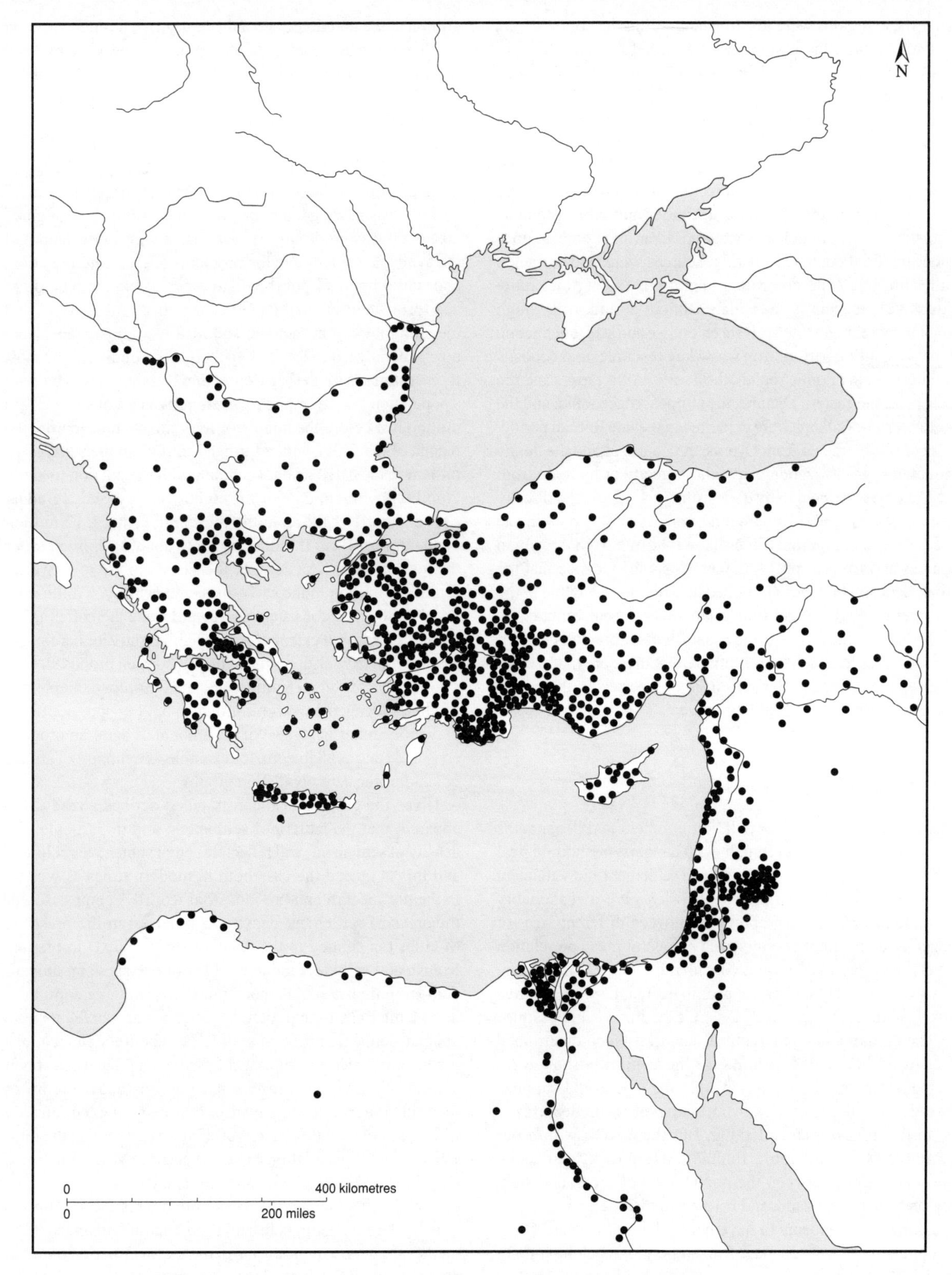

Map 1.5 Major population centres c. 500 CE. (After Jones, *Cities of the Eastern Roman Provinces.*)

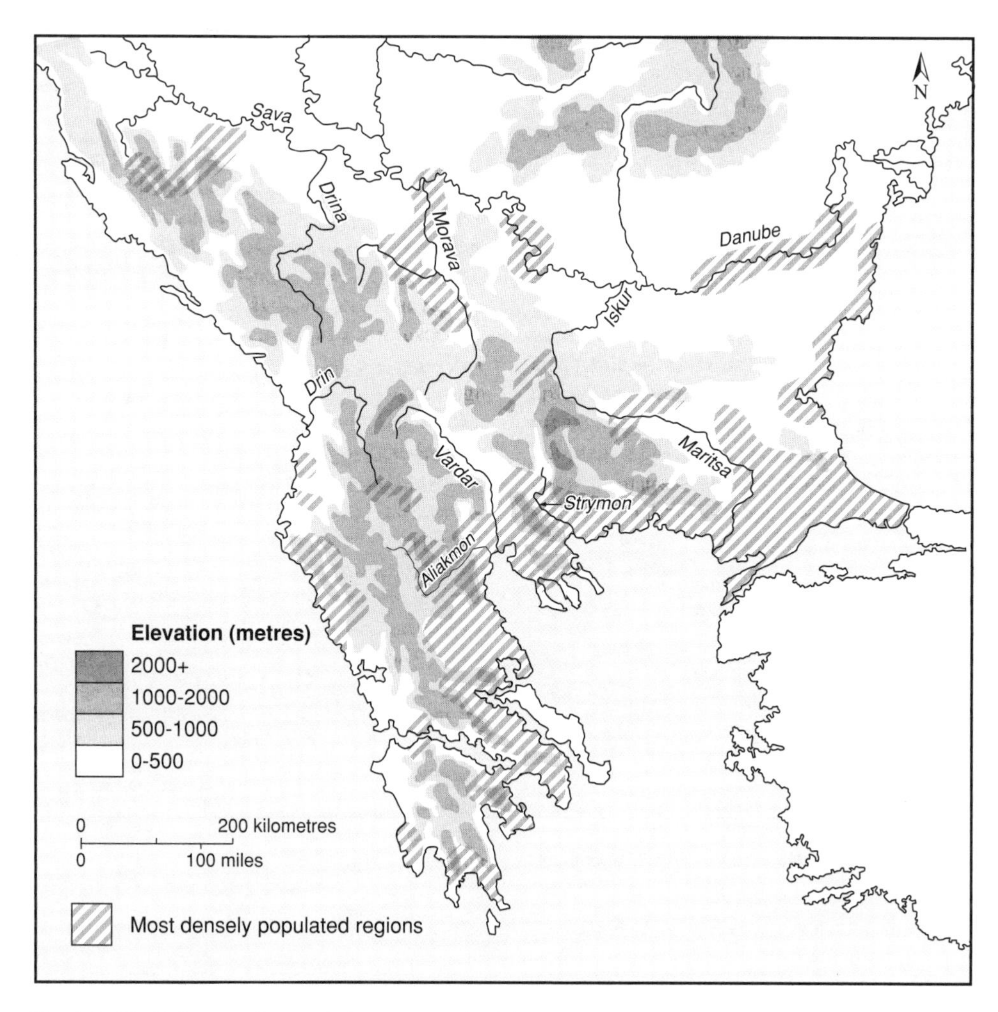

Map 1.6 The Balkans: major population centres, 7th–12th centuries.

modern demographic concentrations in the Balkans, bearing in mind the changes brought about by industrialisation and mechanisation of both industrial and agrarian production.

There were several phases of evolution in the overall settlement pattern of the empire, which will be dealt with in greater detail in the following chapters. But the two most apparent shifts occurred during the fifth and sixth centuries in the Balkan territories of the empire and during the sixth and seventh centuries in Asia Minor, when towns decreased in size, when a larger number of intermediate semi-rural/semi-urban fortified centres evolved, and when village communities came to play a more significant fiscal and political role than they previously had; and in the ninth to twelfth centuries, when relatively peaceful circumstances saw a demographic upswing, an increase in urban consumption and market activity, a growth of local industry and in a closer relationship between supply, demand and consumption in the Byzantine territories and the neighbouring zones, especially with the west and the Islamic world. Both these movements can be related to the changes in general climatic conditions in the period from the later fourth century onwards, and again from the middle of the ninth century on. While it would not be correct to draw too many direct relationships, there can be no doubt of the indirect causal associations which evolved.

Rivers, Roads and Communications

Again, and as we would expect, major communications routes were determined by the geography of the landscape, and for the heartlands of the medieval east Roman empire the inter-regional routes can be identified with some certainty, although their physical traces are not always so readily located. In the Balkans, the major as well as the less important routes pass in several places through relatively narrow and often quite high passes, easily blocked. Winter conditions alone made passage hazardous, as even today in many cases, but human agency might also close access – for example, to an invading army. Political control has always been difficult, and the fragmented geography made for a fragmented political landscape also. The history of the Balkans, the pattern of communications and the degree and depth of Byzantine political control show this especially clearly, for there was no obvious geographical focal point in the south Balkan region – the main cities in

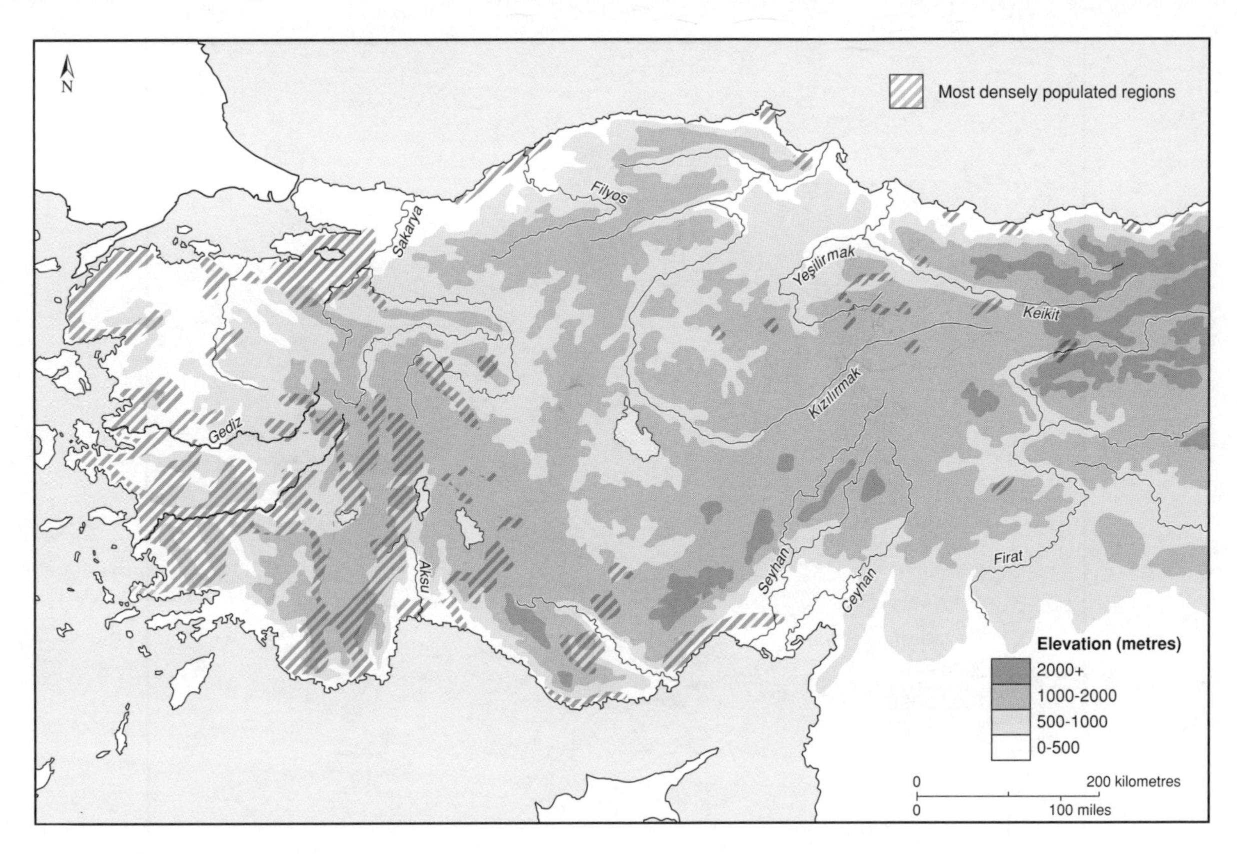

Map 1.7 Asia Minor: major population centres, 7th–12th centuries.

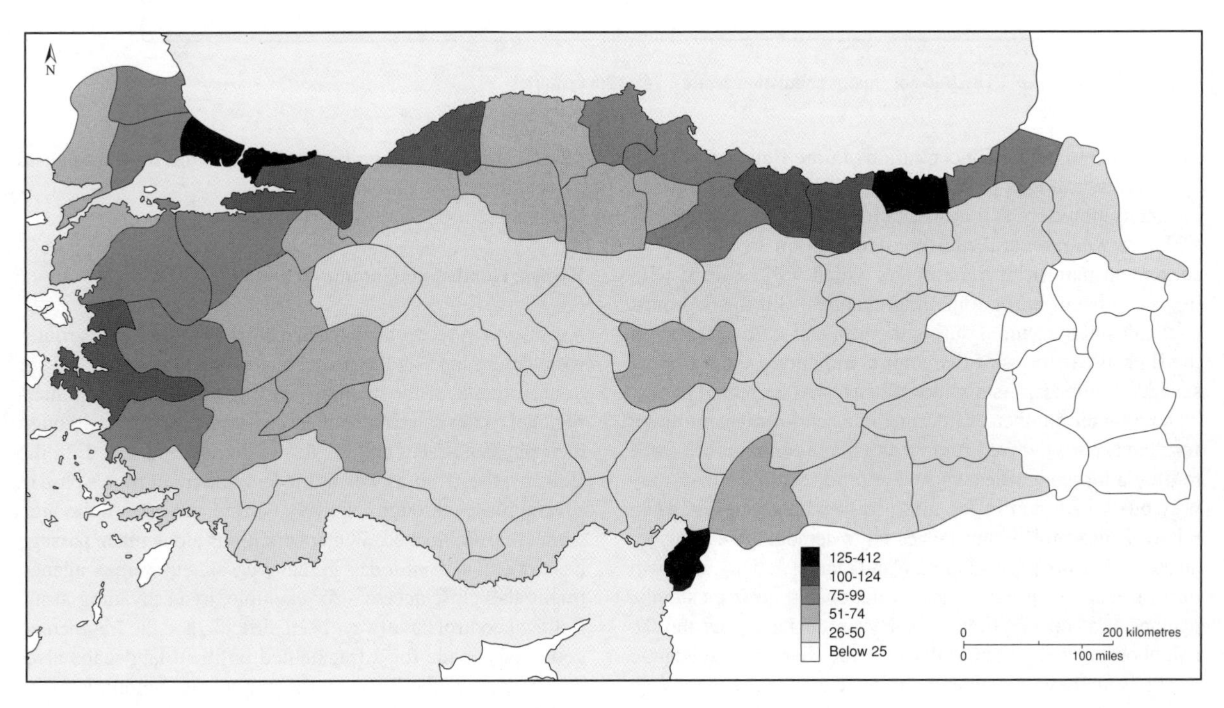

Map 1.8 Turkey in 1935: average population per square mile.

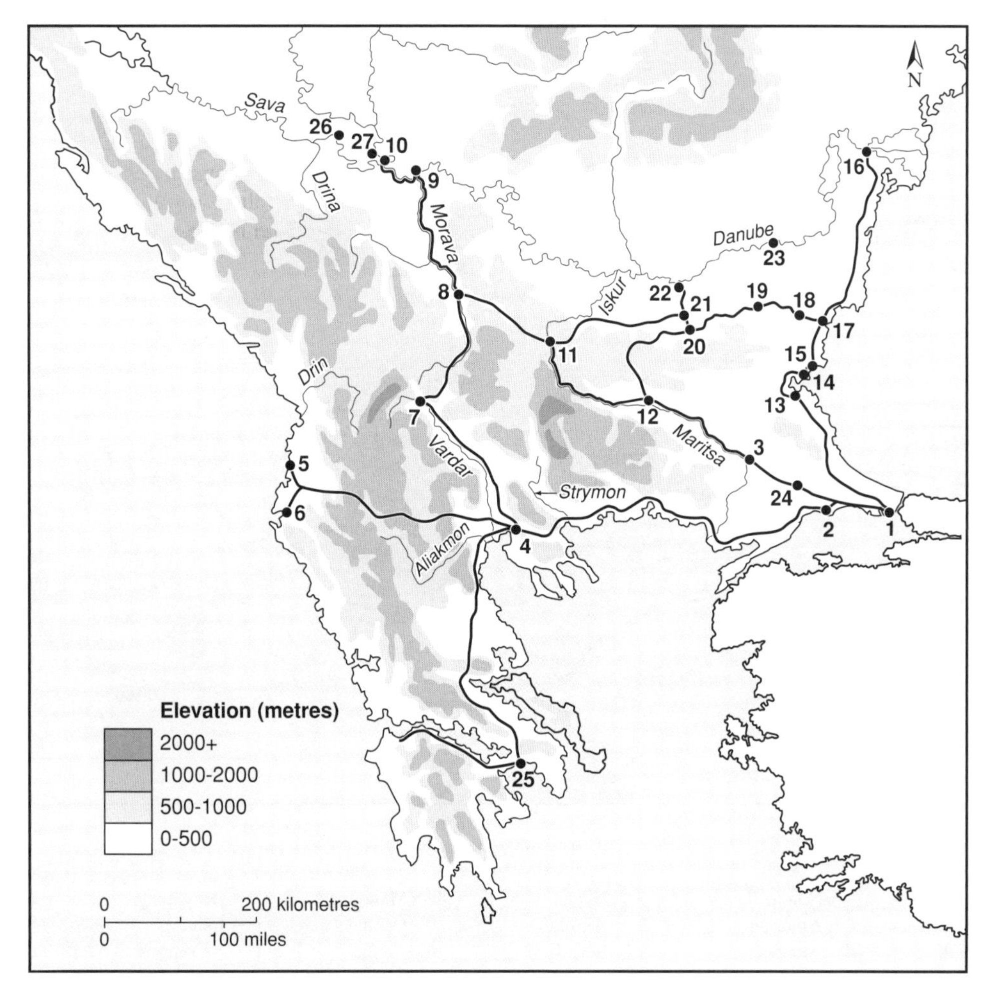

Towns/fortresses

1	Constantinople	2	Herakleia	3	Adrianople	4	Thessalonica
5	Dyrrhachion	6	Avlona	7	Skopje	8	Naissus (Niš)
9	Viminacium	10	Singidunum (Belgrade)	11	Serdica (Sofia)	12	Philippopolis
13	Develtus	14	Anchialos	15	Mesembria	16	Noviodunum
17	Varna	18	Markianoupolis	19	Pliska	20	Trnovo
21	Nikopolis	22	Novae	23	Dorostolon	24	Arkadiopolis
25	Corinth	26	Sirmium	27	Semlin		

Map 1.9 The Balkans: major routes, 7th–12th centuries.

the medieval period were Thessalonica and Constantinople, both peripheral to the interior of the peninsula. In the highland districts, especially the Rhodope and Pindus ranges, government power was always circumscribed by distance and remoteness, regions where paganism and heresy could survive relatively uninterrupted by state or ecclesiastical authority. In Asia Minor the waterless tracts across the central plateau similarly made travel hazardous, while the eastern highlands were particularly difficult to negotiate in the winter season. The narrow mountain passes across the Taurus made that range a natural barrier, and it was successfully employed by the imperial government in this way during much of the eighth, ninth and tenth centuries. Across the Middle Eastern and African provinces of the empire the road system continued to expand into the fifth century as the frontier between Roman and Persian lands shifted and as strategic priorities altered over time. In North Africa again strategic considerations, and in particular the maintenance of communications between key coastal garrisons and ports and the fortresses covering the interior, were important factors, and continued to influence imperial construction into the reign of Justinian.

The eastern Roman empire benefited from the creation of military roads, constructed largely in the period 100 BCE–100 CE by the Roman army – one of the reasons for their success and efficiency on campaign, for this network also eased and aided non-military communications, the movement

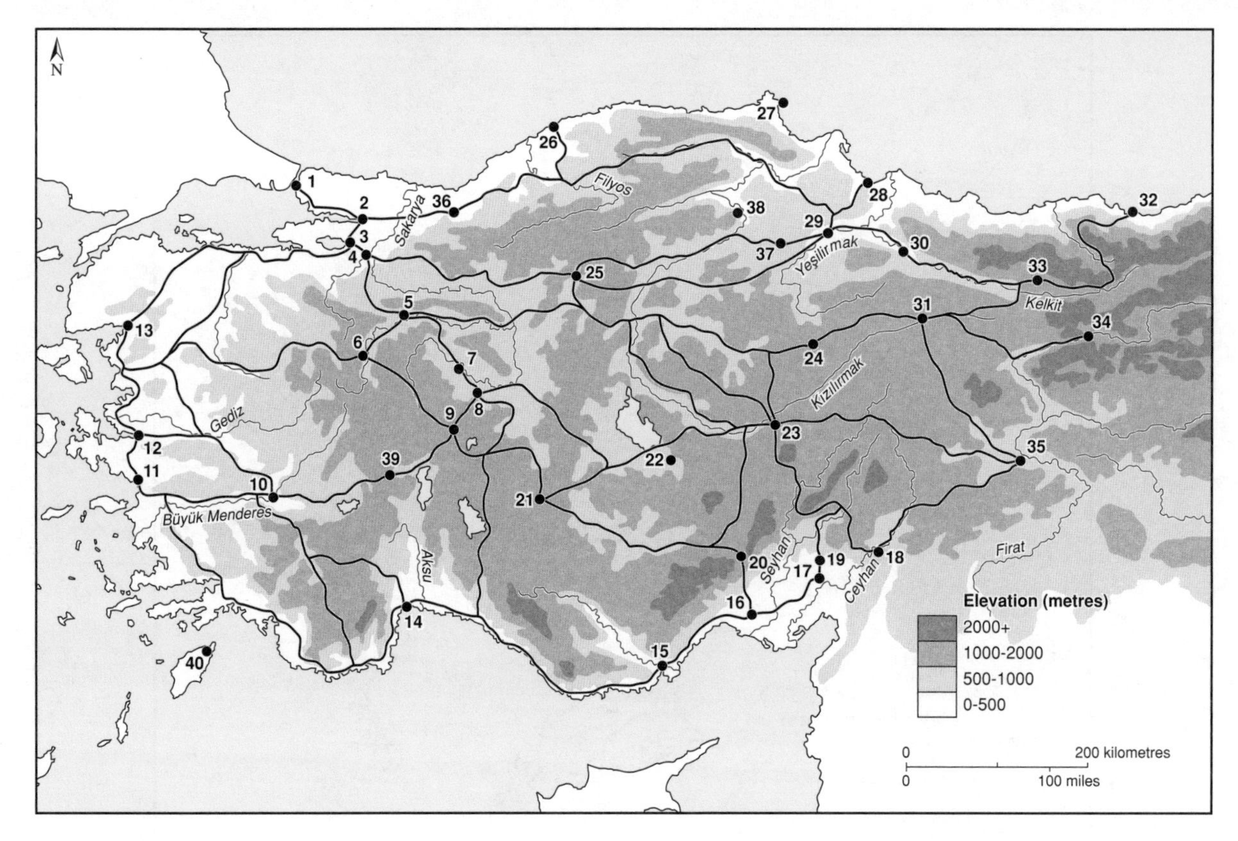

Towns/fortresses

1	Chaledon	2	Nikomedeia	3	Nikaia	4	Malagina	5	Dorylaion	6	Kotyaion
7	Kaborkion	8	Amorion	9	Akroinon	10	Chonai	11	Ephesos	12	Smyrna
13	Adramyttion	14	Attaleia	15	Seleukeia	16	Tarsos	17	Anazarbos	18	Germanikeia
19	Sision	20	Podandos	21	Ikonion	22	Koron	23	Kaisareia	24	Charsianon
25	Ankyra	26	Amastris	27	Sinope	28	Amisos	29	Amaseia	30	Dazimon
31	Sebasteia	32	Trapezous	33	Koloneia	34	Kamacha	35	Melitene	36	Klaudiopolis
37	Euchaita	38	Gangra	39	Sozopolis	40	Rhodes				

Map 1.10 Asia Minor: major routes, 7th–12th centuries.

of goods, people and information. But the regular maintenance of roads, which was a state burden upon towns and which was administered and regulated at the local level, seems during the later Roman period to have suffered somewhat. One significant consequence of this change, and the difficulties it created for the use of wheeled vehicles, was an ever-increasing dependence on pack-animals – horses, ponies, mules, donkeys, camels. Strict regulations were established during the later third and fourth centuries on the size, loads and types of wheeled vehicle employed by the state transport system. This was divided into two branches, the fast post (faster-moving pack-animals, light carts, and horses or ponies) and the slow post (ox-carts and similar heavy vehicles) and although the service was drastically reduced after the sixth century (and cut back already under Justinian), it seems that a unified transport and courier service continued to operate through the Byzantine period.

There were many types and standards of road: wide roads, narrow tracks or paths, paved and unpaved roads, roads suitable or unsuitable for wagons or wheeled vehicles are all mentioned in the sources. Roads of strategic importance were generally more regularly maintained. After the sixth century, it would appear that certain key routes only were kept up, largely by means of compulsory duties imposed on local communities and appropriately skilled craftsmen. The road system from the middle of the seventh century in Anatolia was thus less extensive than in the fifth century or before, but still effective. Similar considerations apply in the Balkans. The maintenance of much of the network became a localised and irregular matter, and the limited evidence suggests that the great majority of non-military routes became little more than paths or tracks suitable only for pack-animals, with paved or hard surfaces only near towns and fortresses.

Transport by water was generally much faster and certainly far cheaper than by land. Long-distance overland movement of bulk goods such as grain was generally prohibitively expensive – the cost of feeding draught-oxen, maintaining drovers and carters, paying local tolls, combined with the extremely slow rate of movement of ox-carts, multiplied the value of the goods being transported beyond the price of anyone who would otherwise have bought them. Although the bulk transport of goods over long distances did sometimes happen, it was really only the state, with some activity funded by wealthy private individuals, which could pay for this. The cost-effectiveness of shipping, entailing the carriage of large quantities of goods in a single vessel handled by a small crew, also gave coastal settlements a great advantage with regard to their access to the wider world.

Balkan Routes

- The *Via Egnatia*: Constantinople – Herakleia in Thrace – Thessalonica – Edessa – Bitola – Achrida (*Ohrid*) – Elbasan – Dyrrachion (*Durrës*) on the Adriatic coast.
- Constantinople – Adrianople (*Edirne*) – along the Maritsa – Philippopolis (*Plovdiv*) – the pass of Succi (guarded at the northern exit by the so-called 'gates of Trajan', and barred by a wall and forts) – the pass of Vakarel – Serdica (*Sofia*) – the Nisava valley – Naissus (*Niš* – key crossroads along the routes southwards to the Aegean and Macedonia, westwards to the Adriatic, south-eastwards to Thrace and Constantinople, and northwards to the Danube) – the valley of the Morava – Viminacium (nr. mod. *Kostolac*) – Singidunum (*Belgrade*). This was a key military route, and it was complemented by a number of spurs to east and west, giving access to the south Danube plain, the Haimos mountains and Black Sea coastal plain, as well as, in the west, the valleys of the west Morava, Ibar and Drin rivers.
- Thessalonica – the Axios (*Vardar*) valley and the pass of Demir Kapija (alternative easterly loop avoiding this defile and leading through another pass, known to the Byzantines as Kleidion – *the key*) – Stoboi (*Stobi)* – Skopia (*Skopje*) – Naissos (*Niš*).
- Constantinople – Anchialos (*Pomorie*) – Mesembria (*Nesebar*) – Odessos (*Varna*) – mouth of the Danube.
- Adrianople – across the Sredna Gora range – over the Shipka pass through the Balkan range itself – Nikopolis (*Veliko Trnovo*) – Novae (*Svistov*) on the Danube.

Anatolian Routes

- Chrysoupolis (opposite Constantinople) – Nikomedeia – Nikaia – Malagina (an important imperial military base) – Dorylaion – (easterly route via Kotyaion/westerly route via Amorion) – Akroinon – Ikonion/ Synnada – Kolossai/ Chonai. There were two options to turn off to the south along this last route, the first down to Kibyra and thence across the mountains to the coast at Attaleia or, farther west, at Myra. Alternatively, the road from Chonai led westwards via Laodikeia and Tralles to Ephesos on the coast.
- Ikonion – Archelais – Tyana/Kaisareia.
- Ikonion – Savatra – Thebasa – Kybistra/Herakleia – Loulon – Podandos – Çakit River gorge (through the Anti-Taurus mountains).
- Kaisareia – Tyana – Loulon – Podandos – 'Cilician Gates' (*Külek Boğazı*) – the Cilician plain – Tarsos/Adana.
- Kaisareia – (i) – Ankara/ Basilika Therma – Tabion – Euchaita/ (ii) – Sebasteia – Dazimon – Amaseia
- Sebasteia – Kamacha/ Koloneia – Satala.
- Dorylaion – valley of the Tembris river (mod. Porsuk Su) – Trikomia – Gorbeous – Saniana – Timios Stavros – Basilika Therma – Charsianon Kastron – Bathys Ryax – Sebasteia – (and on to Kaisareia, north to Dazimon, east to Koloneia and Satala, or south-east to Mélitene).
- Saniana – Mokissos – Ioustinianoupolis – Kaisareia.

Routes across the Taurus Ranges into Byzantine lands

- Cilician Gates – Podandos – Loulon – Herakleia – Ikonion/ Loulon – Tyana – Kaisareia.
- Germanikeia (Mar'aş) – Koukousos – Kaisareia
- Adata – Zapetra – Mélitene – Kaisareia – Lykandos/ Kaisareia – Sebasteia/Mélitene – Arsamosata (Simsat) – Khliat (on L. Van)
- Mopsouestia (al-Massisa) – Anazarba ('Ain Zarba) – Sision – Kaisareia.

Part One:

The Early Period (c. 4th–7th Century)

2 *Historical Development: from Rome to Byzantium*

The Roman Empire c. 400 CE

Following the civil wars of the first decades of the fourth century, the Emperor Constantine I recognised that the empire as a whole could no longer effectively be ruled from Rome. He moved his capital eastwards, to the site of the ancient Megaran colony of Byzantium, and renamed it Konstantinoupolis, the city of Constantine. Its strategic position was attractive, for the emperor could remain in contact with both eastern and western affairs from its site on the Bosphorus. The city was expanded, new walls were constructed and the emperor undertook an expensive building programme. Begun in 326, the city was formally consecrated in 330.

Constantine inaugurated a series of important reforms within both the military and civil establishment of the empire. The fiscal system was overhauled and a new gold coin, the *solidus* introduced in a successful effort to stabilise the monetary economy of the state. Military and civil offices were separated, the central administration was restructured and placed under a series of imperially-chosen senior officers responsible to the emperor directly. The armies were reorganised into two major sections, those based in frontier provinces and along the borders, and several field armies of more mobile troops attached directly to the emperor's court as a field reserve, ready to meet any invader who broke through the outer defences. The provincial administration was reformed, more and smaller provincial and intermediate units being established, the better to permit central control and supervision of fiscal matters. Finally, with the toleration of Christianity and its positive promotion under Constantine at the expense of many of the established non-Christian cults, the church began to evolve into a powerful social and political force which was, in the course of time, to dominateeast Roman society and to vie with the state for authority in many aspects of civil law and justice.

In spite of Constantine's efforts at reform, the size of the empire and the different concerns of west and east resulted in a continuation of a split government, with one ruler in each part, although the tetrarchic system was never revived. Upon Constantine's death in May 337, his three sons succeeded to his authority with the support of the armies. Constantine II, the eldest, was recognised as senior and ruled the west. Constantius ruled in the east and Constans, the youngest, was allotted the central provinces (Africa, Italy, Illyricum). Tension between Constans and Constantine resulted in war in 340 and the defeat and death of the latter, with the result that Constans became ruler of the western regions as well. As a result of popular discontent among both the civilian population and the army in the west, however, Constans was deposed in 350 and his place taken by a certain Magnentius, a high-ranking officer of barbarian origin. Magnentius was not recognised by Constantius, and he invaded Illyricum. But he was defeated in 351, escaping to Italy where, after further defeats, he took his own life. Constantius ruled the empire alone until his death in 360.

In 355 Constantius had appointed his cousin Julian to represent him in Gaul; in 357, he was given the command against the invading Franks and Alamanni and, following a series of victories, he was acclaimed by his soldiers as Augustus. Constantius was campaigning against the Persian king Shapur who had invaded the eastern provinces in 359, and the acclamation may have been stimulated by the emperor's demand that Julian send him his best troops for the Persian war. Julian marched east, but on the way to meet him Constantius died in 361, naming Julian as his successor. Although a competent general and efficient administrator, Julian may have been unpopular with some of his soldiers because of his attempts to revive paganism, often at the financial expense of the church. During the Persian campaign of 363 he was mortally wounded, although it is not clear in what circumstances. The troops acclaimed the commander of Julian's guards, a certain Jovian, as emperor. Having made peace with Shapur, Jovian marched back to Constantinople, dying in Bithynia a mere eight months later.

Jovian's successors were Valentinian and Valens, brothers from Pannonia (roughly modern Austria and Croatia), the former having been elected by the military at Constantinople then appointing his brother as co-emperor. Valentinian ruled in the west and established his capital at Milan, while Valens had to face a rebellion almost immediately, led by the usurper Procopius and caused by the soldiers loyal to Julian, whose favourite Procopius had been. But the rebellion petered out in 366.

The two new emperors each had substantial military challenges to overcome. But Valentinian died in 375 while dealing with the Quadi in Pannonia, and was succeeded by his chosen successor, Gratian. In the east, Valens had to deal with repeated Gothic invasions of Thrace, where in 378 he was disastrously defeated and killed near Adrianople (mod. Edirne) in Thrace.

Gratian appointed as Valens' successor the general Theodosius, son of a successful general of the same name and himself an experienced commander, initially as commander-in-chief and then Augustus; and by a combination of diplomacy and strategy Theodosius was able to make peace with the Goths, permitting them to settle within the empire under their own laws, providing troops for the imperial armies in return for annual food subsidies. Following the death of Gratian in 383 as the result of a coup, and the eventual overthrow of the usurper, Magnus Maximus, by Theodosius in 388, Theodosius became sole ruler. He was, however, the last emperor to hold this position. At his death in 395 his two sons Arcadius (in the east) and Honorius (in the west) ruled jointly.

Migrations and Invasions: Huns, Germans and Slavs

The Roman empire at the end of the fourth century had an enormously long frontier, stretching in the north-west from the Tyne-Solway line followed by Hadrian's Wall in Britain,

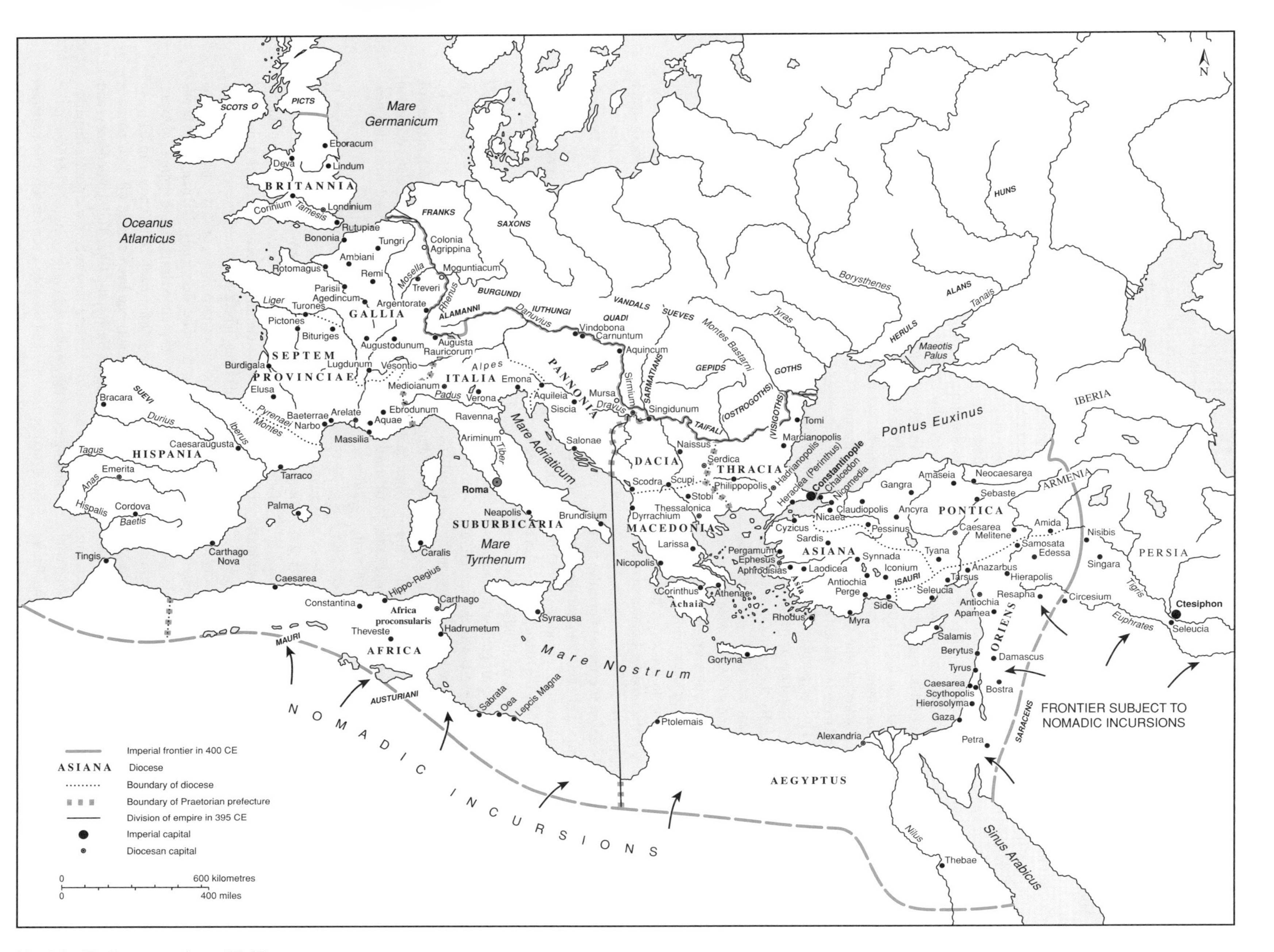

Map 2.1 The Roman empire c. 400 CE.

along the length of the Rhine and Danube rivers to the Black Sea, and in the east from the eastern littoral of the Black Sea near modern Batumi down through the Caucasus into Armenia, across the upper reaches of the Tigris and Euphrates rivers, through the great Syrian desert down to Sinai and across to Egypt, whence it followed the desert fringe across Libya/Tripolitania into modern Tunisia and further west, north of the Atlas mountains, as far as the Atlantic Ocean. Defending such a vast territory was always a formidable task, and with relatively limited resources – an army of perhaps 250,000, including auxiliaries and allied forces, to defend a perimeter over 8,000 miles in length, as well as maintain internal security, combat brigandage and banditry and carry out a range of other less obviously military tasks – necessarily depended less on military power alone than on trade and commerce, diplomacy and cultural influence to avoid constant conflict. It is ironic, therefore, that much of the pressure on the frontier came not from forces who were hostile to the empire, but from those who wished to be part of the Roman state but who found that they were threatened by others behind them or rejected as barbarians by the culture they admired. This is, perhaps, to formulate the issues far too simplistically, but there is nevertheless an important element of truth here. By the same token, wars of conquest and then of containment into the third century CE had familiarised Roman armies and strategists with Germanic peoples and tactics, and Roman diplomacy, power-politics and cultural influence had all worked to maintain a degree of stability. From the late third century in particular, however, a series of developments across the broader Eurasian context destabilised these arrangements.

Germanic peoples had been on the move since the first century BCE, migrating from Scandinavia into north-eastern and central Europe. By the middle of the second century some had arrived in the Pontic steppe north of the Black Sea and others had settled west of the Carpathians. A short-lived stability was reached with the establishment by the Ostrogoths and Visigoths of semi-nomadic pastoralist confederacies which evolved in contact with nomadic groups, such as the Iranian Alans, in the regions north of the Black Sea across to the Caspian. But while the Visigoths occupied a restricted region in what is now the western Ukraine and Romania, the Ostrogoths dominated the whole area from the Crimea up through the Ukraine and north to the shores of the Baltic, whose indigenous, largely Slav populations were made tributary.

Other groups had been under Roman influence for far longer, including the various west Germanic peoples described by Tacitus, for example, and with whom the Romans had had both friendly and hostile relations over the centuries. Some of these had been absorbed into Roman territory; the majority had by the fourth century come to form a series of independent, often competing but still Roman-influenced tribal entities along and behind the Rhine, again exercising tributary authority over many smaller groups, both Slav and Germanic. The two largest groups in the west were the Franks (along the northern and central Rhine), with the Burgundi – an eastern Germanic group – and the Alemanni (to their south). But associated with the latter in particular, and stretching along the upper Rhine and Danube, were the Marcomanni and the Quadi. Behind these groups the Jutes, Angles and Saxons in the north, the Lombards and Thuringi in the centre, and the Vandals, Gepids and Heruls in the south and east were also in frequent conflict with one another and with the dominant tribes. Raids across the frontier, or in the case of the northern groups, across the North Sea into Britannia, became increasingly frequent during the later fourth century, but pressure on the frontier and warfare with the various Germanic groups had always been a factor of Roman imperial existence. Marcus Aurelius had defeated the Marcomanni in the second century, Frankish and Alemannic raids had been common during the third century, and in the 350s and early 360s a Frankish-Alemannic attack was defeated by Julian.

This situation was transformed by the arrival of the Huns, however, who appeared on the borders of the Ostrogothic world in the late 360s CE. A mixed group of Turkic and Mongol tribes which had arisen out of the collapse of the great Hsiung-Nu confederacy on the eastern and central steppe in the first century CE, the Huns split during the fourth century into two major sub-factions, the White Huns, also called the Hephthalites, who invaded Iran from the north-east and caused substantial disruption and devastation, and the Black Huns, who set the Germanic peoples in motion – partly in response to Ostrogothic attempts to extend their control eastwards. The clash resulted in the rapid destruction of the Ostrogothic and Visigothic confederacies and the expansion of the Huns to the Danube by the early fifth century. In turn this set in motion the other Germanic peoples, and the enormous pressure this placed on Roman defences finally led to the collapse of the western frontiers and the occupation of large stretches of the western provinces by Germanic groups, initially as federates granted land and protection in return for military service, then as occupiers and conquerors. The breaching of the Rhine frontier by the Suevi, Vandals and Alans and their move into southern Gaul and then Spain, the Visigothic invasion of the Thracian provinces in the 370s and their subsequent move first into Italy (Rome was sacked in 410), and then on to southern Gaul and Spain, the occupation of the region of Tunisia by the Asding Vandals who had fled the new Visigoth masters of Spain in the 420s, and the Frankish and Burgundian occupation of northern and eastern Gaul, all followed from this new international situation.

In eastern Europe the movement of the Slav peoples is related to, but slightly later than, these developments. By the middle of the sixth century the eastern empire was becoming familiar with the raiding of small bands of Slavs, and during the second half of the century it became clear that many of these bands were intent on permanent settlement wherever they could find suitable unoccupied land, or drive the indigenous population off. But the small, disorganised, if numerous, bands of Slavs were soon overwhelmed by the more aggressive Avars, a Turkic people whose dominant clan (known in Chinese sources as the Juan Juan) had been chased off their pastures by their former subordinates, the Blue Turk confederacy, and had fled westwards. Allying themselves with other disparate nomad groups they appeared on the empire's borders in the 560s, and by the 580s had become a serious threat to imperial power in the Balkans.

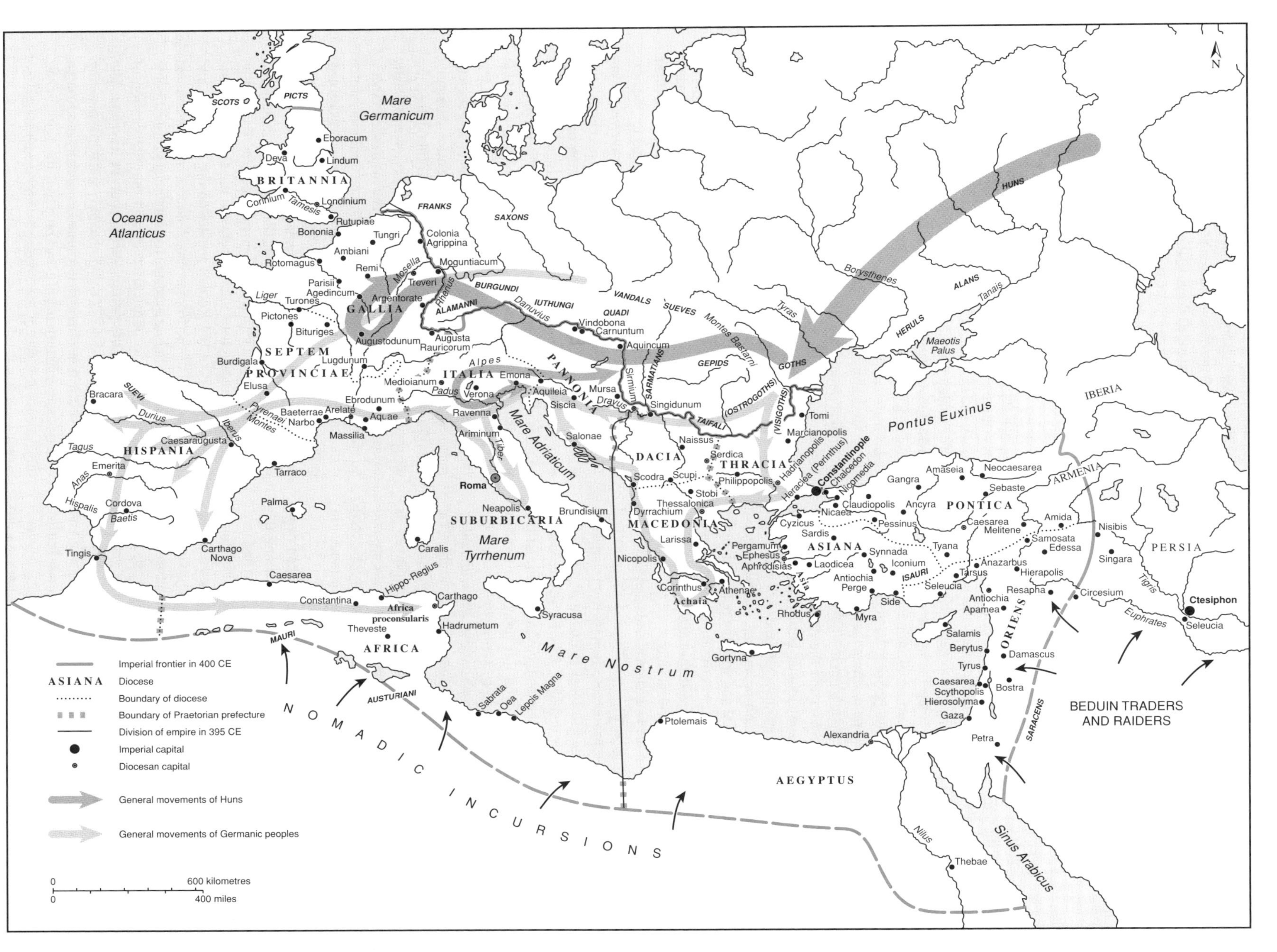

Map 2.2 Migrations and invasions: Huns, Germans and Slavs.

The West and the Rise of the Successor Kingdoms

The polities which succeeded Rome during the fifth and sixth centuries in the west all have their origins in the tumultuous changes which affected the western Eurasian world in the two preceding centuries, yet they had very different histories and outcomes. The loose confederacy of Suevi, Alans and Vandals who crossed the Rhine in 406 spent the next three years extracting tribute and booty from Gaul, before crossing the Pyrenees in 409 and entering Spain. Here, the Suevi established their own kingdom in Galicia. The Visigoths, who had moved from the Balkans into Italy and thence (from 412) into southern Gaul after the sack of Rome and subsequent death of their leader Alaric in 410, had established their own state around Toulouse by about 418 – a move encouraged by the imperial government, which pitted the Visigoths against a rival emperor set up under Frankish and Alemannic influence. In 416 the Visigoths then advanced against the Siling Vandals and Alans in south-western and southern Spain, whom they crushed, before being offered an independent kingdom in south-western Gaul. This saved the remaining Suevi and the Asding Vandals in the north-western regions and Galicia; but by the late 420s the latter were on the move again, crossing in 429 into North Africa. Threatened by the establishment of a Vandal kingdom with naval power at its disposal and with the potential fatally to disrupt the grain supplies of Rome, the imperial government was forced to accept and to recognise formally the King of the Vandals, Gaiseric, as an independent ruler.

Germanic raiders from the Danish peninsula and the North Sea coastlands had meanwhile transformed the political landscape in the British provinces. In 410 Rome appears officially to have conceded to local British authorities the right to organise their own defences, in view of the lack of substantial imperial forces. Although the history of the British provinces is clouded in obscurity at this time, local polities led by Romano-British nobles and by Celtic warlords appear to have evolved, competing with one another and with raiders from Ireland, from the Pictish lands to the north, and from the Saxons, Jutes and Angles in north-west Germany, Denmark and the Low Countries. The latter were also employed as mercenaries, and by the later fifth century certain groups had a firm foothold and would, during the sixth century, succeed in establishing a political dominance in much of the southern and central lowlands.

Both the Salian and Ripuarian Frankish groups had been officially permitted to reside on Roman territory along the Rhine by Honorius in 410 as a result of the pressures he faced elsewhere. Several other Frankish groups remained in Franconia. Those Franks who settled within the empire supplied federate troops to the Roman armies. In central and northern Gaul the Salian Franks, having moved first into the low countries, were then able to establish themselves, precariously at first, in the valleys of the Moselle and Rhine, and by the last years of the century had succeeded in defeating the last remnants of independent Roman rule in the Seine valley, defeating and incorporating into their territory the Ripuarian Franks (settled originally on the right bank of the Rhine but occupying territory on the 'Roman' side during the fifth century), and driving off the Alemanni who threatened them from the south-east in the late 490s. Frankish control was broad – the valleys of the Loire and Seine and the central French plain were the heartland, but Frankish rule extended down to the Visigothic lands stretching across from the Pyrenees into northern Italy, across to the valleys of Main and Rhine in the east, and down to the Burgundian lands about the headwaters of the Rhône in the south-east. The conversion to orthodox Christianity of the Frankish king Clovis in 506 won the Franks the support, or at least disarmed the opposition of, the Gallo-Roman élite and the church, facilitating the consolidation of Frankish power, gaining diplomatic and political support from the eastern emperor Anastasius against the Visigoths and Ostrogoths, as well as the support of the papacy and thus political legitimacy.

The situation in Italy was, if anything, more complex. The general Odovacar (Odoacer), the effective ruler in the Italian provinces, deposed the Emperor Romulus (Augustulus) in 476. In his manifesto to the eastern emperor, Zeno, he claimed that the western army had deposed its commander-in-chief Orestes (Romulus' father) and the emperor, and that he was himself acting on behalf of the senate. And upon sending the western emperor's diadem to Zeno, he asked to be recognised as the emperor's representative in Italy, with the title of *patricius*, on the grounds that one emperor was sufficient. Political circumstances demanded that Zeno concur. But Odovacar styled himself *rex*, king, not simply as senator, magistrate and patrician, and his followers – made up of the eastern Germanic groups of the Scyrii, Rugii and others – clearly saw him as their king and warleader. He ruled Italy from Ravenna for the next 17 years, until – after a conflict that lasted some five years from 488 to 493 – he was defeated by Theoderic and his Ostrogoths, who had been offered the opportunity of acting on behalf of the emperors to re-establish imperial authority in Italy (and as a means of removing the threat they posed to Roman power in the Balkans). While he acted as King of the Goths, Theoderic was a Roman citizen and maintained as far as he was able the structures and fabric of Roman government and society, retaining the framework of Roman administration, hierarchy and offices. As an Arian Christian, of course (although he has also been understood as a 'homoean'), he was viewed by many of his non-German subjects as a heretic. But in all other respects he made a genuine effort to shore up Roman traditions, which – like many other 'barbarian' leaders – he greatly esteemed, and seems to have been held in considerable respect by both the papacy and the mass of the population. His Gothic soldiery replaced the Scyrii and Rugii as the 'Roman' army in Italy, and were settled according to late Roman principles. Eventually, in the 520s, conflicts of interest between the Gothic élite and some elements of the Roman senate, on the one hand, and other members of the Roman establishment in Italy, coupled with Theoderic's failure to secure recognition at Constantinople for his heir, led to political crisis and the intervention of Constantinople in the politics of the court at Ravenna. The result was the invasion of Ostrogothic Italy and the devastating 20-year war which, although it resulted eventually in a Roman victory, both exhausted Italy and prepared the way for the subsequent successes of the Lombards who marched into the Po valley in 568.

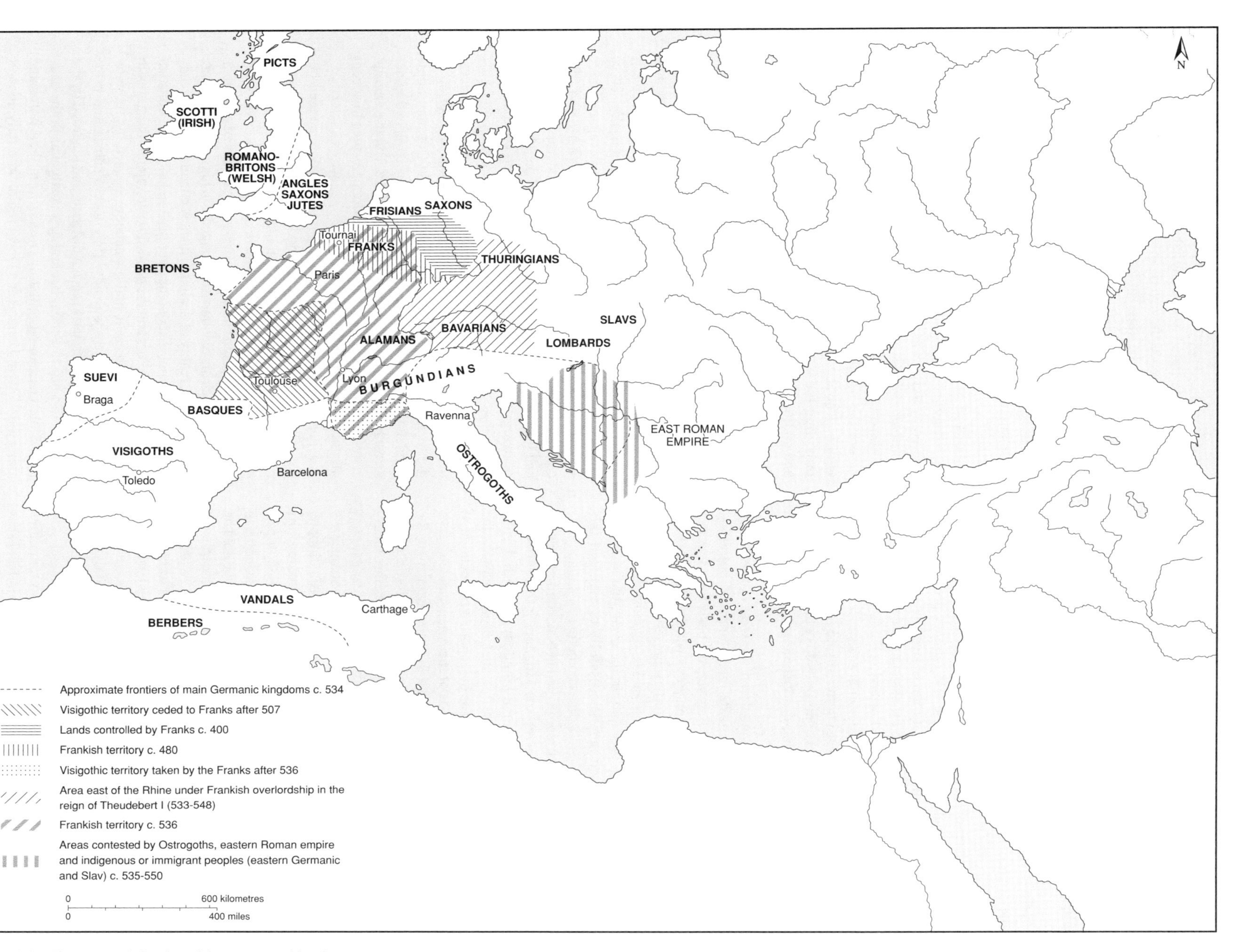

Map 2.3 The west and the rise of the successor kingdoms.

Conflict, Imperial Expansion and Warfare in the 6th Century

The east Roman emperors of the early sixth century faced three problems in terms of military strategy and foreign policy: the constant threat from Persia in the east; the danger posed by the Vandal kingdom in Africa; and the unstable northern frontier along the Danube. The western part of the empire had been transformed into a patchwork of barbarian successor states, but Constantinople continued to view all the lost territories as part of the empire, and in some cases to treat the kings of the successor kingdoms as their legitimate representatives, governing Roman affairs in the provinces in question until Constantinople could re-establish a full administrative and military presence. This is most obviously the case with the Ostrogoths. By the same token, the leader of the Salian Franks in northern Gaul, Clovis, had quite deliberately adopted orthodox Christianity in order to gain papal and imperial recognition and support for his rule, where he also claimed, at least nominally, to represent Roman authority, exploiting the fact of his orthodoxy to justify warfare against his Arian neighbours, the Visigoths in southern Gaul in particular.

Roman emperors thus considered the west not as 'lost', but rather as temporarily outside direct imperial authority. Under Justinian, this point of view was the basis for a series of remarkable, if opportunistic, reconquests which, whatever their original motivation, were certainly represented as restoring the Roman world in its greatness, and re-establishing Rome's power as it had been at its height. In the event, the resources required to achieve and then successfully to maintain this imperial expansionism were exhausted before Justinian died.

When Theoderic the Ostrogoth died in 526, conflict erupted over the succession, throwing the kingdom into confusion. The same occurred in the Vandal kingdom of North Africa. The political conflict and civil strife which broke out upon the death of the Vandal king, as well as the reported persecution of the Roman population at the hands of the Arian Vandals, gave Justinian his chance and, in 533, in a lightning campaign, the general Belisarius was able to land with a small force, defeat two Vandal armies and take the capital, Carthage, before finally eradicating Vandal opposition. The timing of this campaign is perhaps not an accident, for its success redeemed the emperor's reputation in the aftermath of the Nika riot at Constantinople, which had nearly cost him his throne. Encouraged by this success, Sicily and then southern Italy were occupied in 535 on the pretext of intervening in the affairs of the Ostrogoths to stabilise the situation and to restore orthodox Christian rule. The Goths felt they could offer no serious resistance, their capital at Ravenna was handed over, their king Witigis was taken prisoner and sent to Constantinople, and the war appeared to be won. But at this point Justinian, who appears to have harboured suspicions about Belisarius' political ambitions, recalled him, partly because a fresh invasion of the new and dynamic Persian king Chosroes I (Khusru) threatened to cause major problems in the east. In 540 Chosroes was able to attack and capture Antioch, one of the richest and most important cities in Syria, and since the Ostrogoths had shortly beforehand sent an embassy to the Persian capital, it is entirely possible that the Persians were working hand-in-glove with the Goths to exploit the Roman preoccupation in the west and to distract them while the Goths attempted to re-establish their situation. For during Belisarius' absence they were able to do exactly that, under a new war leader, the king Totila. Within a short while, they had recovered Rome, Ravenna and most of the peninsula. It took the Romans another ten years of punishing small-scale warfare throughout Italy finally to destroy Ostrogothic opposition, by which time the land was exhausted and barely able to support the burden of the newly re-established imperial bureaucracy.

Justinian's ambitions did not end there, however. He had further expansionist plans, but in the end only the south-eastern regions of Spain were actually recovered from the kings of the Visigoths, also Arians (Justinian exploited the opportunity offered by a civil war in 554). But arguably his most significant contribution to restoring imperial greatness was the codification of Roman law which he ordered and which was begun well before the military expansionism which began in 533, and which produced the Digests and the *Codex Justinianus*, providing the basis for later Byzantine legal developments and codification. He persecuted the last vestiges of paganism in his efforts to play both Roman and Christian ruler, defender of Orthodoxy and of the church, and he also introduced a large number of administrative reforms and changes in an effort to streamline and bring up-to-date the running of the empire (although in the event many were rescinded within a few years). But his grandiose view of the empire and his own imperial position brought him into conflict with the papacy. In 543, at the beginning of what came to known as the 'Three Chapters' controversy, the emperor issued an edict against three sets of writings (the 'Three Chapters') of the fourth and fifth centuries, by Theodore of Mopsuestia, Theodoret of Cyrrhus and Ibas of Edessa, who had been accused by the Monophysites of being 'pro-Nestorian'. The intention was to conciliate the Monophysites, and required the agreement and support of the Roman Pope Vigilius. The pope did indeed – eventually – accept the edict, but there remained very substantial opposition in the west, and in 553 an ecumenical council at Constantinople condemned the Three Chapters. The pope was placed under arrest by imperial guards and forced to agree. But the attempt at compromise failed to persuade the Monophysites to accept the 'neo-Chalcedonian' position.

Upon his death in 565 Justinian left a vastly expanded but perilously overstretched empire, both in financial as well as in military terms. Justinian had seen himself as the embodiment of Roman imperial power, and there can be no doubt as to the brilliance of his reign and the enormous enhancement of Roman prestige which his reconquests brought. But his successors were faced with the reality of dealing with new enemies, lack of ready cash, and internal discontent over high taxation and constant demands for soldiers and the necessities to support them. Justin II, Justinian's successor and his nephew, opened his reign by cancelling the yearly 'subsidy' (in effect, a substantial bribe paid to keep the Persian king at a distance, and regarded by the latter as tribute) to Persia, beginning a costly war in the east. In 568 the Germanic Lombards crossed from their homeland along the western Danube and Drava region into Italy, in their efforts to flee the approaching Avars, a Turkic nomadic power

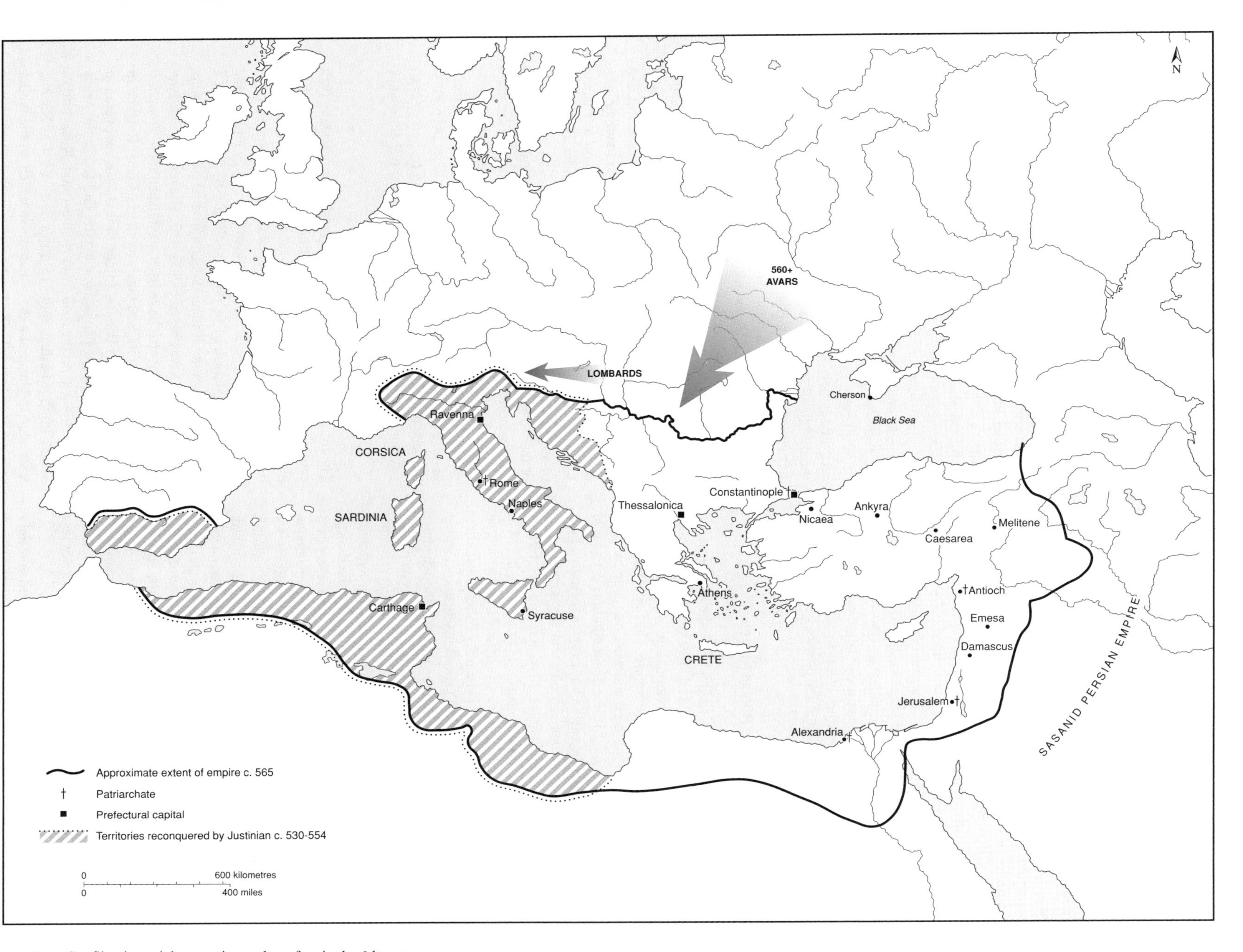

Map 2.4 Conflict, imperial expansion and warfare in the 6th century.

which, like the Huns two centuries earlier, were in the process of establishing a vast steppe empire. While the Lombards rapidly overran Roman defensive positions in the north of the peninsula, soon establishing also a number of independent chiefdoms in the centre and south, the Avars occupied the Lombards' former lands and established themselves as a major challenge to imperial power in the northern Balkan region. Between the mid-570s and the end of the reign of the Emperor Maurice (582–602), the empire was able to re-establish a precarious balance in the east. Although the Romans suffered a number of defeats, they were able to stabilise the Danube frontier in the north, the lands over which the campaigning took place, especially in Italy and the Balkans, were increasingly devastated and unable to support prolonged military activity. Maurice cleverly exploited a civil war in Persia in 590–591 by supporting the young, deposed king Chosroes II. When the war ended, with Roman help in the defeat of Chosroes' enemies, the peace arrangements between the two empires rewarded the Romans with the return of swathes of territory and a number of fortresses which had been lost in the previous conflicts.

Defence and Strategy: Late Roman Structures

At the beginning of Justinian's reign in 527, the armies of the east Roman empire were organised into five mobile field armies and a large number of smaller regional divisions along and behind the frontier regions of the empire. The field army units were referred to as *comitatenses*, and were each commanded by a *magister militum*, or 'Master of the Soldiers'. The five divisions were those of the east (a huge region including both the Armenian and Mesopotamian fronts with Persia, as well as the Egyptian desert front), Thrace, Illyricum, and two further corps 'in the presence' of the emperor (*in praesenti*), based in north-west Asia Minor and in Thrace to defend Constantinople – in the days when emperors had personally commanded their field troops, these had been their divisions. By Justinian's time this tradition of personal command had lapsed, although under Heraclius in the Persian war (622–629) it was revived. The troops making up the frontier divisions and permanent garrisons were known as *limitanei*, mostly composed of older legionary units and associated auxiliary units, backed up by mixed corps of auxiliary and legionary cavalry to provide local reserves.

Justinian undertook several reforms of these arrangements, introducing new commands for the Masters of Soldiers for Africa and Italy after their recovery, and establishing a Master of Soldiers for Armenia out of the older eastern field command. By the end of his reign there were over 25 regional commands behind the frontiers and deeper inland, serving both as military and police force for internal matters, stretching from that for Scythia in the north-west Balkans through the Middle east and Egypt to Mauretania in north-west Africa. The real differences between field troops and garrison units were not always very clear, mainly because of cross-postings from one type of army to the other, and because so many field units were more or less permanently based in and around garrison cities.

Justinian established a strategically very important new field command, known as the *quaestura exercitus* (loosely translated as 'regions allocated to the army'), similar to that of a Master of Soldiers, but whose commander was entitled *quaestor*. This command included the troops based in the Danube frontier zone (the provinces of Scythia and Moesia II), but included in addition the Asia Minor coastal province of Caria along with the Aegean islands. The aim was to supply the Danube divisions by sea from an Aegean hinterland and thus relieve the oppressed local population of the frontier regions and their hinterland from the burdens of supporting a large military force. In addition to the regular corps, the empire maintained substantial numbers of allied forces: Arab clans and tribes were essential to the empire's strategic arrangements in the east, and were subsidised with food, cash, vestments, imperial titles and weaponry.

The emperors had also several guards units based in or near the imperial palace, or in the districts about Constantinople. The most important were the Schools, or *scholae palatinae* and the *excubitores*. The former were organised in seven divisions of 500 heavy cavalry soldiers. Originally élite shock units recruited largely from German peoples, they had become by the middle of the fifth century little more than parade units. In their stead as active guards the Emperor Leo I (457–474) recruited the latter, a much smaller élite unit of a mere 300 men. Imperial naval forces were relatively limited – several small flotillas maintained along the Danube, a fleet was based at Ravenna, and a squadron at Constantinople.

Imperial strategy was based on a first line of defence that consisted of a linear frontier screened by fortified posts, major fortresses and a connecting network of minor fortified positions. This was supported by a second line made up of a reserve of mobile field units scattered in garrison towns and fortresses across the provinces behind the frontier. By the end of Justinian's reign the gap between the different functions of the 'frontier' and 'field' armies had been narrowed, for the reasons noted already, and in the 560s and 570s garrison units seem to have reinforced and fought alongside field army units. In effect, the late Roman army was a relatively expensive force of very variable quality, which consumed a large proportion of the state's fiscal revenue each year, both in respect of cash payments, as well as in terms of equipment and maintenance in kind for troops on campaign.

The frontier was considerably strengthened from the later fifth century into Justinian's reign as political and military priorities evolved and as new threats developed. Typical of such efforts on the eastern front is the fortress of Dara. This fortress (also called Anastasioupolis, mod. Turkish Oğuz) was built by the Emperor Anastasius I in the years 505–507, to serve as a military base on the Roman–Persian frontier, where the *doux* of Mesopotamia was based c. 527–532. The *magister militum per orientem* may also have been established there from 540 to 573, when the city was taken by the Persians. Retaken in 591, it fell again to Persian forces in 604, was recovered at the end of the great Persian war in 628, and fell to the Arabs in 639. Situated on the road from Nisibis (mod. Nusaybin) to Marde (mod. Mardin), some 15 miles north-west of Nisibis, it stood at the head of a dry watercourse which, in the winter season, flows down to the Khabur river farther south. While

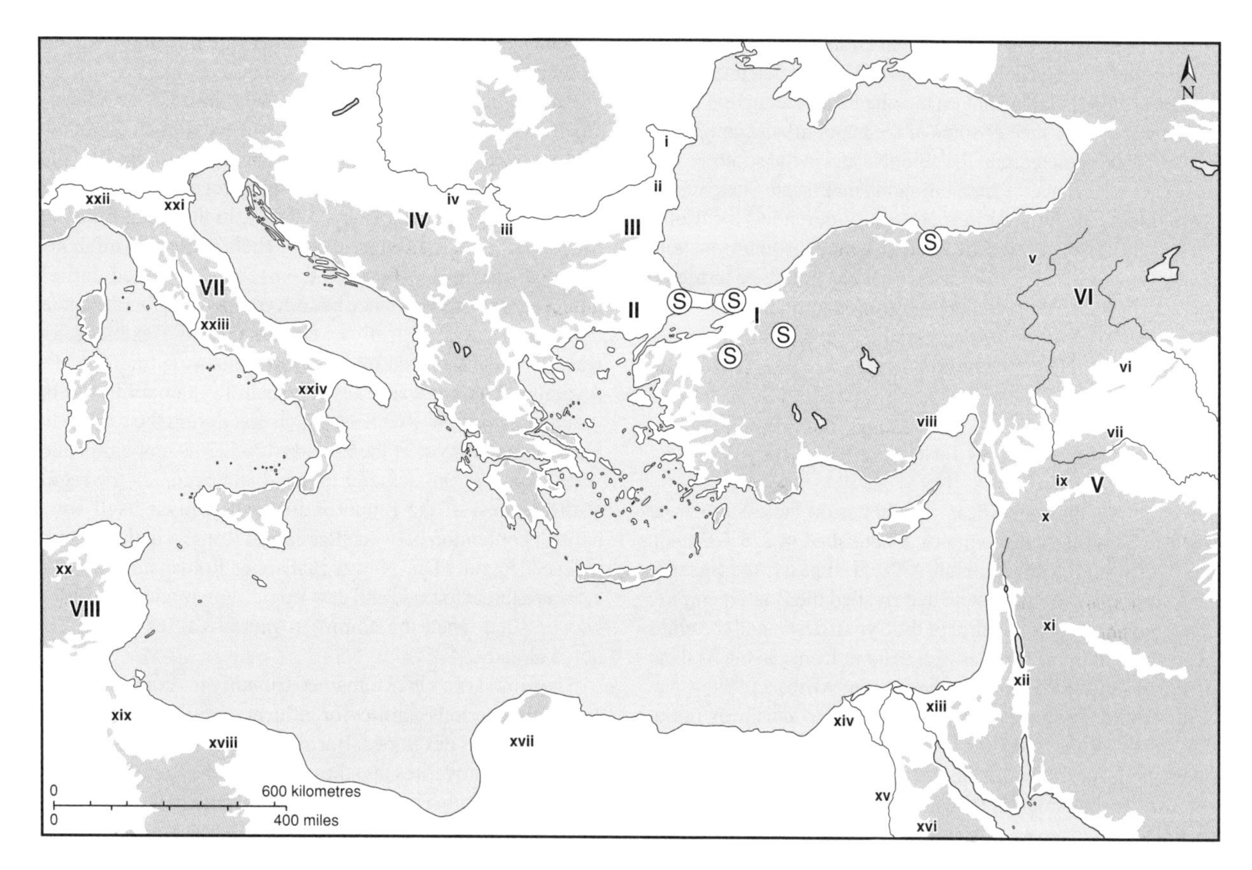

Map 2.5 Defence and strategy: late Roman structures.

Magistri militum

I	*Magister militum praesentalis I*
II	*Magister militum praesentalis II*
III	*Magister militum per Thracias*
IV	*Magister militum per Illyricum*
V	*Magister militum per Orientem*
VI	*Magister militum per Armeniam*
VII	*Magister militum per Italiam (Exarchus Italiae)*
VIII	*Magister militum per Africam (Exarchus Africae)*

Duces

i	Scythia	ix	Syria	xvii	Libya		
ii	Moesia II	x	Phoenice	xviii	Tripolitania		
iii	Dacia	xi	Arabia	xix	Byzacena		
iv	Moesia I	xii	Palaestina	xx	Numidia		
v	Armenia	xiii	Augustamnica	xxi	Ravenna		
vi	Mesopotamia	xiv	Aegyptus	xxii	Liguria		
vii	Osrhoene	xv	Arcadia	xxiii	Roma		
viii	Isauria	xvi	Thebais	xxiv	Neapolis		

Not shown (in West): xxv Mauretania
xxvi Hispania

Ⓢ Units of *scholae palatinae*

the terrain of the region is fairly barren, consisting for the most part of an undulating plain dissected by several shallow dry watercourses and occasional ridges, the strategic importance of Dara was considerable, since it covered a major route into Roman Mesopotamia and beyond into both north Syria or north-westwards into Asia Minor.

Defensive building in the east was characterised by fortresses such as this, and by the maintenance or construction of large numbers of fortlets and defended outposts, linked by a network of military roads, which acted to screen the desert frontier and points of ingress and egress. Under Justinian, and following the recovery of the North African provinces, a similar screen of major fortified cities accompanied by outposts, watchtowers and fortlets was established there, designed to inhibit the depredations of the Berber peoples to the south (or to police their movements within imperial territory). In Italy and the Balkans the pattern was very different. In the Balkans, because of the penetration of the Danube front by Slav groups and other raiders, and in Italy as a reflection first of the long-drawn-out warfare with the Goths and the ensuing fighting with the Lombards, no cohesive linear system was possible. Instead, the government, in the form of the local military and civil authorities, seems to have promoted the development of a dense pattern of small fortified sites which could support both local military defence and defend the interests of agrarian production and local trade. This meant on the one hand a reduction in the importance to the state of some of the major urban centres, but on the other an increase in the numbers and in the strategic and economic relevance of medium- and small-scale sites, which were generally situated in more easily-defended locations, served as local centres of exchange and production, and possibly also fiscal administration. What might be termed a medieval pattern of small, highly-localised fortified centres was beginning to evolve in response to the changed circumstances and military needs of the times.

Imperial Neighbours: the East

The empire's most significant neighbour on the eastern front was the Sasanid Persian empire. Established in 226 CE upon the overthrow of the Parthian Arsacid dynasty, the Sasanid kings created a powerful state that rivalled the Roman empire, challenging Rome for control of the Syrian desert and the cities of northern Iraq and Syria, as well as for influence in the Arabian peninsula and as far afield as the Horn of Africa and Ethiopia. With commercial and political contacts in northern India, the Indian Ocean and China, the Sasanid realm was a major international power with a vibrant culture, dominated politically by Zoroastrianism, but in which Nestorian Christianity in particular came to play a key role after the middle of the fifth century. Relations with Rome throughout the history of the Persian empire were tense – while there were substantial periods without actual fighting, the Romans were always aware of the threat from the east and substantial resources were devoted to holding it in check. By the same token, the Persians likewise spent considerable sums in protecting their own frontier and in challenging Rome for control over the mountainous Armenian principalities and lesser Caucasian states, both because of their strategic significance and because of the wealth of their mineral resources, in particular gold and other precious metals.

The central and southern sectors of the eastern Roman frontier were protected, or at least covered, by the great Syrian desert which stretched down into the Hejaz, and by the Sinai desert. These were by no means impassable, but nevertheless meant that major routes into the empire were few and well-used. The desert was reinforced as a barrier and frontier by the garrisons of the *limes diocletianus*. The semi-nomadic populations of the northern Arabian peninsula occasionally posed a threat as small-scale raiders, but were also a source of mercenary and allied soldiers who could be employed against the Persians; while the commercial centres of the southern coast or the north, such as Medina and Mecca, maintained regular trading contacts between the cities of Syria and Palestine, the Indian Ocean, the east African littoral and the Axumite kingdom of Ethiopia. The frontier with the Arabs remained stable from the fourth to the seventh centuries. A system of allied tribes or clans organised as *foederati*, or federates, under a paramount group (the Tanukhids in the fourth century, the Salihids in the fifth, and the Ghassanids in the sixth and seventh centuries) served to defend Roman interests, and by the time of Justinian the Ghassanid kings were thoroughly integrated into the east Roman system of precedence.

The Arabian peninsula, and especially the organised states of Aden and the Yemen, were also a focus for Roman diplomatic activity, especially in view of their closeness to the Christian state of Ethiopia. The east Roman rulers regarded the kingdom of Axum (named after the capital city, in the north Ethiopian highlands) as a legitimate part of their sphere of influence, although the Axumite rulers themselves remained entirely independent. Christian since its conversion in the fourth century, it depended ecclesiastically on the patriarch of Alexandria, and was heavily influenced by the Syrian monastic tradition. The Axumite kingdom was a key player in Roman politics in the Arabian Peninsula–Red Sea region, and during the wars in the kingdom of Himyar in the period 517–537 its emperor, Kaleb 'Ella 'Asbeha, had actually invaded and occupied the region at the request of the Emperor Justin I. Himyar itself was a bone of contention between Persia and Rome – in the 520s the independent ruler Dhu Nuwas challenged Roman influence and, seen as a threat to trade and east Rome's international position, was crushed when the empire's Axumite allies invaded, as noted already.

The Himyarite kings remained tributary to Axum, but under Justinian, a serious conflict for influence between the Sasanids and the Romans developed. But although the emperor attempted to bring the Himyarites into the conflict on the Roman side, they played for the most part a neutral game until, in the early 570s, a Persian force was invited in by some of the subordinate chiefs, the king was slain in battle, and Persia became the pre-eminent power in the region. Thereafter relations between the Jewish communities and the (monophysite) Christians in Himyar were stabilised until, in the 620s, Mohammed dispatched his initial call to the people of the region to embrace Islam, which quickly became the dominant belief system in the region.

The northern sector of the east Roman frontier in the east was occupied by the Armenians, Georgians and other, minor, Caucasian principalities. Georgia (Greek Iberia, Georgian K'art'li) was divided into two zones, the eastern (Iberia proper) and the western (also called Lazica, later divided into a northern section – Abkhazia – and a southern – Lazica). Converted to Christianity in the fourth century, Georgia remained closely tied to the east Roman and Byzantine worlds, politically and ecclesiastically, breaking with the orthodox (Chalcedonian) tradition at the Council of Dvin in 505, but returning to the east Roman fold in the early seventh century. A source of minerals as well as of soldiers, but in particular a region of strategic significance, west Georgia – Lazica and the coastal region of Suania – became the object of hard-fought campaigns between Roman and Persian armies during Justinian's reign, in a conflict which lasted from 542 until 556. East Georgia – Iberia – was similarly fought over, and occupied by the Persians in 522–523, who installed their own military governor or *marzban*.

Armenia, which had likewise become Christianised during the fourth century, had been divided into three sectors: west of the Euphrates an area under direct Roman control (*Armenia Minor*), to the east the kingdom of Greater Armenia, and in the south the so-called Satrapies. From the 390s this arrangement was altered: the Romans retained control of the western segment

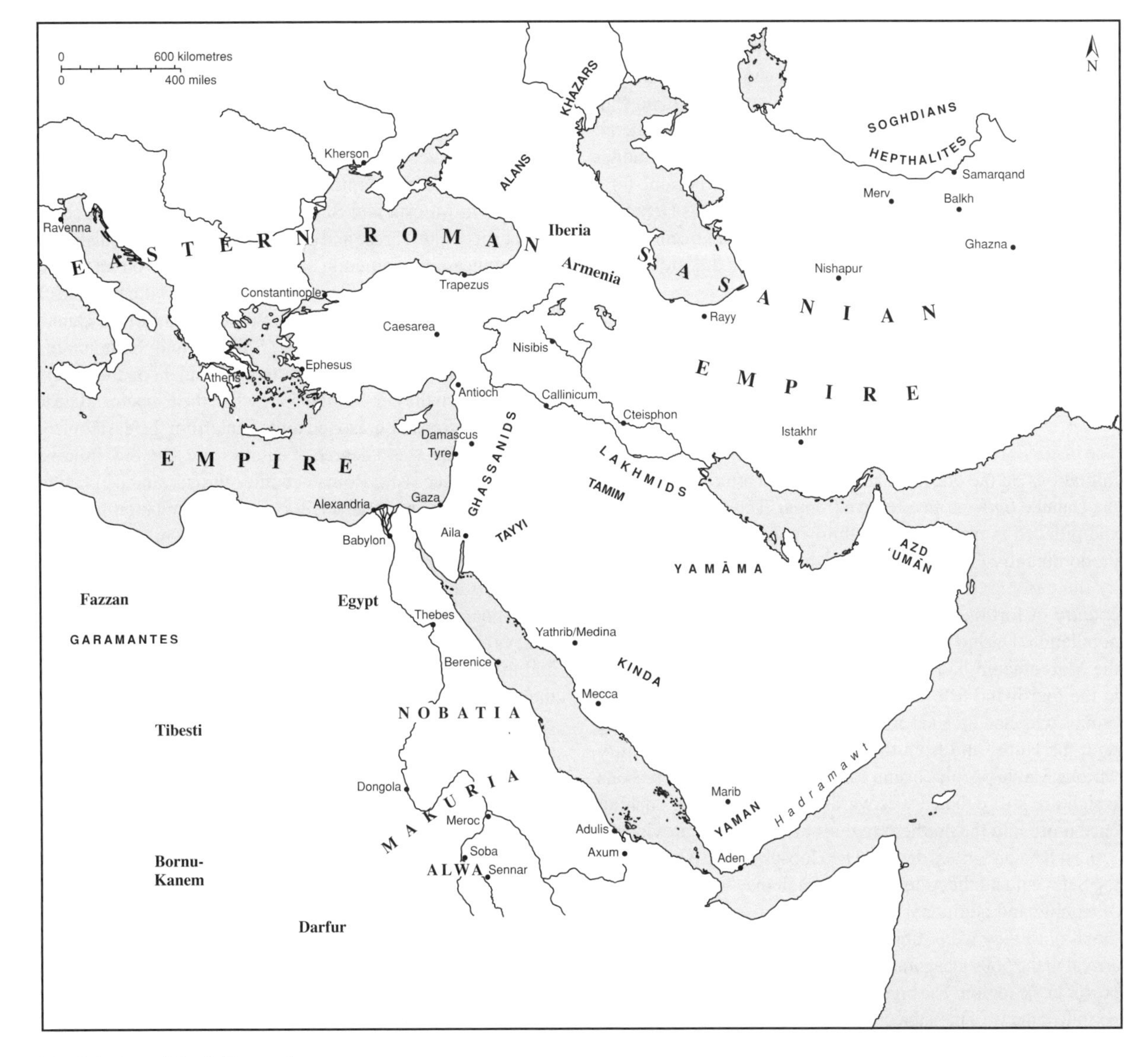

Map 2.6 Imperial neighbours: the east

and were able to extend their control across the westernmost districts of the kingdom, while the remaining regions of Greater Armenia became part of a Persian-dominated zone referred to as Persarmenia, governed by a military governor appointed by the Persian Great King. Under Justinian, the Armenian regions were reorganised and associated with districts separated from adjacent provinces to form four imperial provinces, Armenia I–IV. The successful Roman intervention in the Persian civil war in 590–591 gave the Emperor Maurice the opportunity to negotiate an expansion of Roman control in the region; but Roman attempts to force the Armenian (monophysite) church back into the Constantinopolitan orbit met with resistance at all levels, and made Roman domination extremely unpopular with the majority.

While Maurice had become a close ally of the restored Great King Chosroes II, Maurice's murder in 602 gave the Persian ruler the opportunity to intervene in Roman affairs. Ostensibly intending to restore the dynasty of his former ally, Chosroes' armies invaded the eastern provinces of the empire on a grand scale, harrying imperial provinces in Anatolia and permanently occupying Syria, Palestine and Egypt. Jerusalem fell in 614, Roman Mesopotamia had been conquered by 615, Egypt by 616. A permanent Sasanid administration was set up with military governors for the different provinces. Heraclius' overtures to restore the *status quo ante*, on the grounds that he had defeated and killed Phocas, fell on deaf ears: Khusru aimed at nothing less than the re-establishment of the ancient Achaemenid realm of the days of Darius and Xerxes. In spite of the threat from the Avars in Europe and the trouble with the Lombards in Italy, Heraclius launched a brilliant campaign against the Persians in 622. Refusing to be distracted by the great siege of Constantinople in 626,when Persian forces occupied much of north-western Asia Minor in support of the Avar Khagan, Heraclius was intent on taking the war directly onto Persian territory, where he outmanoeuvred the enemy commanders, sacked key Persian fortresses and cult centres and forced Chosroes to flee his capital at Ctesiphon. With Chosroes' assassination the war came quickly to an end. By 629 the Persian forces had withdrawn from the conquered territories, a new Persian ruler had been put in place under Roman protection, and the great Sasanid empire became for a short while tributary to Constantinople. But even as Heraclius triumphed, unexpected new developments were occurring in the Arabian peninsula which were to transform the late ancient world and its empires for ever.

Imperial Neighbours: Italy, the Slavs, the Balkans and the North

The history of the Balkan region is overshadowed by the great migrations on the one hand and by Roman efforts to maintain the Danube *limes* as an effective frontier. Traversed, occupied and pillaged at times from the third century on by peoples of predominantly Germanic culture, the Balkan landscape had by the early sixth century already been transformed into a country of fortified settlements, military bases and declining population. During the third century Germanic groups such as the Marcomanni had continued to push against the frontier; in the fourth and fifth century it had been the various Gothic groups who had invaded and occupied Roman territory, along with the Huns; and from the beginning of the sixth century various Slavic peoples began to appear, migrating for reasons which are still debated westwards and southwards, pushing once more into the northern regions of the Balkan provinces.

East Roman writers describe two loosely-organised groups, the Sclaveni and the Antes, and although they were the source of trouble and additional expense, they posed no substantial threat until they were subjugated by the Turkic Avars, whose arrival in the 560s inaugurated a period of real decline in Roman power in the region. The break-up of the empire of the Huns had permitted many of the subject peoples or lesser clans to establish an autonomous existence. But in the middle years of the sixth century a new Eurasian empire, formed by the so-called 'Blue Turks' (Kök Türük, mod. Turkish Gök Türk), had extended its power as far west as the Volga. Byzantine–Turk relations were at first cordial, but came to nothing when the Byzantines entered into negotiations with the Avars, the sworn enemies of the Turks, and their former masters. Known by the Chinese term Juan Juan, they had been overthrown earlier in the sixth century and chased west. Meeting a stout opposition from the Franks their westward expansion was halted, although they were drawn, partly through the intervention of Constantinople, into a war between the Germanic Lombards and their neighbours the Gepids, the result of which was the effective disappearance of the Gepids and the decision of the Lombards to move into Italy. The slow process of reconstruction and economic recovery in the peninsula was thus fatally compromised. At the same time the relaxation of pressure on the Avars from their former tributaries the Blue Turks (whose khanate was divided in 582) enabled the Avars to consolidate their hold over central eastern Europe. By the end of the decade the Avar Khagan exercised hegemony over a large swathe of territory focused in the Pannonian plain and stretching east as far as the Crimea and the Don, subjugating the various Turkic groups who made up the residue of the former Hunnic empire, in particular the Kutrigurs, Utigurs and Sabiri.

With the successful use of Slav groups along the Danube, he was able to move into Roman territory along the eastern Danube in Moesia and Scythia. From there Avar horsemen swept south into Thrace and as far as Constantinople, disrupting communications, inflicting substantial damage on an already strained economy, and threatening the imperial capital itself. Several key fortresses along the middle reaches of the Danube fell over the same period, notably Sirmium and Singidunum, a serious blow to the defensive system dependent on the riverine *limes*. Only in the period after 591, when the Emperor Maurice was able to transfer seasoned units back from the eastern front to the Balkans, was a degree of equilibrium restored, followed by several successful Roman counter-thrusts aimed at driving the Avars out of imperial territory and reducing their hold over the Slav immigrant groups, who had meanwhile been able to settle as far south as the Peloponnese in southern Greece.

Unfortunately, Maurice's fiscal policies and his strict disciplinarianism led to a mutiny of the Balkan field army in 602, as a result of which the emperor and his family were slain and the usurper Phocas, a lower-ranking officer, succeeded to the throne. Phocas' reign did not see a collapse of east Roman power in the Balkans, although Maurice's offensive was stopped. But in the brief civil war which followed Heraclius' seizure of power in 610 and the ensuing Persian invasion of the empire's eastern and Anatolian provinces, the empire lost the initiative. Slav immigration continued unchecked, and the Avars were able to reassert their control in the region. Only with the defeat of the great siege of Constantinople in 626 by a combined Avaro-Slav army (launched in conjunction with the Persians on the other side of the Bosphorus) was Avar power reduced as the Khagan's vassals among the Slavs and elsewhere began to challenge his authority. The Kutrigurs and Utigurs joined forces to establish an independent khanate between the lower reaches of the Dniepr and the Don, under the new name of 'Bulgars'.

While Heraclius was able to defeat the Persians in the east by the late 620s and recover all the lost territory in the east, in the Balkans the imperial position was fatally undermined by the flow of Slav immigrants and the establishment of a number of autonomous '*sklaviniai*', ostensibly tributary to Constantinople but effectively independent when no imperial army was present. At some point in the early seventh century, perhaps under Heraclius, two Irano-Slavic groups, the Croats and Serbs, had thrown off Avar control and migrated from north of the Pannonian plain into 'Roman' territory in the north-west and central Balkan region, where they established loose confederacies incorporating the indigenous and migrant populations of the region, again nominally under imperial authority, but effectively quite independent. By the end of Heraclius' reign in 642 the empire could exercise its authority in the inland regions only through military force.

In Italy, the invasion of the Lombards in 568 under their leader Alboin had led to a fairly rapid collapse of imperial

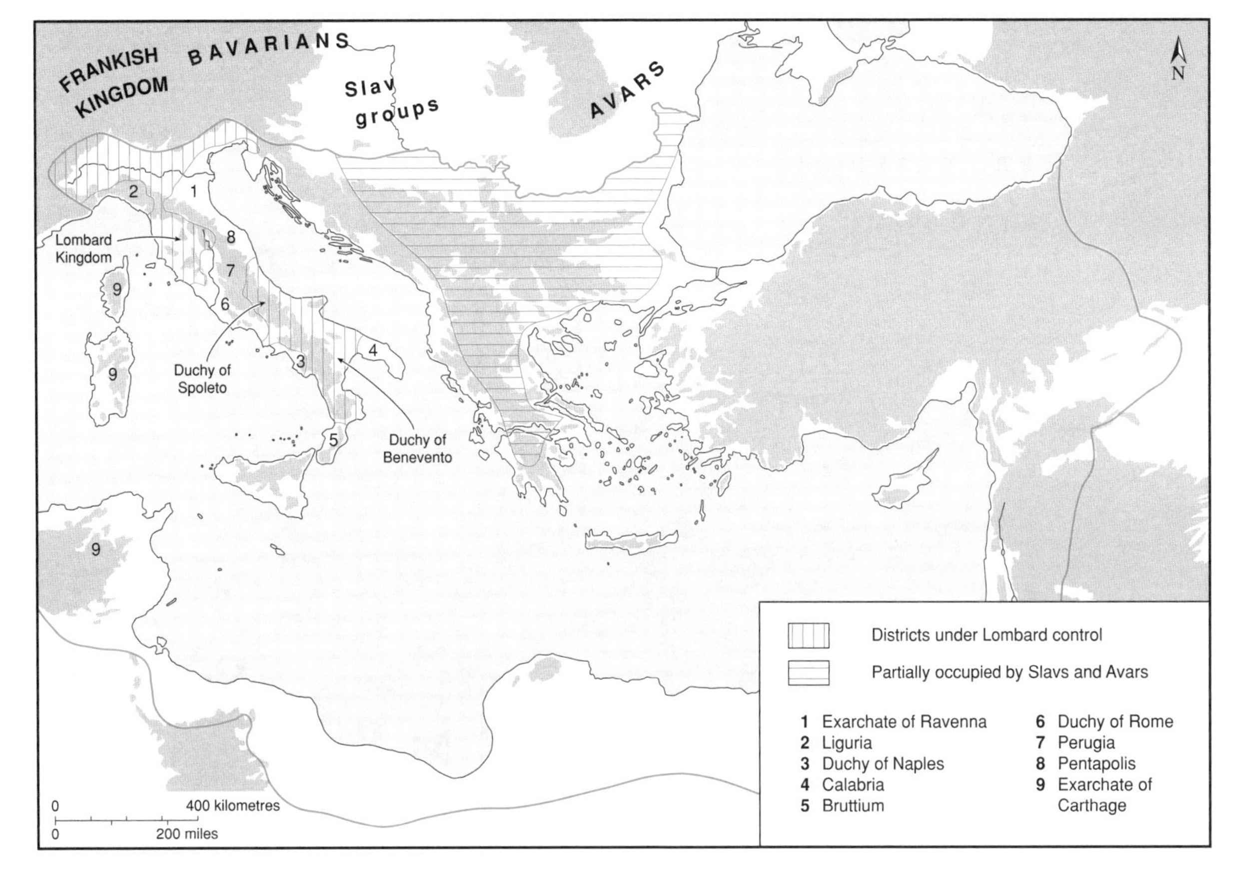

Map 2.7 Imperial neighbours: Italy, the Slavs, the Balkans and the north in 600.

defences and the establishment of a Lombard kingdom in the Po valley, centred on Pavia, and the two principalities of Spoleto and Benevento in the centre and south. It also led to the militarisation of Italian provincial government and the creation of the exarchate based at Ravenna, a military supreme commander responsible for co-ordinating the defence of all the imperial territories in Italy.

The Rise of Islam and the Beginnings of a 'Byzantine' Empire

The origins of Islam lie in the northern Arabian peninsula, where different forms of Christianity and Judaism had competed and co-existed for centuries with indigenous beliefs, in particular in the much-travelled trading and caravan communities of Mecca and Medina – Mohammed was himself a respected and established merchant who had several times accompanied the trade caravans north to Roman Syria. Syria and Palestine already had substantial populations of Arabs, both farmers and herdsmen, as well as mercenary soldiers serving the empire as a buffer against the Persians. Although Mohammed met initially with stiff resistance from his own clan, the Quraysh, who dominated Mecca and its trade (as well as the holy Ka'ba), by 628–629 he had established his authority over most of the peninsula. On his death in 632 there followed a brief period of warfare during which his immediate successors had to fight hard to reassert Islamic authority; and there is little doubt that both religious zeal combined with the desire for glory, booty and new lands motivated the attacks into both the Persian and Roman lands. A combination of incompetence and apathy, disaffected soldiers and inadequate defensive arrangements resulted in a series of disastrous Roman defeats and the loss of Syria, Palestine, Mesopotamia and Egypt within the short span of ten years, so that by 642 the empire was reduced to a rump of its former self. The Persian empire was completely overrun and destroyed. The Arab Islamic empire was born.

The most important loss for the Romans was Egypt. Already during Heraclius' Persian wars Egypt had been lost to the Persians, albeit briefly, with serious results for the empire, since it was from Egypt that the grain for Constantinople and other cities was drawn. It was a rich source of revenue; and along with Syria and the other eastern provinces had provided the bulk of the empire's tax revenue. Constantinople was forced radically to restructure its fiscal apparatus and its priorities, including the way the army was recruited and supported; and the result was, by the later seventh century, an administratively very different state from that which existed a century earlier. We will examine these changes in the following chapters.

The reduced and impoverished east Roman or Byzantine empire now had to contend not only with an aggressive and extremely successful new foe in the east; it had far fewer

Map 2.8 The rise of Islam and the beginnings of a 'Byzantine' empire. (After Kennedy, *Historical Atlas of Islam*.)

resources at its disposal, it had lost effective control in the Balkans, and had no real power in Italy, where the military governor or exarch, based at Ravenna, struggled against increasingly difficult odds to maintain the imperial position. The insistence of the imperial government during the reign of Constans II on enforcing the official Monothelete policy reflected the government's need to maintain imperial authority and the views of those in power that the Romans were being punished for their failure to deal with the divisions within the church. But it also brought the empire into conflict with the papacy and the western church, as well as provoking opposition within the empire, bringing a further degree of political and ideological isolation with it. In Italy the exarchs and the local *duces* in charge of the defence of the various east Roman enclaves fought a long-term war of raid and counter-raid with the Lombards, while the papacy did its best both to support this effort, to encourage the emperors to commit more resources to the struggle (largely without success), and to fight on the diplomatic level to maintain its own position and a degree of equilibrium. In the long term, the balance was slowly tilting against the imperial interest, in particular because ideological conflicts such as monotheletism could only damage the chances for a constructive co-operation between Constantinople and Rome.

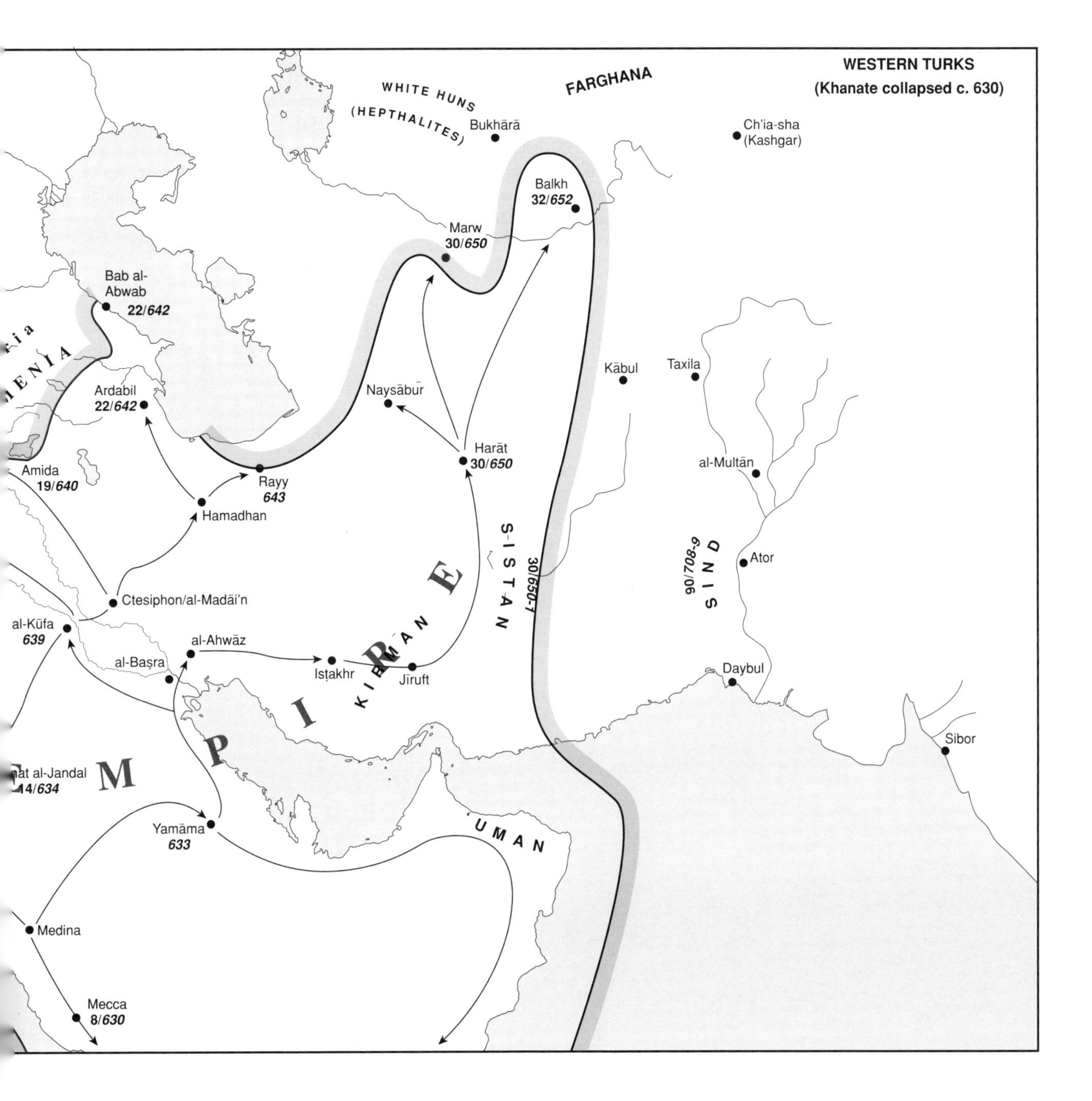

Arab strategy can be followed through several phases – until the defeat of the siege of 717–718 Byzantine resistance was relatively passive, limited to defending fortified centres and avoiding open contact. On the few occasions when imperial troops did mark up successes, this was due to the appointment of particularly able commanders, but was unusual. During the Arab civil wars of the late 680s and early 690s the Emperor Justinian II was able to stabilise the situation for a short while; but it was only during the 720s that the empire was able effectively to begin meeting Arab armies in the field and reasserting imperial military control. In the meantime the Byzantine resistance, focused on fortified key points and a strategy of harassment and avoidance, had at least prevented a permanent Arab presence in Asia Minor, aided of course also by the geography of the region: the Taurus and Anti-Taurus ranges acted as an effective physical barrier, with only a few well-marked passes allowing access and egress; while the climate was in general unsuitable to the sort of economic activity preferred by the invaders.

The Balkan front was also a concern for Constantinople. Technically, the Danube remained the border even in the 660s and 670s, but in practice Constantinople exercised very little real control. In 679 the situation was transformed by the arrival of the Turkic Bulgars, a nomadic confederation made up of the Kutrigur and Utigur Huns and other groups who had been forced out of their homelands and pastures around the Volga by the encroachments of the Khazars from the east. Petitioning the Emperor Constantine IV for permission to seek refuge and protection south of the Danube, on 'Roman' territory (the Danube river itself remained in fact largely under Byzantine control because it was navigable, and the imperial fleet could patrol it), they were refused. They nevertheless did succeed in crossing over, where they were met by an imperial army under Constantine himself. But the imperial army fell into panic (poor discipline, misunderstood signals and a lack of cohesion all contributed), and was defeated by the Bulgars, who over the next 20 years consolidated their hold over the region and established a loose hegemony over the indigenous Slav and other peoples in the region. By 700 the Bulgar Khanate was an important political and military power threatening Byzantine Thrace, and was to remain so for the next three centuries.

3 *Cities, Provinces and Administration*

Imperial Administration

In many respects civil government was effectively the same as fiscal administration: the main function of the administration was the assessment, collection and redistribution of fiscal resources, in whatever form, towards the maintenance of the state. The amount of tax required by the government varied year by year according to the international political situation and according to internal requirements. By the time of Constantine I state finances had come to be controlled and administered through three departments, the praetorian prefectures, the 'sacred largesses' (*sacrae largitiones*) and the 'private finance department' (*res privata*).

The most important was the praetorian prefecture, through which the land-tax assessment was calculated, collected and redistributed. Each prefecture comprised a specific territory, although they were reorganised and redistributed on several occasions: at the beginning of Constantine's reign there were three major prefectures: Oriens (stretching from Moesia and Thrace in the Balkans around to Upper Libya in Africa); Illyricum, Italy and Africa; and the Gauls, including Britain and Tingitana in North Africa. By the 440s these had been rearranged into four prefectures: the Gauls, Italy, with North Africa and parts of Illyricum; and the east (Oriens). The Gallic, Italian and much of the North African prefectures were lost during the middle and later fifth century, leaving Illyricum and Oriens only, but with Justinian's reconquests new prefectures for Italy and for Africa were established. Each prefecture was subdivided into dioceses (*dioecesae*, 'directorates'), under a deputy – *vicarius* – of the praetorian prefect; and each diocese was divided into provinces under provincial governors. The lowest unit of administration was then the city – *civitas* or *polis* – each with its district – *territorium* – upon which the assessment and collection of taxes ultimately devolved.

Taxes were raised in a variety of forms, but the most important regular tax was the land-tax. This could be raised in money, and traditionally had been so; but during the financial crisis which the state suffered in the later third century, and as a result of the restructuring of finances and military arrangements under Diocletian and Constantine, much of it was actually raised in kind – grains, other foodstuffs and so forth – and deposited in a vast network of state warehouses, where it could be drawn on by both soldiers and civil administrators, who received a large portion of their salaries in the form of rations, *annonae*. As the financial situation of the government improved during the fourth and into the fifth century, so these rations could be commuted once again for cash – assuming the producers were able to obtain it – but the government always kept available the option of raising revenues in kind, especially when military requirements demanded it. The prefectures, through their diocesan and more particularly their provincial levels of administration were also responsible for the administration of justice, the maintenance of the public post, the state weapons and arms factories, and provincial public works.

The two remaining finance departments had evolved out of earlier Roman palatine departments, and had more limited functions. The sacred largesses were responsible for bullion from mines, minting coin, state-run clothing workshops, and the issue of military donatives – regular and irregular gifts of coin to the troops for particular occasions such as an imperial birthday, accession celebration and so forth. It had local branches in each diocese, and representatives in the cities and provinces to administer the revenues drawn from civic lands (which it administered after the middle of the fifth century) and from other income, such as the cash for the commutation of military service or the provision of horses for the army. The *res privata*, under its *comes*, was essentially responsible for the income derived as rents from imperial lands, whatever their origin (from confiscation, for example, or by bequest or escheat). It was as complex as that of the *comes sacrarum largitionum*, with different sections responsible for its various tasks. During the sixth century its responsibilities were divided between income destined for state purposes and that employed to maintain the imperial household, and a new department, the *patrimonium*, was established.

During the course of the sixth century the sacred largesses and the private finance department continued to evolve: the various estates administered by the latter were organised into five sections, each independent (including the original *res privata*), responsible for different types of estate and expenditure; while the diocesan level of the activities of the sacred largesses was gradually subsumed by the provincial level of the praetorian prefectures. Under Heraclius, mint production was centralised – mints at Ravenna, Carthage, Alexandria and Constantinople continued to function, while the rest were closed down. In the same period, and over the following 20 or so years, the *sacrae largitiones* disappears as a separate department, while the praetorian prefecture of the east (that of Illyricum disappeared as imperial control over most of the Balkans was lost) was broken up, so that each of its subsections becomes an independent bureau, mostly under its own *logothetês*, or accountant, placed directly under the emperor and a senior officer at court, often the *sakellarios*.

The role of the *sakellarios*, as superintendant of the imperial household finances, is illustrative of the sorts of changes which occurred. His close association with the emperor and the imperial household shows that a process of centralisation was taking place in which the emperors played a much more active managerial role, a reflection of the crisis in the empire's financial and political situation in the years from 640 on.

Cities and Urban Life

The city – *polis* or *civitas* – occupied a central role in the social and economic structure of the empire and in its administration. Cities could be centres of market-exchange, of regional

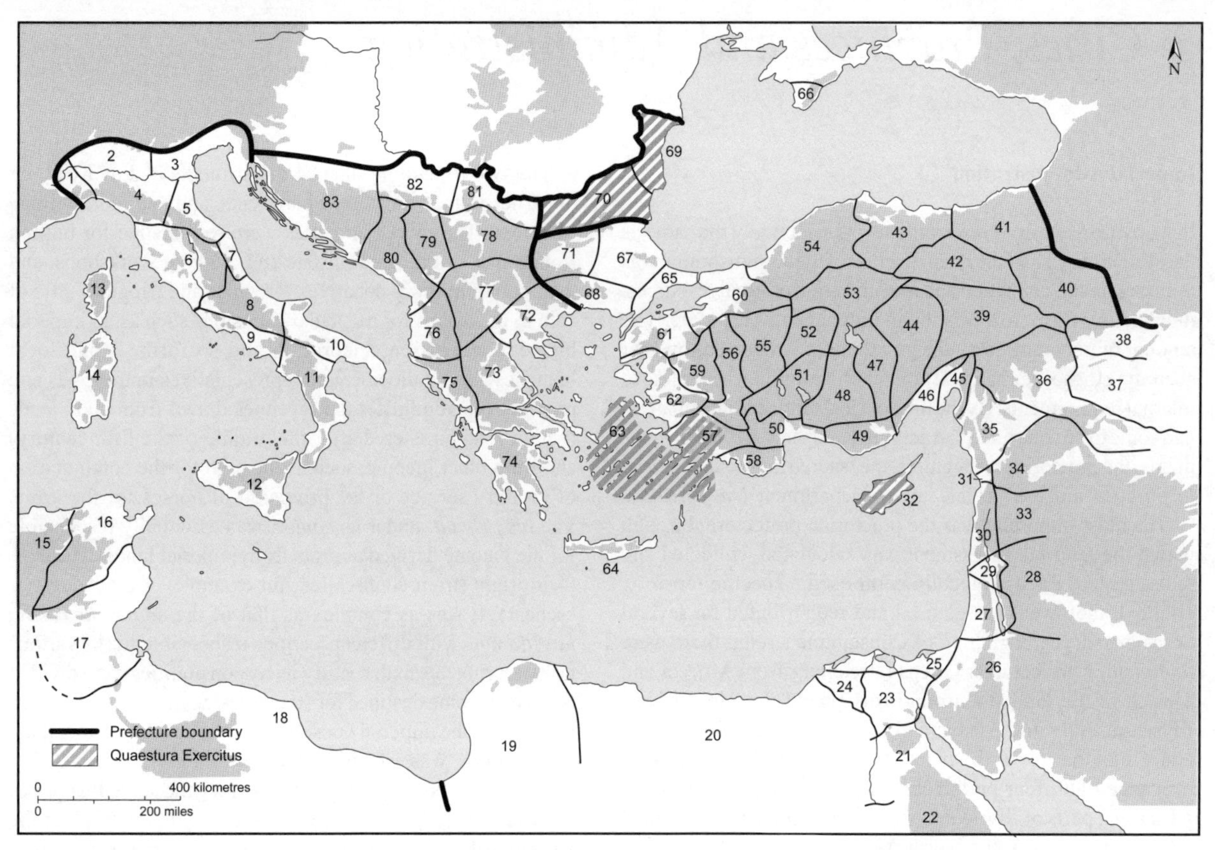

Prefecture of Italy
1 Alpes Cottiae
2 Aemilia
3 Venetia
4 Liguria
5 Flaminia
6 Tuscia et Umbria
7 Picenum
8 Samnium
9 Campania
10 Apulia et Calabria
11 Lucania et Bruttium
12 Sicilia

Prefecture of Africa
13 Corsica
14 Sardinia
15 Numidia
16 Zeugitania
17 Byzacena
18 Tripolitania

Prefecture of Oriens
19 Libya Pentapolis
20 Libya Inferior
21 Arcadia
22 Thebais Inferior
23 Augustamnica II
24 Aegyptus I and II
25 Augustamnica I
26 Palaestina III
27 Palaestina I
28 Arabia
29 Palaestina II
30 Phoenice
31 Theodorias
32 Cyprus (in *quaestura exercitus*)
33 Phoenice Libanensis
34 Syria II
35 Syria I
36 Euphratensis
37 Osrhoene
38 Mesopotamia
39 Armenia III
40 Armenia IV
41 Armenia I
42 Armenia II
43 Helenopontus
44 Cappadocia I
45 Cilicia II
46 Cilicia I
47 Cappadocia II
48 Lycaonia
49 Isauria
50 Pamphylia
51 Pisidia
52 Galatia Salutaris
53 Galatia I
54 Paphlagonia
55 Phrygia Salutaris
56 Phrygia Pacatiana
57 Caria (in *quaestura exercitus*)
58 Lycia
59 Lydia
60 Bithynia
61 Hellespontus
62 Asia
63 Insulae (in *quaestura exercitus*)
64 Creta
65 Europa
66 Bosporus
67 Haemimontus
68 Rhodope
69 Scythia (in *quaestura exercitus*)
70 Moesia II (in *quaestura exercitus*)
71 Thracia

Prefecture of Illyricum
72 Macedonia I
73 Thessalia
74 Achaea
75 Epirus vetus
76 Epirus nova
77 Macedonia II
78 Dacia Mediterranea
79 Dardania
80 Praevalitana
81 Dacia ripensis
82 Moesia I
83 Dalmatia

Map 3.1 Imperial administration: Justinianic prefectures and provinces c. 565. (After Jones, *Later Roman Empire*.)

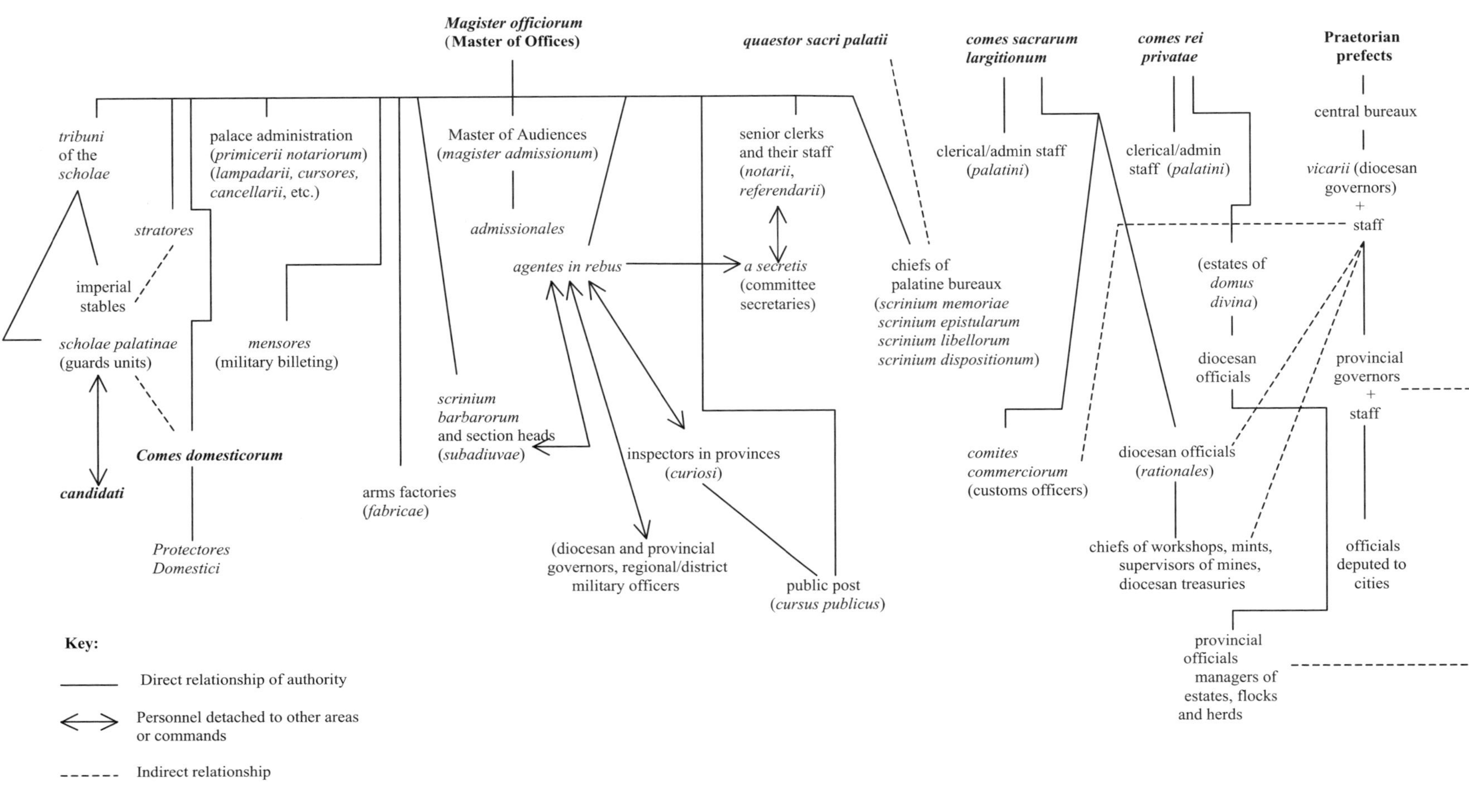

Figure 3.1 The imperial civil and fiscal administration c. 560.

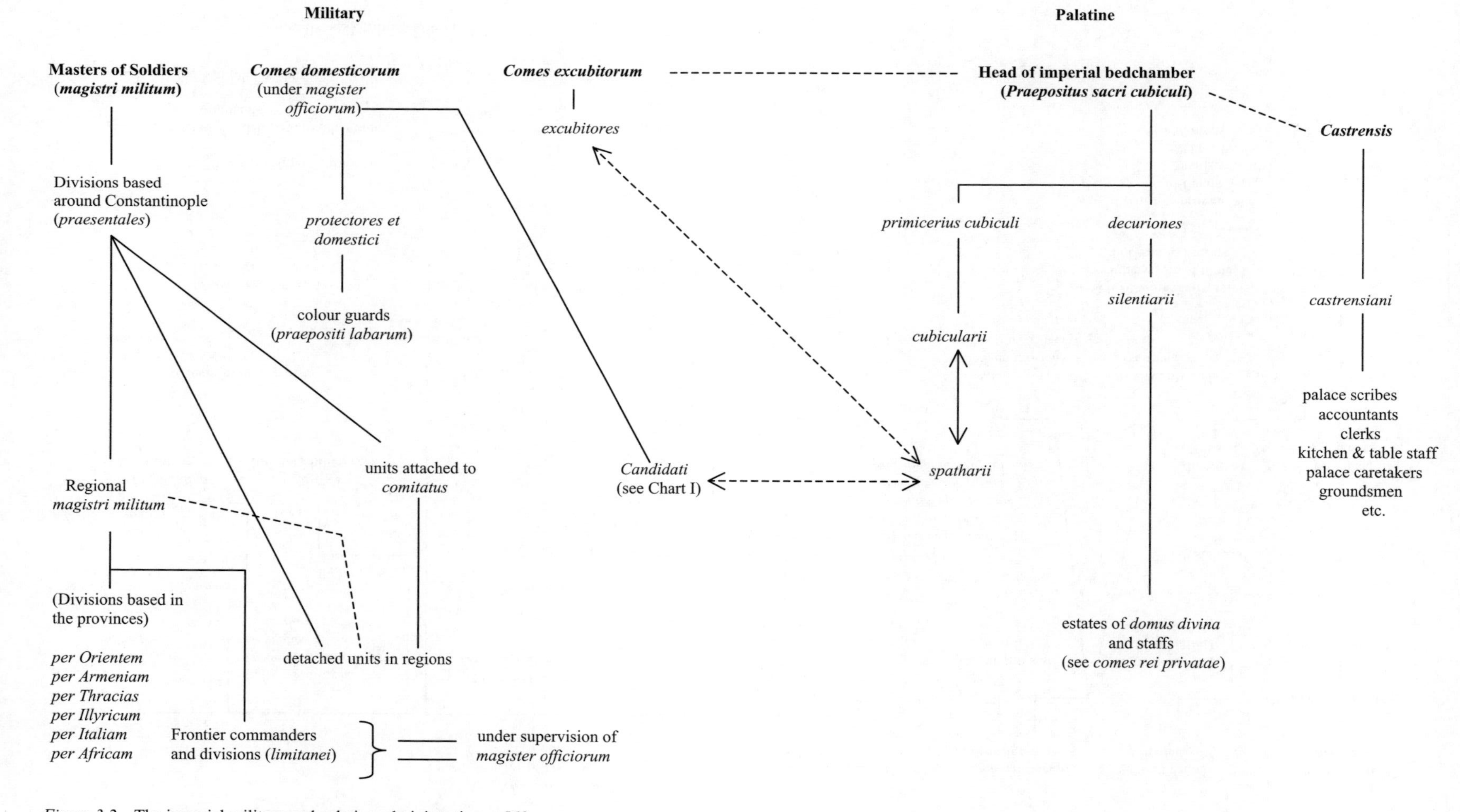

Figure 3.2 The imperial military and palatine administration c. 560.

agricultural activity, occasionally of small-scale commodity production or, where ports were concerned, major foci of long-distance commerce. Some fulfilled all these roles, others remained merely administrative centres created by the state for its own fiscal administrative purposes. Cities were also self-governing districts with their own lands, and were responsible to the government for the return of taxes. Where such cities did not exist, the Roman state created them, sometimes *ab initio*, sometimes transforming pre-existing settlements and endowing them with the institutional structure and corporate identity of a *civitas*. The great majority of towns and cities were dependent on their agrarian hinterlands for their local market and industrial functions and for the foodstuffs from which the urban populace lived.

The role of cities changed as the empire evolved, so that over the period from the third and fourth centuries the appearance of the classical Roman city also changed. The changes themselves were complex and reflected the growing tensions between state, cities and private landowners over the extraction of surpluses from the producing population. Cities failed to weather the contradictions between their municipal independence and the demands of the state. The vested interests of the wealthier civic landowners further complicated the picture. Although there is evidence well into the first half of the seventh century in the east that many *curiales* – the city councillors, members of the *curia* – continued to honour their obligations to both state and city, it is clear by the later fourth century that many did or could not. These *curiales*, who were the chief landowners and leading citizens, had been responsible both for the upkeep of their cities by voluntary subscription, and for the local assessment, collection and forwarding of the revenues demanded by the state. But as many were able to obtain senatorial status (in other words, they became members of the *curia* of either Rome or Constantinople), which freed them from such duties, so the burden fell more and more upon the less wealthy and privileged, who were in consequence less able to extract all the revenues demanded (especially as tax evasion among the wealthy, through bribery as well as physical resistance, was endemic).

The process was both very complex and regionally nuanced, according to traditional pattern of landholding and urbanism, but these were the most obvious features. As a result, and over the period from the later fourth to the later fifth century (in the west until the empire disappears as well as in the east), the government intervened more and more directly to ensure the extraction of its revenues. This it did both by appointing supervisors imposed upon the city administrators, as well as through the confiscation of city lands, the rents from which were now the guarantee that the state's fiscal income was at least to some extent assured, and eventually through the appointment of tax-farmers for each municipal district. The *curiales* seem still to have done the actual work of collecting, but the burden of fiscal accountability seems to have been removed during the reign of Anastasius (491–518). While this certainly relieved the pressure, and possibly helped promote the brief renaissance in urban fortunes that took place in some eastern provinces in the sixth century, it did nothing to re-establish the traditional independence of cities.

By the early years of the seventh century cities as corporate bodies were less well-off than they had been before about the middle of the sixth century. This is not because urban life declined in any absolute sense, or that there was less wealth available to the local élite and landowning class, or even that cities no longer fulfilled their role as centres of exchange and production. There was as much wealth circulating in urban environments as before, but towns as institutional bodies now had only very limited access to it, and the archaeological data shows a reduction in the area of many cities. Most of their lands and the income from those lands had already been taken from them. From at least the second half of the sixth century the local wealthy tended in addition to invest their wealth in religious building or related objects – thus it was changing patterns of investment as much as a decline in investment which affected urban building and maintenance of structures. In particular, the church was from the fourth century a competitor with the city for the consumption of resources.

The needs of warfare and the pressures imposed by threats from enemies also played a role in the changing face of the city – by the fifth century most urban centres had walls and towers for defence, which further transformed their character and the ways in which they could spend their resources. In the Balkans during the fourth to sixth centuries and in Italy from the sixth century the changing relationship between different types of settlement and the functions they fulfilled in relation to the military needs of the state and to their districts stimulated the development of new types of fortified urban centre (see Map 2.5). The state also played an important role because the government followed a policy of 'rationalising' patterns of distribution of cities for its own administrative ends. Towns in over-densely occupied regions were sometimes deprived of the status and privileges of city (many of the 'cities' which were suppressed in this process had been little more than villages), while others, which were of importance to the state in its fiscal-administrative structure, received city status for the first time. This was entirely associated with the need to maintain a network of centres adequate to the demands of the fiscal system. The ideological and symbolic importance of cities and urban culture in the Roman world, expressed through imperial involvement in urban building and renewal throughout the period, meant that they continued to play an important role culturally. In addition, cities particularly associated with a local saint's cult or fulfilling some other cult function within the Christian perspective enhanced their chances of flourishing where they did not already possess a primary economic character. The combination of all these tendencies – military, administrative and cultural – generated a very different urban landscape from that which had prevailed at an earlier period. Yet whatever changes they underwent in respect of buildings and urban planning, cities and towns in the eastern provinces – in Syria and Palestine especially – continued to flourish right through the seventh century and the Islamic conquests, presenting thereafter a very different picture indeed from the towns remaining within imperial territory.

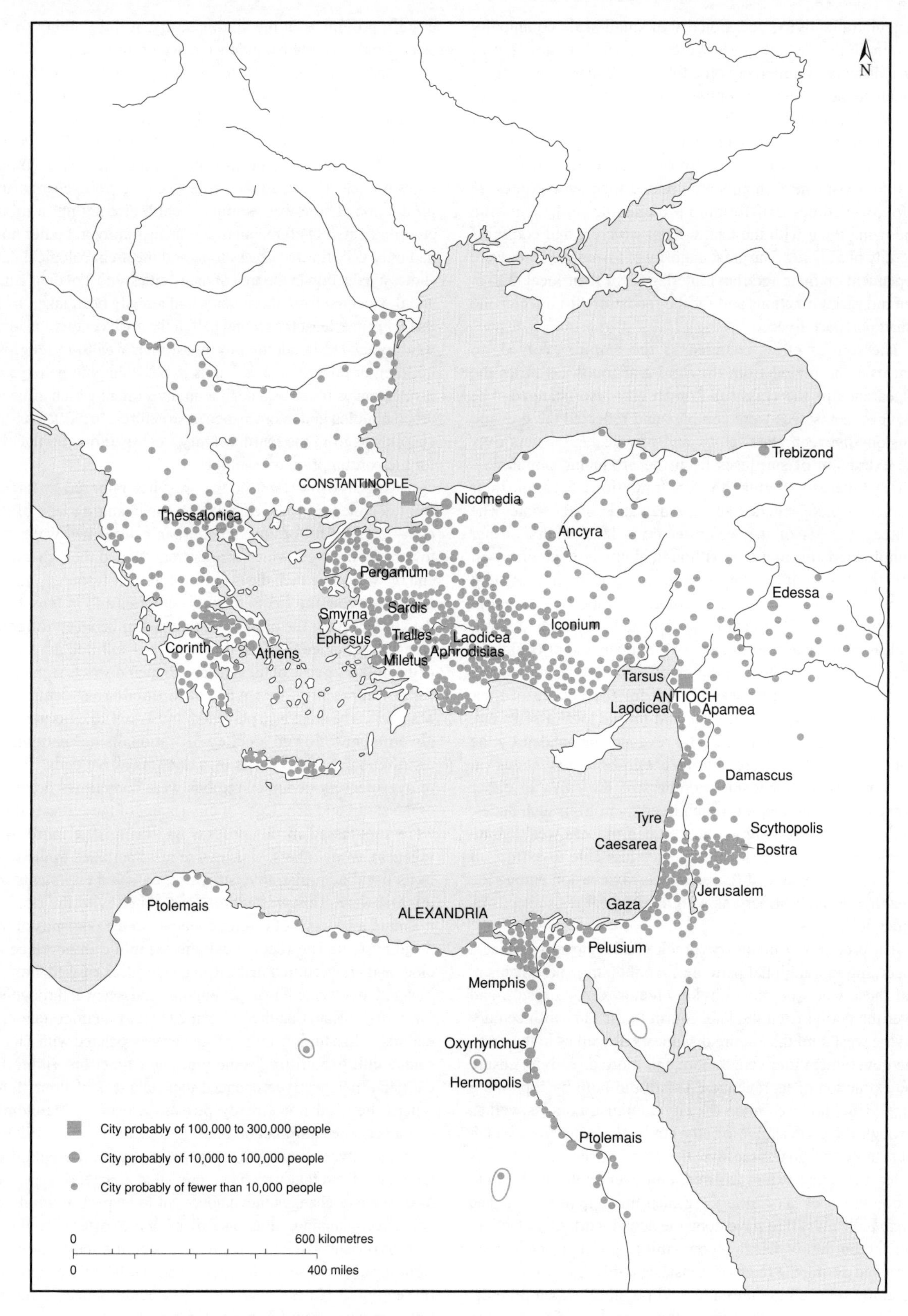

Map 3.2 Cities of the eastern Roman empire in the 5th century. (After Jones, *Later Roman Empire*.)

Constantinople: Evolution of an Imperial Capital

The establishment by the Emperor Constantine I on the site of the ancient city of Byzantion of a new imperial capital in the year 324, and its formal consecration in 330, had far-reaching consequences for the pattern of exchange and movement of goods in the Aegean and east Mediterranean basin, as well as for the politics of the late Roman world. With the imperial court, a senate, and all the social, economic and administrative consequences of a major city, Constantinople soon grew to be the dominant city of the eastern Mediterranean region, rivalling Alexandria and Antioch in wealth, prestige, population and cultural influence. But the foundation of the new capital was not the radical break with tradition it is sometimes suggested: Diocletian had some years previously established his own residence and court at Nicomedia, not far to the south; while the establishment of regional locations for the emperors was a reflection of the needs of the tetrarchy. Constantine's choice was probably based on strategic considerations, since his new capital was located where two major land routes met, both of strategic value: the *via Egnatia*, which crossed via Thessalonica to the Adriatic coast, and the military road from Chalcedon, opposite Constantinople, via Nicomedia to the east.

That the founder of the new imperial city envisaged a substantial population is evident from the fact of his arranging an annual grain supply from Egypt amounting to some 80,000 rations. Rapid growth certainly followed, with greatly expanded water-supply and accompanying structures (aqueducts, cisterns and so forth), grain-storage facilities, and residential areas. The pipes, channels and aqueducts bringing the city's water reached over 80 miles out into the Thracian hinterland, and have still not been fully traced. The imperial headquarters was established with the construction of a palace complex, placed in the south-eastern corner of the original city, accompanied by a substantial hippodrome and a new city wall encompassing an expanded urban area. The major thoroughfare began at the palace in a colonnaded route constructed under the Emperor Septimius Severus (who rebuilt parts of the city after the destruction which occurred in the civil war of 195–196) and led through the circular Forum of Constantine across the city to the Golden Gate, a triumphal entry to the city in the southern section of Constantine's new land wall. Successive emperors then embellished the city with their own monuments, including, for example, stoas, colonnaded streets, baths and other public amenities. In the period from the fourth to the seventh centuries some 40 public bath-houses were built, supplied by a series of vast cisterns, mostly open air constructions. The cistern of Aetius was among the largest and could hold some 160,000 m^3 of water. By the same token the number of imperial and private mansions increased, so that by the early fifth century there were at least five imperial palaces of varying size and function, while the great palace itself continued to be added to and grew into an immensely complex labyrinth of buildings.

In later years the city was famed for its churches, although it seems that Constantine built only three (St Irene, which functioned as the city's cathedral church, and the two churches dedicated to local saints, of St Acacius and St Mocius). But by the 420s there were some 14 churches, and the numbers increased in the following century. Just as they added to the secular ornamentation of the city, later emperors added to this number, and the most famous was built by Justinian in the mid-sixth century, the church of the Holy Wisdom, or Hagia Sophia (on the site of an earlier church of the same name destroyed in rioting).

The defensive walls destroyed by Septimius Severus were rebuilt during the later third century. Constantine began a new circuit further to the west enclosing an area twice as big again as the original city. Completed under his successor Constantius II, the absence of any substantial threat from the sea meant that no sea defences were constructed. The Gothic threat in the 370s and afterwards, the increasing exposure of the city to raids from beyond Thrace, and the rapid expansion of the city population and the needs of the imperial government changed this situation, and during the reign of Theodosius II the prefect Anthemius enclosed more land within the city and built the land walls which can be seen today, a massive three-level system with a moat, stretching for some 6 km from the Sea of Marmara to the Golden Horn. While the land-walls were begun in 412–413, the seaward defences were not begun until the late 430s, but proved their worth in subsequent centuries.

The rapid expansion of Constantinople ground to a halt in the period from the mid-seventh to late eighth century as the empire lived through its centuries of crisis. But from the early ninth century on it began once more to expand both in terms of population and in respect of building activity. In the 530s the total population may have been as many as 500,000 (some estimates are even higher); by the middle of the eighth century, following a major plague in the 540s and endemic pestilence throughout the period up to the 750s, culminating in another major plague in the later 740s, the population may have been reduced to a low of as few as 30,000–40,000 (although all these figures are contentious). Thereafter it gradually rose again, until in the later eleventh and twelfth centuries it may have reached the levels of the early sixth century. The city saw several sieges – successful resistance to the Avaro-Slav siege of 626, the Arab sieges of 674–678 and 717–718, the Bulgar attacks of the early tenth century and attacks from Russian sea-raiders in the tenth and eleventh centuries proved the effectiveness of its defences. In 1204 the city fell by treachery to the forces of the Fourth Crusade, ostensibly en route to attack Islamic Egypt; and the sack that followed witnessed the removal or destruction of great numbers of monuments, as well as the burning of buildings and other forms of destruction which accompany such events. Recovered by the Byzantines in 1261, it remained in imperial hands as the empire shrank to the city and its immediate hinterland in the later fourteenth and fifteenth centuries, to fall in May 1453 to the Ottoman army under Mehmet II after a siege of over two months.

Styles and fashion in building affected Constantinople as they affected any other built environment, and given the length of the city's imperial history it is not surprising that a number of shifts can be seen in this respect. Most obviously, the secular aspect of imperial building diminished as emperors and members of the imperial family and court invested their wealth in churches, palaces and philanthropic establishments, many supported by generous endowments in land and property. Basil

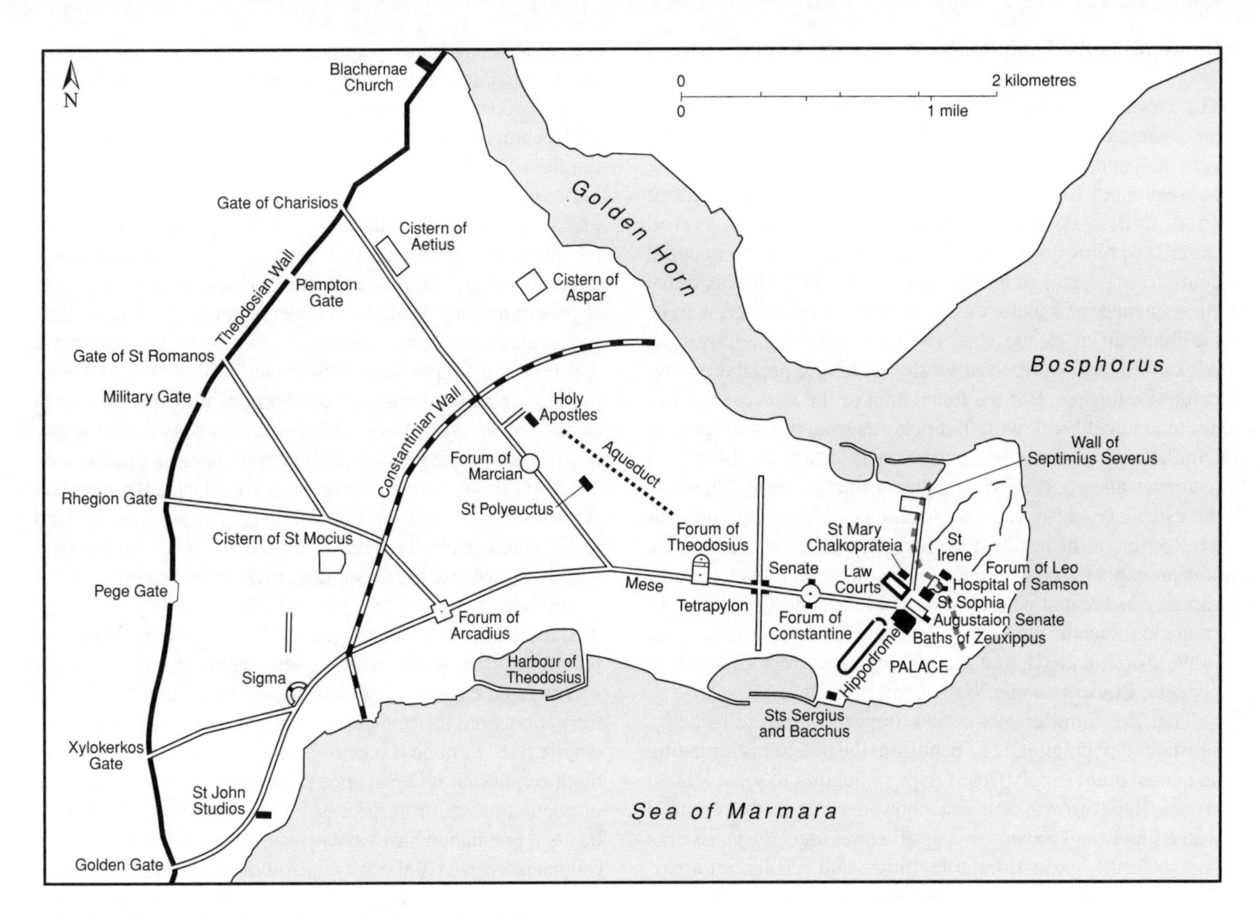

Map 3.3 Constantinople: evolution of an imperial capital.

I is supposed to have rebuilt or restored over 30 churches in the period 867–886. During the eleventh and twelfth centuries many members of the imperial élite donated funds for the construction of monasteries or philanthropic establishments in the city, some of them vast, such as the great Orphanage of St Paul built and endowed by the emperors and reportedly requiring a whole day to inspect. But as well as such buildings the city also contained residential quarters, mansions for the wealthy, a vast range of trades and crafts and the accommodation required to produce and sell their wares, covered and open-air marketplaces, as well as imperial armaments workshops, docks and harbours for both military and civil purposes. From its nadir in the eighth century, when only one commercial harbour seems to have functioned, in the twelfth century the city could boast some four harbours and a larger number of quays along the shores of the Golden Horn for the various merchant communities who had commercial rights in the city, and in its later years was credited with 365 churches, an exaggeration, certainly, but indicative of its image and reputation.

The city was the site of imperial ceremonial events throughout the year, and many were staged as city-wide events. The prefect of the city had the authority to order individual households along any ceremonial route to contribute by hanging out precious fabrics and tapestries, for example, and the streets would be perfumed and decorated to honour the emperors. Ceremonial processions were generally of a religious character and marked key festivals of the church; but military parades, triumphs and the processing of captives and booty were also common. There were several favoured routes, but the most important was that which led to or from the imperial palace and the church of the Holy Wisdom, along the Mese or a parallel major street, to the Golden Gate.

The Evolution of Late Roman Cities

The transformations in urban life that took place over the three centuries or so of the late Roman period are well-attested in the archaeological record. This shows an almost universal tendency for cities to lose by neglect many of the features familiar from their classical structure. Major public buildings fall into disrepair, systems of water-supply are often abandoned (suggesting a drop in population), rubbish is dumped in abandoned buildings, major thoroughfares are built on, and so on. The undoubted decline in the maintenance of public structures or amenities – baths, aqueducts, drains, street-surfaces, walls – is suggestive of a major shift in the modes of urban living: of finance and administration in particular. As we have seen (see pages 36–37) the causes were many and complex, but it is clear that what used to be thought of as a 'medieval' or even 'middle eastern' street plan and arrangement of public and private space was already beginning to appear in the towns of the Roman world long before the conquests of Islam or the end of the late ancient period during the seventh century.

Scores of excavated examples can illustrate these transformations. A good example is the town of Apamaea in the province of Syria II. The city was struck by serious earthquakes in the second half of the fifth century, and again in 526 and 528, and was then besieged and sacked by the Persians in 573. But the archaeological evidence illustrates a more gradual process of increasing impoverishment, alleviated on occasion by imperial and some private largesse. After 573, for example, the *agora* was abandoned, and the scale of rebuilding was much more limited than after earlier disasters. A major transformation occurred between about 625 and the middle of the century, marked by a functional subdivision of the larger town houses and a ruralisation of the activities carried on in the city. Antioch on the Orontes underwent similar changes, exacerbated by a Persian sack in 540, and many other towns show the sorts of transformations in function described above.

Like one or two other cities which suffered a similar fate, however, the history of Apamaea may also reflect local and regional economic change as much as any general tendency in urbanism; for it is also clear that while they may have undergone considerable change in internal structure, use of space, architectural style and street plan, many more cities continued to flourish, to be the site of intense commercial and industrial activity, and to support substantial ecclesiastical and government administrative activities. The changes in the way the state operated, the shifts in the relationship between civic élites and the government and imperial establishment, the continuing process of Christianisation, and changing lifestyles and patterns of investment among the social élite all contributed to changes in the way towns worked socially and economically, and this also had a direct impact on the ways in which space was used in cities. One version of such change in respect of street plan, for example, has been presented in a 'model' form which, while it does not do justice to the very considerable regional variations across the different provinces of the empire, nor to the different timescale which this differentiation implies, nevertheless illustrates the process whereby the classical street plan was changed to suit changing patterns of social and economic activity and identity from late Roman into early medieval times, with streets encroached upon by shops and artisans' workshops, being also built upon and divided, tending towards the creation out of the regular Roman street plan of a much more complex, sectional arrangement (Figure 3.4).

One of the most obvious changes in the internal structure of cities was the abandonment of pagan temples and the frequent reuse of the building materials for the construction of churches or related Christian structures. This was often done very carefully, even down to the numbering of the blocks of masonry as they were taken down. The fifth and sixth centuries also saw a vast amount of church building across the eastern provinces of the empire – small towns like Anemurium (on the southern coast of Asia Minor) were endowed with some nine churches within and without the walls in a period of less than 200 years, for example. Such activity reflected both new patterns of élite investment and, just as importantly, the increasing importance in the cities of the local bishop and the church, a major corporate landowner. The pattern of change varies from city to city. Baths which had fallen into disrepair during the third or fourth centuries were sometimes repaired and brought back into use. Theatres often remained in use, although employed for functions approved by the church. Gymnasia and related structures such as *stadia* were sometimes retained, but just as often were built in or over and turned into artisanal or residential structures, occasionally also being rebuilt to include small churches. Large public spaces such as *agorai* were frequently built on – by churches (which in the context of Christianity came to fulfil several of the community functions of an *agora* anyway) or by shops, workshops, pottery kilns or houses, sometimes as well by rubbish dumps. At the same time, extensive private residences, often with substantial associated outbuildings, monumental entrance porticoes, internal courtyards, dining halls and administrative spaces, and also with associated churches, continued to be built well into the sixth century, for senior ecclesiastical or government officials as well as private persons, particularly in the suburbs. But the structure and plan changed – the traditional Roman peristyle house begins to be replaced by buildings with more than a single level, and with some of the key reception rooms on the second storey.

Many of these trends are common to the whole Roman world, east and west. Substantial wealth was invested in many major and large numbers of smaller towns until the later sixth century and in some cases, especially in Syria and neighbouring regions, beyond, but the form of that investment reflected new social, administrative and cultural priorities which gave to cities a very different physical appearance from their classical Roman forebears. Some of the key differences were already clearly marked out in the form of the few new cities which were constructed, all at imperial command, during the sixth century. Their characteristics have been summarised as small, fortified, imperial and Christian. Many older provincial cities, where they played a role in imperial civil or military structures, also changed to conform to this pattern – from the later fourth and fifth century in the Balkans, somewhat later in less exposed parts of the eastern empire. Their evolution in Asia Minor into the typical middle Byzantine *kastron* is not difficult to follow. But the path which urban development would take thereafter is determined also by the political histories of the areas in question: while they share a common late Roman heritage in respect of the developments already described, the fate of towns in territories remaining to the empire after the middle of the seventh century was very different from that of the towns and cities which were in Islamic territory, for example, a reflection of the beleaguered and impoverished situation of the eastern Roman or Byzantine empire in the seventh and eighth centuries.

Economic life: Production, Trade and Circulation of Goods

There were in several respects two 'economies' in the Roman world. On the one hand, there was the economy of the state and government, representing a fairly straightforward relationship between producers, taxation, and the redistribution of the

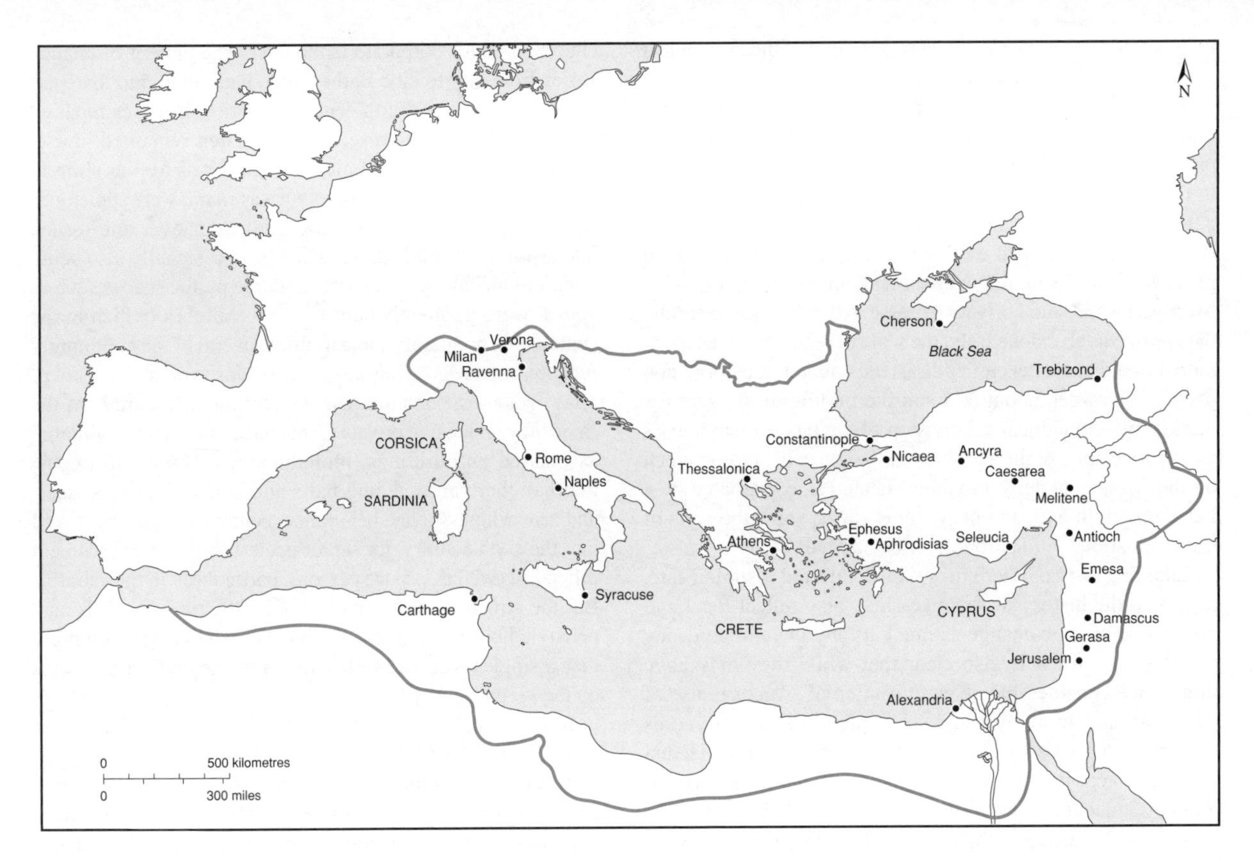

Map 3.4 Major cities of the 6th century.

resources thus extracted to the army, court, government and civil administration. The coinage issued by the government through its mints facilitated the smooth operation of this system. On the other hand, there was the 'economy' of ordinary society, a vast web of intersecting pools of activity and resources tying town to countryside and vice versa, pools of activity that sucked in the state's coinage and circulated it around entirely different patterns of transaction and exchange from those of the state's fiscal system. Both were inter-related, yet both were in several respects autonomous, operating on different principles and responding to substantial shifts in context and circumstance in very different ways.

Long-distance trade displayed particular patterns that reflected the prosperity of the regions where certain goods were produced. Oil and grain were exported, along with the pottery in which they were transported, from North Africa throughout the Mediterranean world; oil and wine travelled from Syria to the western Aegean and south Balkan coastlands, whence they were re-exported inland or further west – to Italy and southern Gaul. The distribution of ceramic types provides much information about these movements: many types of produce were carried in pottery containers of one sort or another – amphorae, for example, often very large vessels, were used for transporting and storing liquids such as wine and oil, as well as solids such as grains, the shape varying according to need. They were moved around chiefly by sea, and were often accompanied by other ceramic products, including fine tablewares, which were exported alongside the bulk goods. Finds of such pottery, in conjunction with knowledge of their centres of production, offer a fairly detailed picture of such trade in this period. Some of these goods travelled considerable distances – in the later sixth century ships were still sailing from Egypt into the Atlantic and around to south-west England, trading corn for tin. This Mediterranean trade did not reflect private, market-led demand alone: much of the commerce in fine wares travelled in the ships of the great grain convoys from North Africa to Italy and from Egypt to Constantinople, the captains of the ships being permitted to carry a certain quantity of goods on their own account for private sale in return for a percentage of the price obtained.

A marked difference existed between those centres of population and production with access to the sea, and inland towns and villages. The economy of the late Roman world was intensely local and regionalised, and this was reflected also in the attitudes and outlook of most of the population: the expense of transport, especially by land, meant that patterns of settlement and the demographic structure of the empire conformed to the limits of the resources that were locally available. And here, it was chiefly the major urban centres that were involved.

Pottery is crucial in revealing patterns of production, exchange and consumption in the late Roman world. Indicative of the strongly market-orientated nature of estate and smallholder production in North Africa, for example, is the fact that until the late fifth and early sixth century imports from this region were strongly represented throughout the eastern Mediterranean and Aegean regions. Commerce and exchange seem in no way

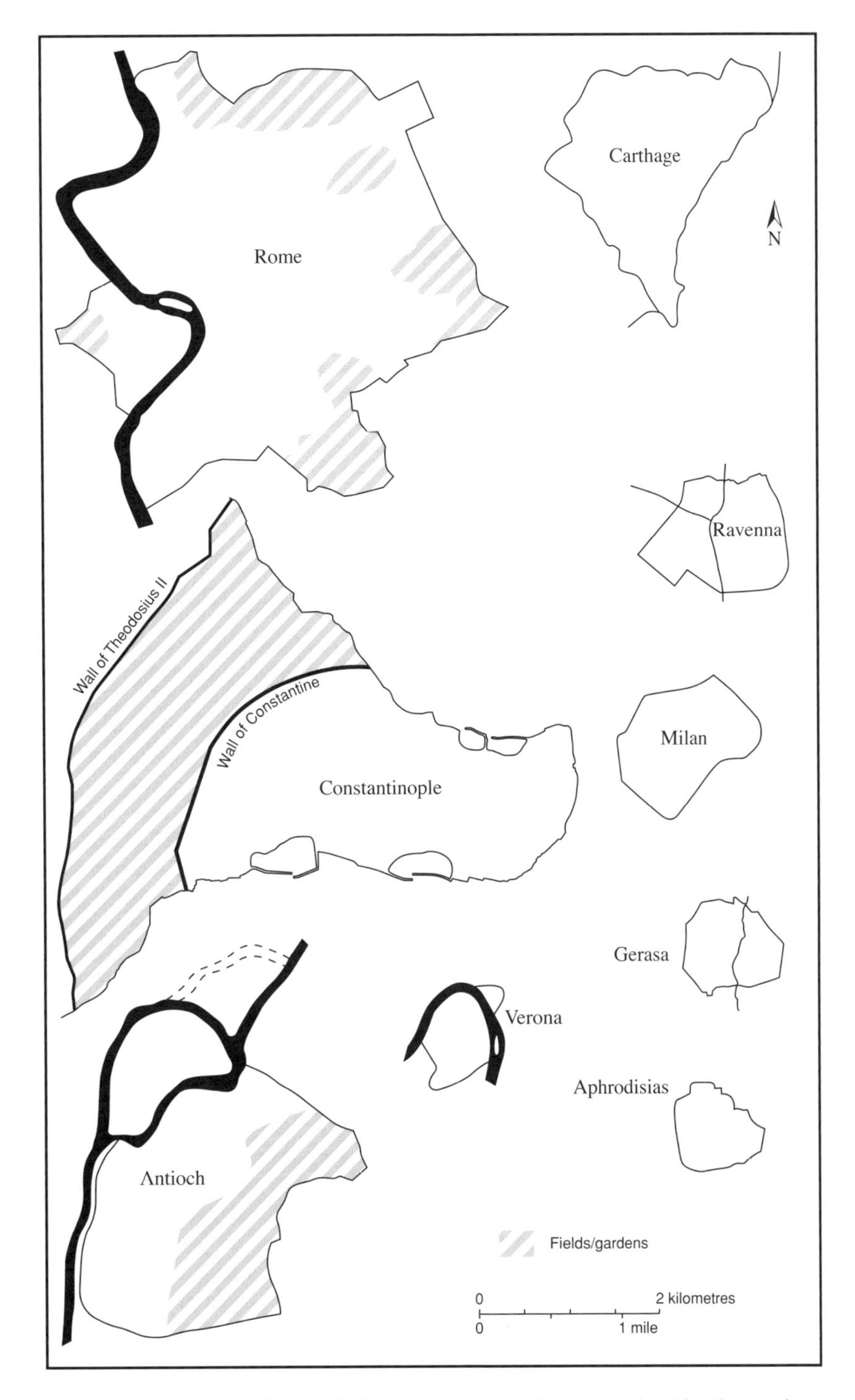

Figure 3.3 Comparative size of walled towns in the eastern Roman empire, 5th–7th centuries.

to have been hindered by the political boundaries of the period. From the later fifth century, there was a reduction in regional North African ceramic production, a reduction in the variety and sometimes the quality of forms and types, especially of amphorae, and a corresponding increase of eastern exports to the west. African imports to the east Mediterranean, both fine and coarse wares, decline sharply from about 480–490 on, to achieve a limited recovery after Justinian's destruction of the Vandal kingdom in the 530s. The incidence of wares produced in the Aegean region and connected with the development of Constantinople as an imperial centre during the fourth century increases in proportion as that of African wares decreases; while over the same period the importance of imported fine wares from the Middle East, especially Syria and Cilicia, also

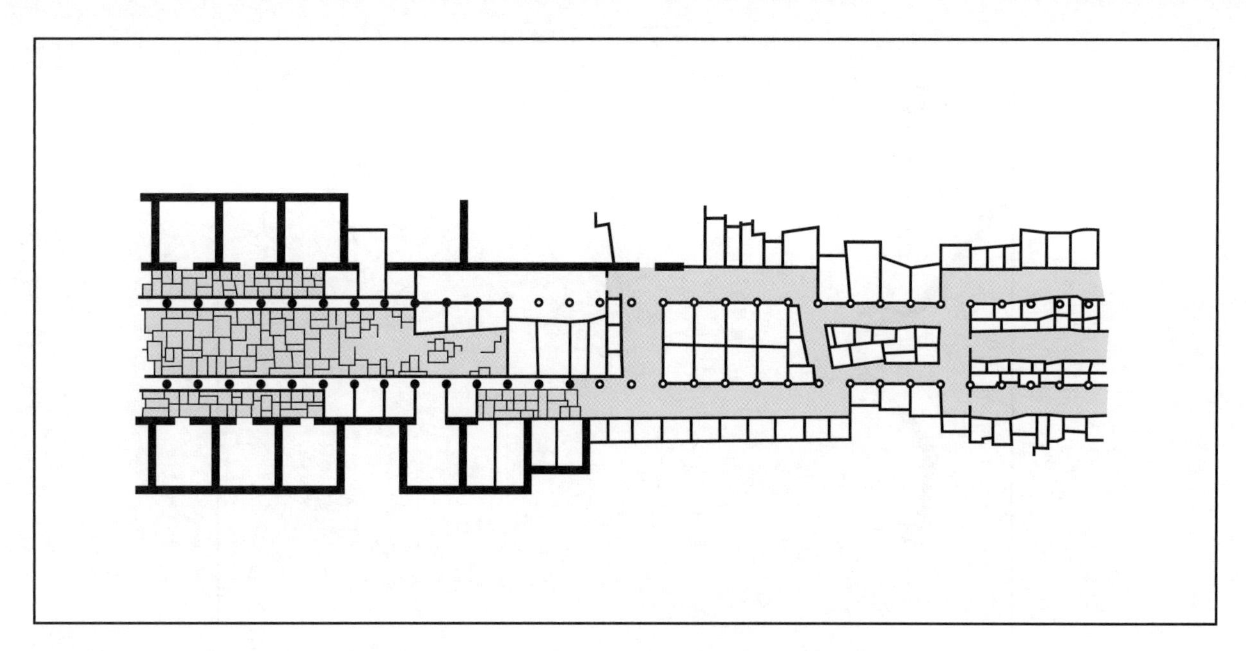

Figure 3.4 The transformation in eastern towns. (After Sauvaget, 'Le plan antique de Damas'.) From left to right the plan shows the process of change over time from the late Roman colonnaded street to medieval alleys and shops/artisanal quarters.

increases. Aegean coarse wares – transport vessels such as amphorae, and cooking vessels, in particular – begin to compete with the western imports during the sixth century and finally to dominate from the decades around 600. The pattern of ceramic distribution thus reflects a variety of factors, including highly localised economic sub-systems. Amphorae from both Palestine and north Syria are found in quantity in the Peloponnese and in Constantinople from the middle of the sixth century, for example, complemented by amphorae from western Asia Minor, presumably representing imports of olive oil and wine. From the later sixth century a greater localisation of fine-ware production can be observed. The economic implications of these patterns is that several overlapping networks of ceramic production and exchange complemented one another, and these were accompanied by similarly overlapping exchange-patterns for the products which accompanied or were transported in these containers.

The ceramic record thus reflects both facets of economic activity: the movements of the state-controlled grain convoys and the 'piggy-back' commerce which accompanied it; and the networks of private enterprise and commerce reflecting the operations of large estate-based cultivation – especially in North Africa – as well as of smaller-scale cash-crop operations. Beneath this lay the vast substratum of simple peasant subsistence activities, producing enough for the government to take in taxation as well as for landlords (whether individual or corporate – such as the church), to extract in rent.

The picture that emerges of late Roman commerce and trade is of an immensely complex pattern of intersecting local, regional and supra-regional pools or networks of exchange, focused around the shores and major ports of the Mediterranean, Adriatic and Aegean seas. Disruption of one area would thus affect comparable exchange zones elsewhere in terms of reduction of supply or market demand, with all the consequences for local production, employment and income which this entailed. Since these networks operated at different levels, and were certainly fragile in certain respects, they were also flexible: the dramatic shrinkage of one area of activity did not necessarily bring with it similar consequences throughout the whole system. The results of the Vandal occupation of North Africa in the first half of the fifth century, for example, were a series of relatively minor adjustments to markets and sources of supply, although with no doubt more or less drastic effects on particular communities or even households.

Mints and Coinage

Coins were an essential element of Roman state administration and always played a key role in the process of extraction of resources, maintaining the government and its apparatus, and maintaining and supplying the armies. Coins provide information about prices and values, and they are also political objects, bearing symbolic imagery and inscriptions which reflect political values, official ideology and the propaganda and claims of a state or ruler. They cast light on methods and technologies of production, imperial fiscal policy, and on the relationship between centre and provinces, and between taxation and economic life in general.

The fiscal and economic crisis which beset the Roman empire during the third century was resolved in respect of monetary policy by the reforms of Diocletian and Constantine I. The older gold and silver coinages, along with the minor bronze and copper coinages of account, had become unmanageable. In the 280s, Diocletian inaugurated an important reform. A new gold coin, the *aureus*, worth 1/60 of a Roman pound, was introduced,

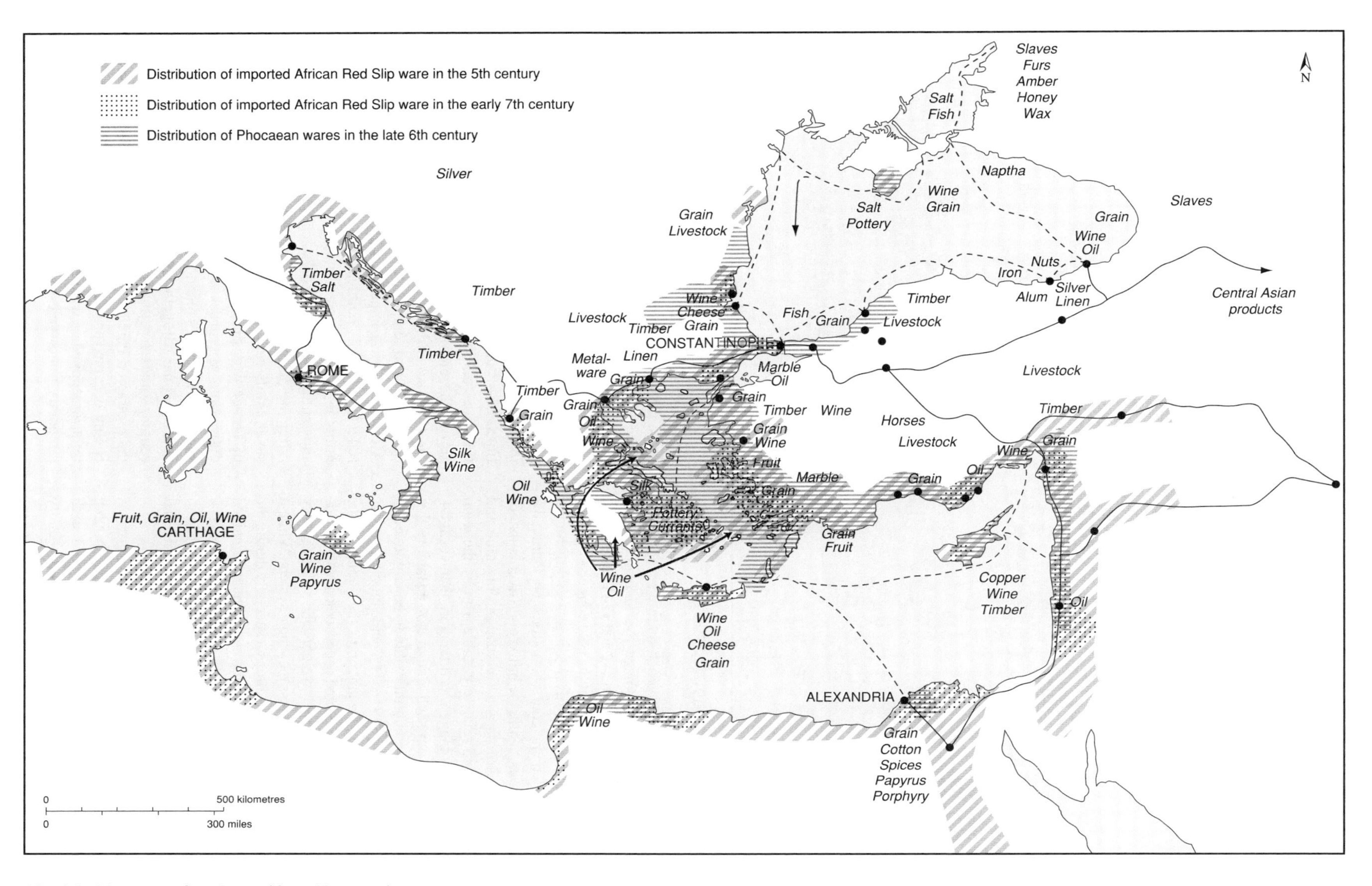

Map 3.5 Movement of goods as evidenced by ceramics.

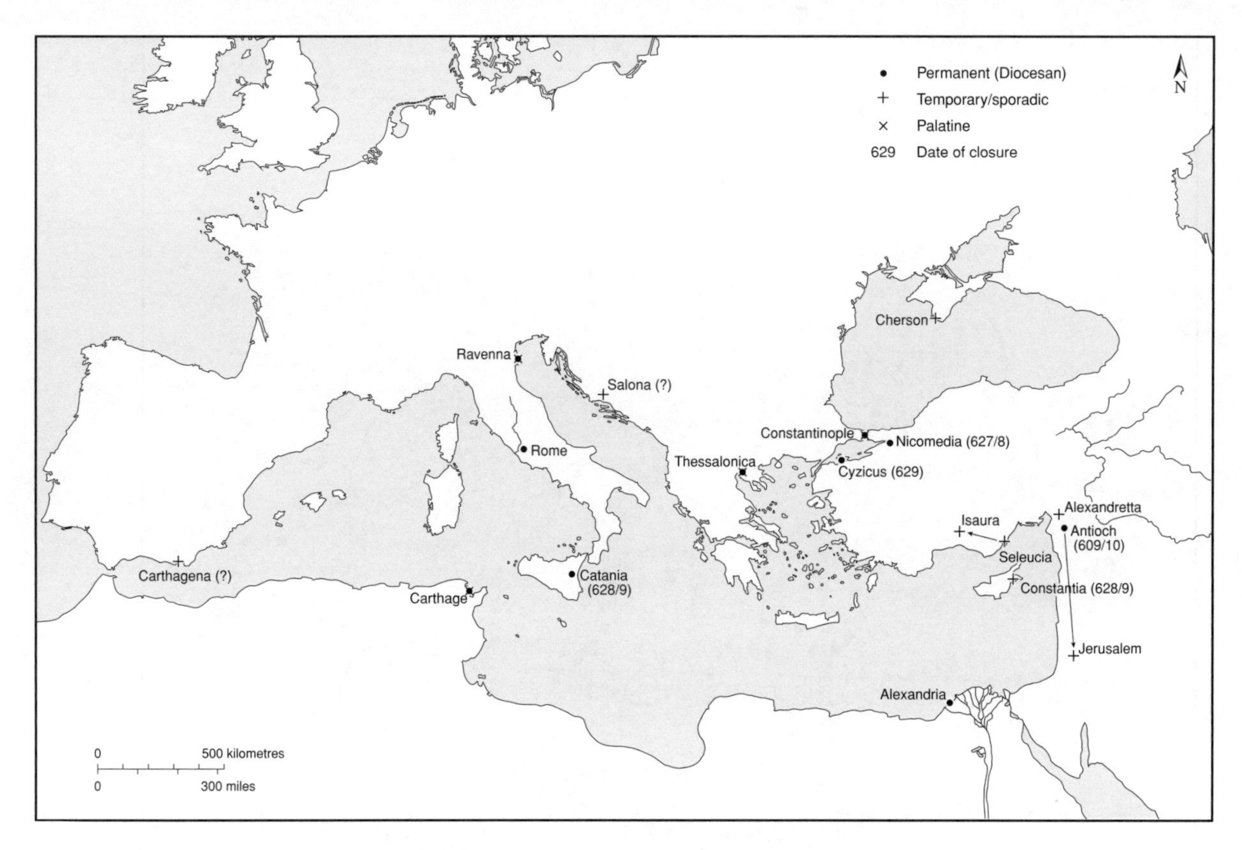

Map 3.6 Mints, c. 527–628/9 (After Hendy, *Studies*).

accompanied by a silver coin, of which there were 96 to the pound, and by a reformed billon coinage, the *nummus* (copper with a small silver content). Between 312 and 324 Constantine I transformed this system: the value of the gold coin was changed to 1/72 of a pound, and a second silver coinage, slightly higher in value than the Diocletianic coin, was introduced. During the fourth and fifth centuries, while the billon and silver coinages suffered a series of reforms and fluctuations in value and weight, the gold remained relatively stable. But by the reign of Anastasius the silver coinage was little more than vestigial, and the billon suffered from such a degree of instability that it became too cumbersome and inflexible to be employed in normal exchange. Anastasius (491–518), while modifying only slightly the gold:silver ratio and maintaining the stability of the gold, introduced a radically reformed copper coinage to replace the older base-metal coinage, with weights and values clearly marked, facilitating exchange across the whole system. While it did suffer from considerable fluctuations, especially during the seventh and eighth centuries, the reformed coinage remained the basis for copper coin until the later eleventh century. The reforms of Anastasius mark a convenient historical point from which the establishment of a specifically east Roman imperial coinage can be said to have taken place.

Silver, especially in the form of the *miliarensis* (Hellenised as *miliaresion*), a heavy coin struck at the rate of 72 to the pound, played a relatively minor role during the later fifth and sixth centuries, except in the empire's western regions (especially those reconquered from the Vandals and Ostrogoths) until the reign of Heraclius, when the hexagram was introduced, a silver coin worth 1/12 of a gold *solidus*.

The administration of coin production and circulation was complex. From the fourth century into the early seventh century mints were under the authority of the *comes sacrarum largitionum*, the count of the sacred largesses, one of the leading financial officers of the empire, who also had responsibility for mines and bullion. There were a large number of mints across the empire producing gold and bronze coins until the early seventh century. Between 296 and 450 some 17 permanent and temporary mints stretching from London to Alexandria struck imperial coinage. The location of the mints reflected primarily political and military concerns – provincial mints in particular were mostly located in regions that had substantial military needs, such as garrisons and frontiers. After the disappearance of the western half of the empire, seven mints continued to operate, although that at Heraclea in Thrace was shut down due to barbarian pressure in the 490s. Gold was struck at Constantinople and Thessalonica, and with Justinian's reconquests it was also struck at Carthage and Ravenna. Altogether there were at the end of the sixth century some ten permanent and two temporary mints. During the protracted Persian wars, from 603 to 626, minting in the enemy-occupied provinces and in other regions affected by the fighting stopped or was severely disrupted, and a number of temporary mints were set up with the specific aim of coining money with which to pay the army. As a result of some of these developments, the Emperor Heraclius undertook a major reform of the mints, probably in 628–629, a reflection both of a straitened state budget and ongoing changes within the state fiscal administration, in particular the gradual absorption into the various levels of the praetorian prefecture, on the one hand, and of the imperial *vestiarion*, the household administration, on the other, of many of the key activities and functions of the *sacrae largitiones*. The result was the closure of a number of mints and the concentration of the striking of gold at Constantinople, Carthage and Ravenna, the rest being permitted to strike copper only. While this structure changed slightly later in the seventh century (see below), it sets the pattern for the highly centralised medieval Byzantine production of imperial coinage in the Balkans and Asia Minor.

4 The Church

Politics, Religion and Heresy 4th–6th Centuries

The Christian Church and the development of Christianity were fundamental to the cultural and political evolution of the empire. For Constantine I the Church had been a valued political ally in his effort to stabilise the empire and to consolidate his own power. For that reason it had been essential that the church remained united: discord and disagreement was politically threatening for an emperor who privileged the Christian Church in terms of landed property and formal recognition in his political plans. But Constantine had to deal with a major split within the church, brought about by the appearance of Arianism, a heresy about the Trinity and the status of Christ. Arius (250–336) was a deacon of the church at Alexandria. Trained in Greek philosophy, he became an ascetic, and in his attempts to clarify the nature of the Trinity, produced a creed that was for many contemporaries heretical. He could not accept the notion that God could become man: he taught that Jesus was not eternal and co-equal with the father, but created by Him. Arius was excommunicated in 320 by the Bishop of Alexandria, and in 325 at the Council of Nicaea he was condemned and exiled. Unfortunately, Constantine himself began later to favour the Arian position, and after his death in 337 his son and heir Constantius adopted it in the eastern part of the empire. The Emperor Constans in the west supported the Nicene position. Many synods were held to debate the issue, until in 350 Constans died and the Nicenes were persecuted. But the Arians were themselves split in three factions: those who argued that Father and Son are *unlike*; those who believed that Father and Son are alike, but not consubstantial; and those who thought that Father and Son were of *almost* one substance – a group which eventually accepted the Nicene position. Constantius died in 361, in 362 the Council of Alexandria restored Orthodoxy, and in 381 the ecumenical Council of Constantinople reaffirmed Nicaea.

The early fifth century saw a further Christological split in the form of Nestorianism, which took its name from Nestorius, a monk of Antioch who had studied under Theodore of Mopsuestia. In 428 he was appointed Bishop of Constantinople by Theodosius II, but aroused considerable hostility in Constantinople when he publicly supported the preaching of his chaplain, that Mary should not be referred to as the '*Theotokos*' – the God-Bearer. The Nestorians developed a theology in which the divine and human aspects of Christ were seen not as unified in a single person, but operated in conjunction. Their position was condemned in 431 (Council of Ephesus), and they then seceded, establishing a separate church at their own council at Seleucia-Ctesiphon in Persia in 486, where they established a firm foothold and carried out successful missionary activity across northern India and central Asia as far as China during the following centuries. Nestorianism survives today, particularly in northern Iraq, as the Assyrian orthodox church.

The debates thus contributed to the evolution of a much more significant split within Christianity in the form of the Monophysite movement, which – although only referred to under this name from the seventh century – represented a reaction to some of the Nestorian views. The key problem revolved around the ways in which the divine and the human were combined in the person of Christ, and two 'schools' of Monophysitism evolved. The most extreme version argued that the divine was prior to and dominated the human element – hence the description 'monophysite': *mono* – 'single' and *physis* – 'nature'. A council held at Ephesus in 449 found in favour of the Monophysite position. But at the Council of Chelcedon in 451 a larger council rejected it and redefined the traditional creed of Nicaea to make the Christological position clear. The political results can be seen in the politics of the court at Constantinople and in the regional identities of different regions of the empire. In Egypt and Syria in particular Monophysitism became established in the rural populations. At court, in contrast, imperial policy varied from reign to reign leaving some confusion within the church as a whole, and involving persecutions by both sides. The Emperor Zeno (474–491) issued a decree of unity, the *Henotikon*, which attempted to paper over the divisions. Anastasius supported a Monophysite position, Justin I was 'Chalcedonian', and Justinian, partly influenced by the Empress Theodora (d. 548) swung between the two. Theodora lent her support to the Syrian Monophysites by funding the movement led by the bishop Jacob Baradaeus (whose name was afterwards taken to refer to the Syrian 'Jacobite' church); a similar 'shadow' church evolved in Egypt, and the Armenian church also adopted a Monophysite view. In each of these cases the form of traditional belief may have been one of the most important factors, but alienation from the imperial regime also played a role.

These were not the only heretical movements to affect the church and directly involve the emperors. The 'Donatist' movement was a strictly North African heresy, led by a puritan sect claiming that the tradition of consecration of bishops of Carthage was improper. Because the church authorities were supported from Rome, African regional feeling was inflamed, and the heresy flourished, although as a small minority until the seventh century. Other regional heresies included Messalianism, a Syrian monastic heresy that spread from Mesopotamia to Syria in the fourth century, but was condemned by the Council of Ephesus in 431. Pelagianism was a largely western heresy, begun by a British or Irish monk, Pelagius, during the later fourth century, condemned repeatedly – in 411 and again in 416–448, and finally – because its chief spokesman Celestius associated himself with Nestorianism – at Ephesus in 431. These local heresies had few longer-term results, but directly involved the emperors on every occasion and cemented the association between the interests of the church and those of the imperial government.

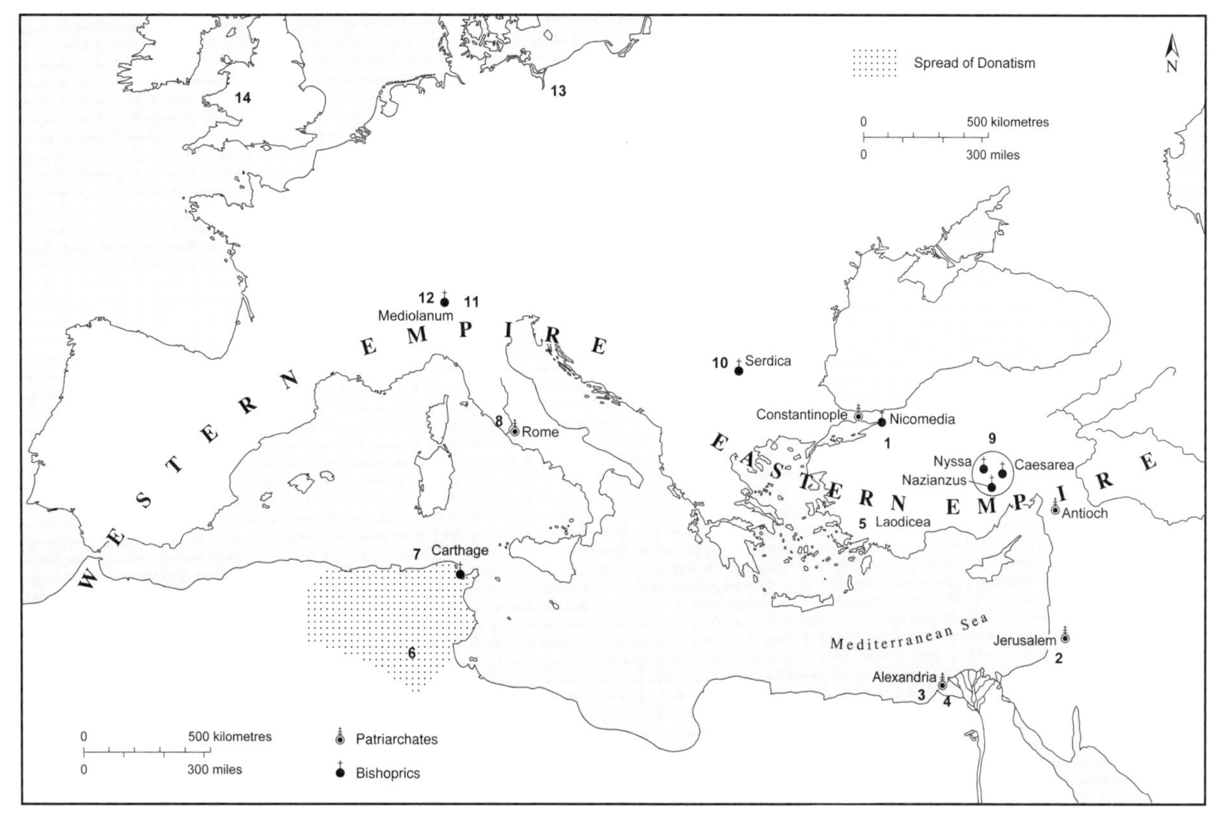

1 *Eusebius of Nicomedia (- c. 342), studies with Arius; from c. 320 - Arian leader, makes Arianism an ecumenical controversy, but in 325 he signs anti-Arian creed of Nicaea.*

2 *Eusebius (c. 260-c. 340); c. 315 - Bishop of Caesarea; 325 - attempts reconciliation with Arianism at Council of Nicaea; later accepts anti-Arian creed of Nicaea.*

3 *Arius (c. 250-336) ascetic and preacher, whose teachings split orthodox church from c. 319 - leading to Council of Nicaea in 325. He was excommunicated and his teachings condemned.*

4 *Athanasius (c. 296-373) - Bishop of Alexandria: greatest opponent of Arianism; attempts reconciliation with Arianism at Council of Alexandria (365) defeats Arianism within orthodox church at Council of Constantinople (381).*

5 *Apollinaris (c. 310-c. 90), Bishop of Laodicea, friend of Athanasius and opponent of Arianism - teachings condemned (347-481); secedes from orthodox church c. 375.*

6 *Donatist schism divides part of African church away from Rome c. 311.*

7 *Augustine (354-430). Bishop of Hippo, defines orthodox beliefs in response to Manichaeism, Donatist and Pelagian heresies.*

8 *Jerome (c. 342-420), 382-385 secretary to Pope Damasus (later ascetic in Palestine) opposed Arian and Pelagian heresies.*

9 *Basil, Gregory of Nazianzus and Gregory of Nyssa oppose Arianism.*

10 *343 - Council of Serdica restores Athanasius; condemns Arian views.*

11 *355 - Council of Milan. Arianism re-established under Constantius.*

12 *Ambrose (c. 339-97), Bishop of Milan, opponent of Arianism.*

13 *Ulfilas (c. 311-83), preaches among Goths, translates Bible into Gothic.*

14 *Pelagius, British or Irish monk, goes to Rome, Africa and Palestine early in 5th century; opposes teachings of St Augustine; excommunicated in 417.*

Map 4.1 Politics, religion and heresy, 4th–5th centuries.

Ecclesiastical Administration

The patriarchate of Constantinople was one of the five major administrative units into which the Christian Church had organised itself territorially within the Roman world, the others being Rome, Jerusalem, Antioch and Alexandria. Constantinople was the last to be raised to ecumenical status, which it first claimed formally at the council held there in the year 381, but always contested by Rome. Each see had an apostolic tradition, and the history of their relations with one another are heavily inflected by claims and counter-claims about precedence and rights. The position of the patriarchate of Constantinople was enhanced by the fact that the city was also an imperial capital – and after 476, the imperial capital – but the Roman see never admitted Constantinopolitan equivalence. Whereas the development of the see of Constantinople thus took place in an environment which transformed it slowly but surely into an 'imperial' church, Rome was, with a few exceptions, in effect independent of direct political influence from the emperors. One of the results was that Rome appeared at times as an independent authority, and conflicts within the Byzantine Church could be referred to Rome, with the result that the tensions between Rome and Constantinople, on the one hand, and within the Byzantine world, between different factions within the church, as well as between the secular power and one or another of these factions, were further heightened.

Bishops typified the growing importance of the church both spiritually and institutionally within the empire. Bishops were by the sixth century recognised as key members of city governing councils; they were, as managers of church lands in their sees, major controllers of economic resources, a position they maintained throughout the history of the empire and

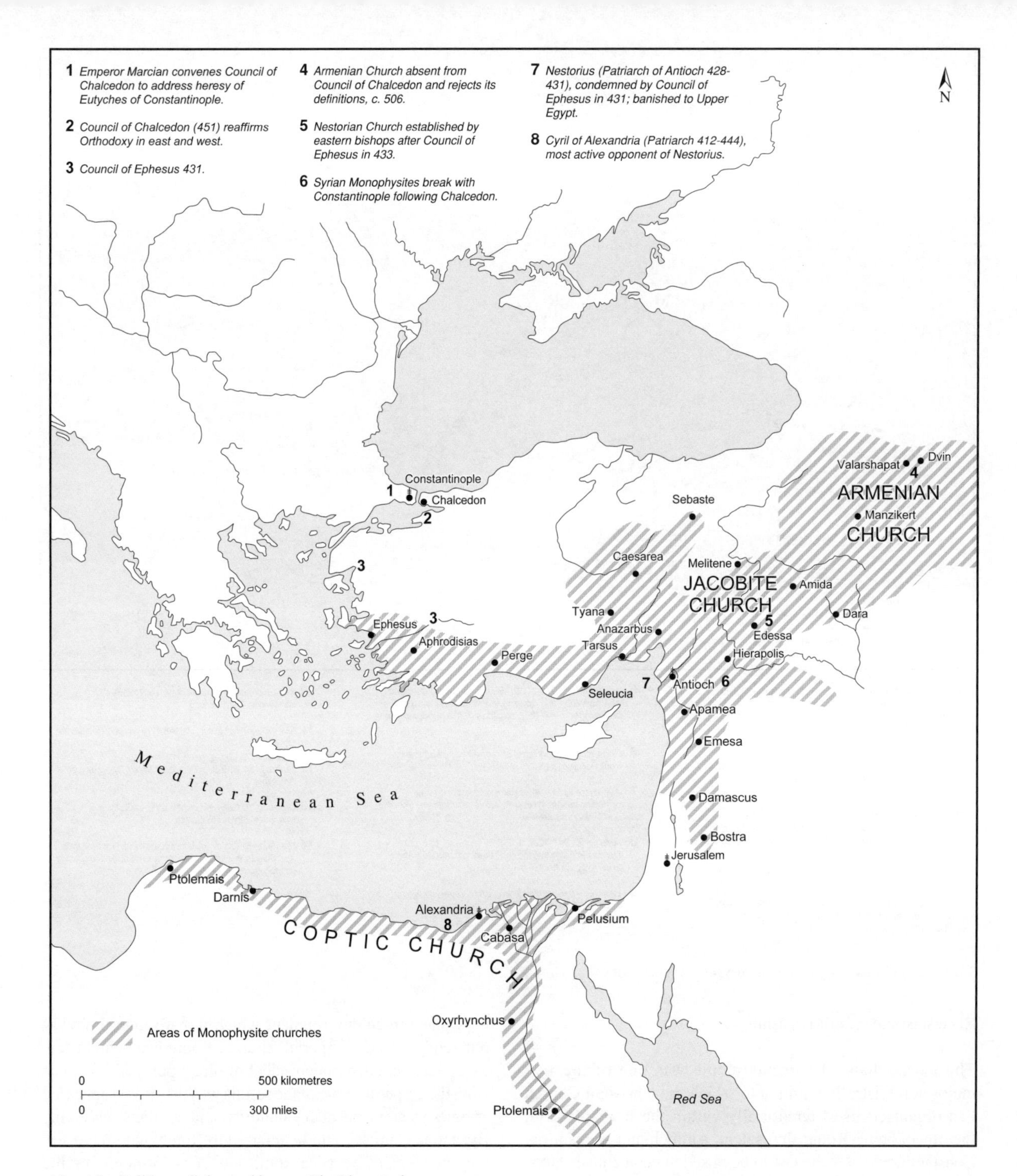

Map 4.2 Politics, religion and heresy, 5th–6th centuries.

beyond. Such was the wealth accumulated by the church by the later sixth century that one estimate has suggested that the resources it consumed to maintain its charitable institutions, its clergy and episcopate, its public ceremonial and its public buildings were greater even than those of the state – excluding the army, of course. This wealth was mostly in land, although substantial amounts were invested in gold and silver or in buildings. The church derived a large income from rents from its numerous estates, whose importance was clearly understood – the canons, or laws, of the general council held at Constantinople in 692, known as the Quinisextum, included prescriptions about bishops staying in areas subject to hostile attacks in order to look after their flocks – and, presumably, church lands and property.

Charitable foundations were a major focus for the church's philanthropic activities, but they had to be supported, and were usually funded by the rents from large estates. Orphanages, hospitals, almshouses, for example, were built and maintained, some by private persons after they had decided to endow their lands, sometimes by the church itself – the importance of the church in the state is reflected in the establishment by the emperors of various imperial charitable houses, illustrating the ideological importance of such commitments. Vast amounts of land were given to the church over the centuries, sometimes in large bequests from wealthy persons, just as often in the form of tiny parcels of land willed to the church by the less well-off. church land could not in theory be alienated – once devoted to God, always devoted to God – so that the accumulation could only grow. But in fact, the use of emphyteutic leases, by which life-long and often hereditary lease contracts were agreed, meant that land frequently did in practical terms fall outside church control and supervision.

Church administration followed the secular organisational pattern of the late Roman empire. Below the patriarch were metropolitan bishops, autocephalous archbishops, and bishops. The first was the senior cleric in each province, and was personally appointed by the patriarch. Bishops were elected by the provincial synods attended by all bishops, and until the sixth century the ordinary clergy and congregation of the region also had some influence, at least in nominating candidates. The bishop was the highest-ranking church official in his region, with both the regular clergy as well as any monastic communities present under his authority. Bishops were leaders of their communities spiritually and politically: many bishops had been martyred during the great persecutions of the later third and early fourth century, and the position of bishop was one to which considerable esteem was attached – but of which expectations were accordingly high. Bishops were based in cities, and insofar as the bishop's jurisdiction extended over the *territorium* of his city, the ecclesiastical administration of the empire preserved the late Roman secular administrative pattern of the Roman world.

The task of the bishop was threefold: seek out and combat heresy, ensure that orthodoxy prevailed; and impose and apply church law. He was also the chief manager of church lands, giver of charity to the poor and needy; as well as the leading ecclesiastical authority and judge in his district. He presided over the church courts regarding clerical mores and correctness; and he arbitrated in cases involving conflicts between laypersons and the church – indeed, from the early seventh century the clergy had an especially-privileged legal status, to prevent their being mistreated at the hands of secular authorities. Bishops were generally, although by no means exclusively, drawn from the more privileged and the best-educated sections of society, at least from the middle and later fourth century.

While responsible for the physical as well as spiritual welfare of their flock, bishops tended also to share the views of the privileged élite from which they were drawn, so that those based in the provinces in particular have left letters bewailing their fate, relegated as they felt themselves to be to regions of cultural darkness and barbarism. Of course, such images are often deliberately overdrawn for effect, and many bishops and senior clergy in the provinces were thoroughly committed to the care of their congregations. Corrupt and dishonest clergy there certainly were, but there is little clear evidence of the degree of these problems.

A major concern of the church throughout the empire, and an aspect in which bishops played a fundamental role, was in public welfare: bishops were responsible within their sees for the organisation of charitable activities, in particular the relief of the poor. Such activities depended, naturally, on the commitment and involvement of individuals, so that there is no uniformity of practice across the empire. Many bishops did very little, but many also funded the construction of almshouses, hospitals, refuges for lepers, and so on. Bishops and senior clergy were also involved in politics, both secular and ecclesiastical. Bishops were often asked to speak out for particular individuals, or for certain groups in their jurisdiction – the poor peasants suffering from heavy taxation, for example, at one extreme, or the imperial official seeking redress from the emperor for political victimisation.

The extent to which bishops were successful depended largely on their contacts – their own network of connections at court and in the government – as well as on their ability to present the case convincingly, and on the context and personalities concerned. Senior clergy, bishops especially, had the right to exercise *parrhesia*, 'freedom of expression', which they often did, particular in respect of powerful secular figures – emperors, imperial officials and so on, on behalf of their flock or in order to speak out against what they perceived as unorthodox beliefs or behaviour. This often got them into trouble, especially where a patriarch, for example, attempted to oppose a strong-willed or determined emperor in respect of some issue of state affairs or high church politics.

Monasteries, Pilgrims and Holy Places

Monasteries were a significant feature of the late Roman social, demographic and cultural landscape. Monasticism represented an alternative both to the secular church and to life in 'the world'. Its origins – in late third-century Egypt and in Syria and Palestine during the fourth century – lie in the particular conditions of that time, but it rapidly assumed a universal relevance for the late Roman world, and underwent a dramatic expansion during the period up to the sixth century. In spite of efforts on the part of the church and the state to exercise some control over monasticism – embodied in particular in the acts of the Council of Chalcedon in 451, and repeated in Justinian's legislation – its diversity and its anti-authoritarianism and spiritual utopianism represented a source of independence from the established structures of Christian life.

The founders of monasticism, as far as the sources show, were Pachomius (c. 290–346) and Anthony (c. 251–356), both of whom developed their own forms of the monastic life in Egypt. While Anthony seems to have been the first to practise desert asceticism, Pachomius established a number of cenobitic monasteries. They inspired many followers in later generations, among whom Jerome (c. 342–420), Euthymius (376–423) Theodosius (423–529) and Sabas (439–532) in Judaea and

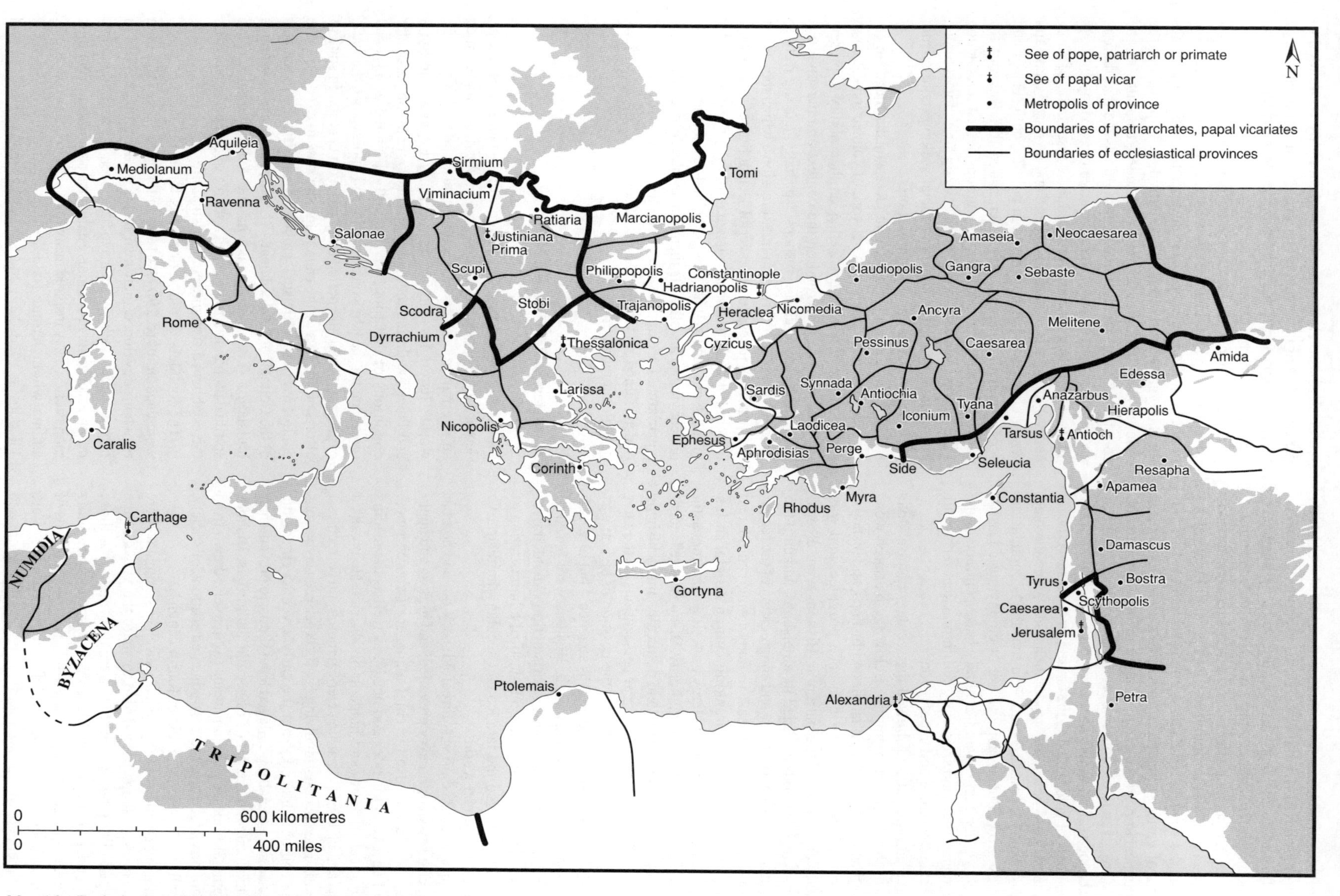

Map 4.3 Ecclesiastical administration. (After Jones, *Later Roman Empire*.)

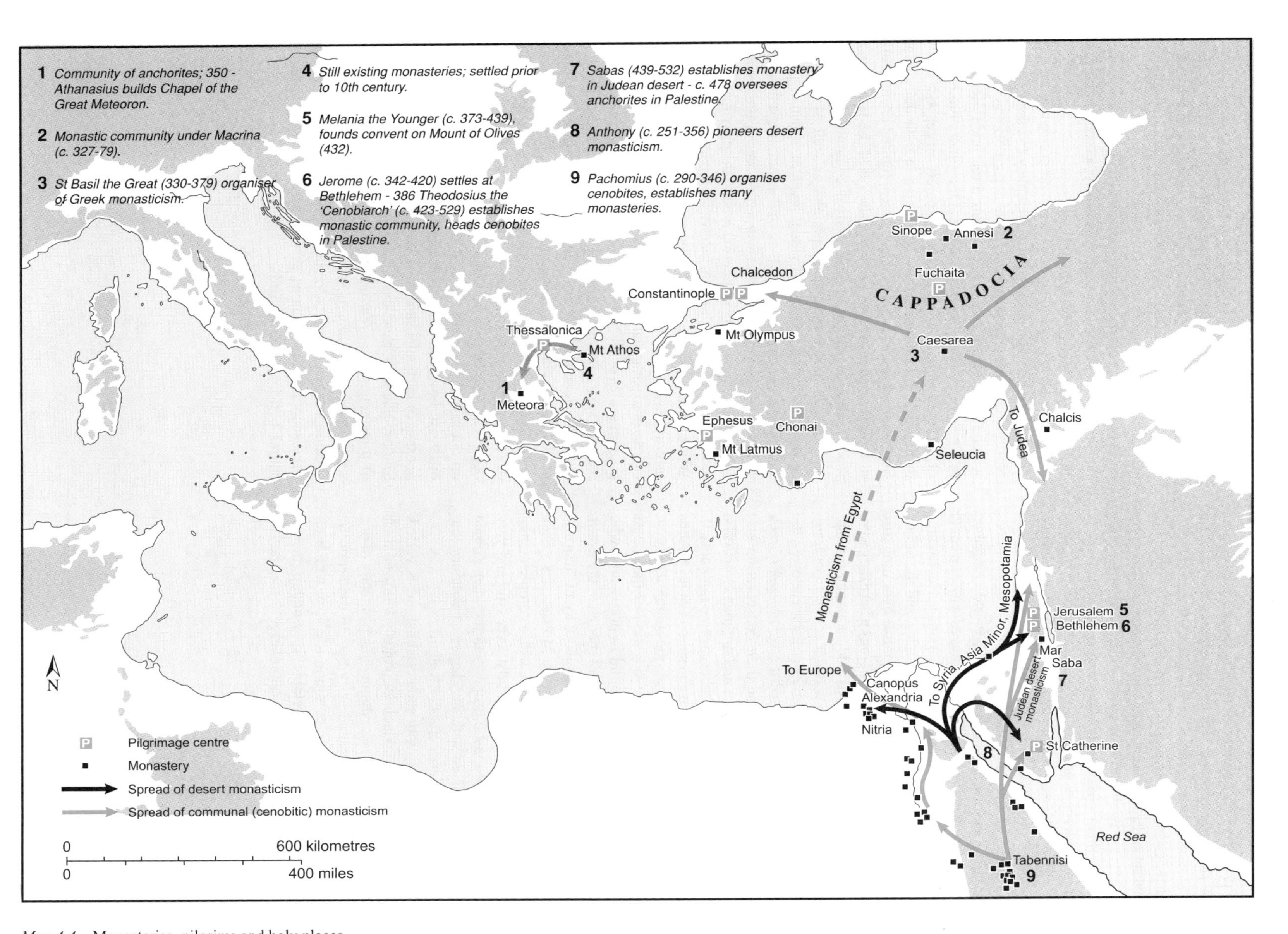

Map 4.4 Monasteries, pilgrims and holy places.

Palestine became among the best-known in their own lifetimes. From Egypt and Syria-Palestine monasticism quickly established itself in Asia Minor, especially in Cappadocia, where St Basil 'the Great' (330–379) became the leading figure, establishing a rule for monastic communities which remains the basis for much orthodox monasticism.

Monasteries and monks, and the even less-readily controlled individual hermits and 'holy men' who separated themselves from society by dwelling in the countryside away from other human settlement (frequently choosing quite deliberately a region of wilderness or desert in order to emphasise their withdrawal), often exercised a spiritual and moral authority gained through their lifestyle and their struggle with the forces of evil, which the regular church could not. But this does not mean that they turned their back on the world altogether. Monasteries, like the church, could be endowed with and could own property, in land or other forms. Some monasteries became substantial landlords with considerable estates, and the efforts of the church to retain some degree of control over monastic establishments gave the bishops of the regions where they were to be found an important role. It also led to tensions and conflict: the agreement of the local bishop was necessary in order to establish a monastery, and each such establishment owed the local episcopate a regular tax, called the *canonicum* (Greek *kanonikon*). Elections of abbots and their consecration was also the responsibility of the bishop, and arguments between monastic communities and the local church were not unusual.

Monasteries attracted the attention and support of laypersons from all ranks in society, from emperors to the humblest peasant, who endowed them with land for the salvation of their souls – for which the monks would be bound to pray – and who sometimes retired there when they had had enough of the world. Indeed, entering a monastery was a popular way of 'retiring' in the late Roman and Byzantine world. Monasteries varied enormously. Some remained very small, with a complement of a few monks. Others became immensely successful and very wealthy. The extent of monastic property in the later Roman period cannot be calculated with any degree of accuracy, but some monastic houses were certainly well off.

The lifestyles associated with such establishments varied, from the 'idiorrhythmic', in which individuals followed their own daily rhythm, eating and worshiping independently of one another and with few communal activities, to the cenobitic (from the Greek *koinos bios*, 'common life'), in which all followed the same communal timetable for worship, meals, work and meditation. The former was relatively rare before the fourteenth century. In the period from the later fourth to the early seventh century monastic centres flourished in Constantinople, in Cappadocia, Syria, Palestine and Egypt, although there were monasteries in many other regions too. In some regions there were almost as many monasteries as there were villages, all involved in a closely interlinked network of agricultural and pastoral exploitation and often tied together in market and commercial relationships.

In many respects monks and individual holy men constituted an alternative source of spiritual authority, a type of authority that challenged implicitly the formally-endowed authority of the regular clergy and the church. Such authority was won by men and women who demonstrated their piety and their spiritual worth by enduring hardships, physical and emotional, through which they were thought to gain a more direct and fuller access to God, and thus at the same time met the superstitious needs of ordinary people at all levels of society in respect of the difficulties and problems they confronted in their daily lives. Such individuals, men as well as women, figured prominently in the political as well as the religious life of the late Roman world, acting as patrons of rural or urban communities in local conflicts, as representatives of popular as well as marginal opinion, as well as, at an individual level, as spiritual guides and advisers to persons of high as well as humble status.

Many of the holy men and women of whom the sources speak spent many years wandering around the provinces of the empire before establishing themselves at a particular location. While their wanderings were often random, the journeys of the pilgrims who travelled to the Holy Land, to established cult centres or to particular holy men were not, and several well-worn pilgrim routes evolved over this period. The pilgrims' journey represented a particular form of Christian piety and endeavour, but it also generated a considerable 'pilgrim industry', with the production of clay pots containing holy water from the Jordan or the Sea of Galilee, for example, as well as the production of reliquaries for items from the bodies, vestments or other objects associated with holy figures. Pilgrimage centres sprang up not just in the Holy Land, at sites associated with the life of Christ, however, but also at many sites associated with the apostles or particular saints – in the eastern empire at Ephesus, Sinope, Euchaita, Seleucia, Chalcedon, for example. Centres of pilgrimage also grew up around particular holy men, too – the site of the column of St Symeon the Stylite in Syria soon became a major tourist attraction with all the appurtenances: guest accommodation, shops, as well as a monastic community, church buildings and so forth.

Part Two:

The Middle Period (c. 7th–11th Century)

5 *Historical Development: the Rise of the Medieval East Roman World*

The East Roman Empire c. 650–717

Through the reigns of Constans II (641–668), Constantine IV (668–685) and Justinian II (685–695) Asia Minor was subjected to constant raiding, with substantial tracts of territory devastated on a yearly basis from the early 640s well into the first half of the eighth century. This devastated the population, the economy of the regions affected, especially the border zones, and urban life, which was reduced effectively to fortified garrison towns. A series of sieges and attempts to break Constantinopolitan resistance between 674 and 678 was finally driven back; and a major siege in 717–718 was defeated with great losses on the Arab side. But the situation appeared desperate enough for Constans II to move the imperial court to Sicily in 662. His assassination in 668 brought the experiment to an end, but illustrates the nature of the situation. Justinian II was deposed in 695; a series of short-lived usurpers followed until Justinian II himself recovered his throne in 705. But he was again deposed and killed in 711, and the situation of internal political and military confusion lasted until the seizure of power by the general Leo, who became Leo III (717–741) and, having defeated the Arab besiegers in 717–718, finally re-established some political order. The wars were not entirely one-sided. Although the empire was largely on the defensive, being forced to give up attempts to face Muslim armies on the field and adopt a strategy of avoidance and hit-and-run raids, internecine strife within the Umayyad family or between different elements in the newly-established Arab Islamic empire helped the Byzantines to survive these most adverse of circumstances. A civil war over the succession to the Caliphate followed the death of the third Caliph, Uthman, in 656. Although his successor, Ali, was the son-in-law of the prophet Mohammed, this did not prevent the powerful Umayyad clan, to which Uthman had belonged, from challenging his authority. He was defeated in 661 and, under the new Caliph, Mu'āwiya, the Umayyad dynasty established itself firmly in power, where it remained until deposed in 750.

The changes that accompanied the enormous loss in territory and in revenue were considerable and sometimes drastic. In the period between the later years of Heraclius and the end of the century the whole fiscal apparatus began to be remodelled; the organisation of the imperial field armies underwent dramatic alterations to cope with the changed circumstances in which they had to operate, both in respect of strategic geography and in terms of resources (or lack of them). The political ideology of the empire regenerated itself in an increasingly exclusive orthodoxy that rejected heterodox belief and was suspicious of anything not 'Roman', even though at the same time that epithet applied to anyone who spoke Greek, was orthodox and accepted the emperor as God's representative on earth – whether Armenian, Slav or Arab.

The eastern empire was fortunate in its strategic geographical situation. For although the new power of Islam was a major threat to the continued existence of the empire, the peoples to the north and west offered a far less systematic and organised, and thus much less effective, challenge. They could be destructive, and they certainly forced the empire onto the defensive in Italy, for example, or along the Balkan front, but they were unable or uninterested in challenging Constantinople, in part because the imperial capital and the Roman empire still attracted their admiration and envy in a way which was quite irrelevant to the Umayyad power in the east. The Khazars were sufficiently distant to serve as imperial allies (as they had done briefly during the Persian war in the time of Heraclius), acting as a threat to both the Bulgars and the more distant Avars. From the 680s the Bulgars along the eastern reaches of the Danube, while a potential threat to the empire, also acted as a buffer between the declining Avar power to their north and west; while in the southern and central Balkans the various Slav peoples and groups formed a series of competing and disunited groups – the autonomous *Sklaviniai* nominally under Byzantine authority, the Serbs and Croats in the west, and other more amorphous groups, some of whom soon intermingled with the indigenous population, elsewhere. In Italy the complicated territorial situation, with imperial, local and Lombard forces constantly at war, was not improved by direct imperial intervention in local politics when the emperors felt their interests or authority were challenged. Thus in the course of imperial efforts to banish discussion on the issue of the emperor's monothelete policy (see page 65), for example, the pope, Martin, was arrested in the 650s and imprisoned on charges of treason.

For a while, the empire hung on to its North African territories – apart from Italy, the last Latin-speaking regions under its control. But the cities of Africa were already disenchanted with imperial rule as a result of the monothelete disputes of the middle of the century and imperial intervention in local politics. Early Arab raiders pushed in 642 into Tripolitania, and from the 650s and 660s onwards further westwards, and by the 680s Roman rule was seriously compromised. By the early 690s Carthage had fallen, never to be recovered, and by 696 Islamic rule – which, however, faced the same turbulent Berber clans as the Byzantines had had to contend with – was firmly established. By the end of the first decade of the eighth century Islamic raiders were on the Atlantic coast and in 711 Berber troops crossed over into Spain to challenge the Visigothic kingdom there.

In the west the Lombards had continued to put pressure on the fragmented imperial possessions, seizing Genoa, for example, in 640; while the Visigoths had defeated and absorbed the Suevi in 584 and by 631 had seized the last strip of Byzantine-controlled south-eastern Spain. The Frankish kingdom had expanded to become the dominant power, having

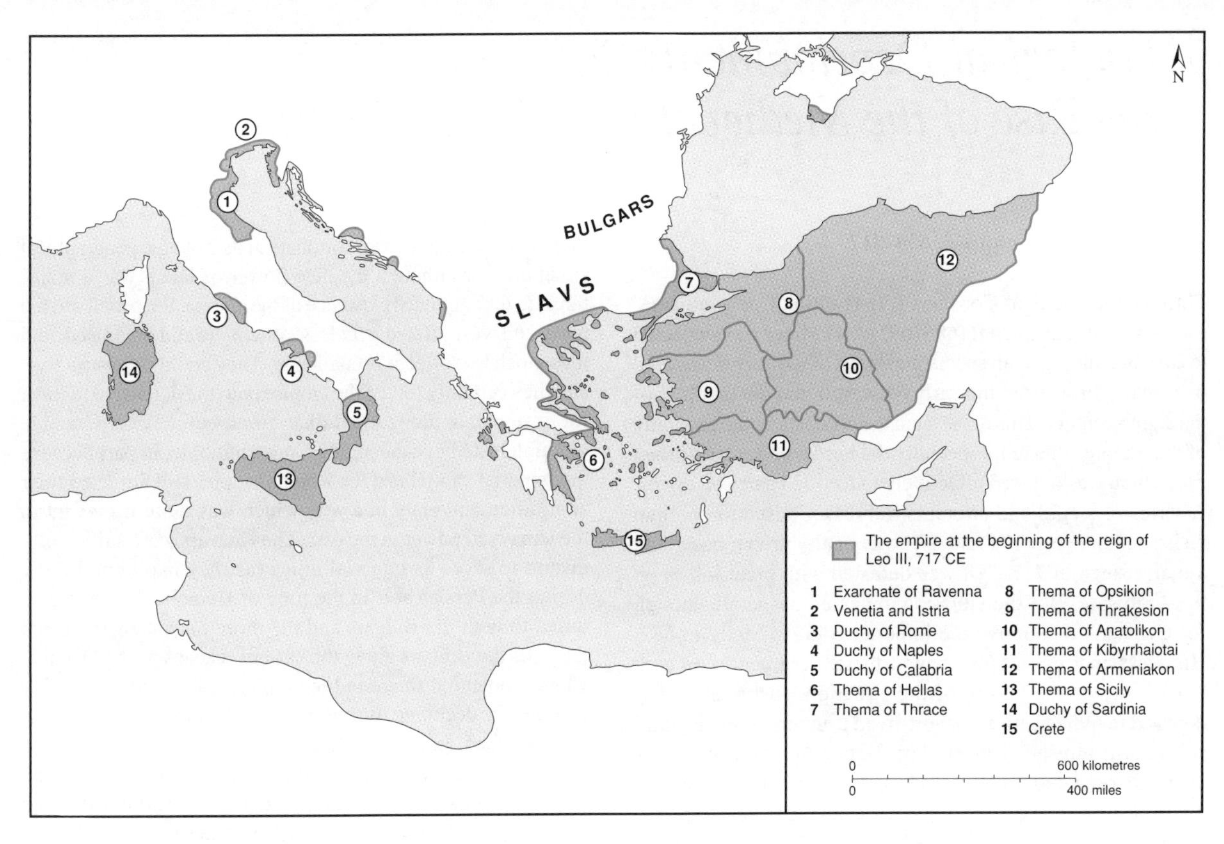

Map 5.1 The east Roman empire c. 650–717.

driven the Visigoths out of southern Gaul before the middle of the sixth century and subjugated the majority of the remaining independent Germanic tribes along its eastern margins by the end of the sixth century. The Baiuvari (Bavarians) accepted Frankish overlordship, and Frankish authority extended as far east as the Elbe. Only the Saxons and Frisians in the north-east remained independent. But the kingdom was divided, initially by the sons of Clovis, then again in 561, and after a series of fierce internecine wars settled permanently into the two kingdoms of Neustria and Austrasia.

The East Roman Empire 632–1050: Transformation and Recovery

The defeats and territorial contraction which resulted from the expansion of Islam from the 640s in the east, on the one hand, and the arrival of the Bulgars and establishment of a permanent Bulgar Khanate in the Balkans from the 680s, on the other, radically altered the political conditions of existence of the east Roman state, and established a new international political context. The evolution of this context was characterised by the political, cultural and economic relations between the empire and its neighbours, on the one hand; and by the fluctuations in imperial political ideology and awareness of these relations, on the other. At the same time, the cultural imperialism of Byzantium, and the powerful results of this in the Balkans and Russia, had results which have influenced, and continue to influence, the Balkans and eastern Europe until the present day.

Under the Emperor Leo III (717–741) and his son and successor Constantine V (741–775), the period of contraction and defeat begins to change. Leo, who was from a military background and had come to power through a coup d'état, seems to have been an able military and fiscal administrator; Constantine proved to be a campaigning emperor who introduced a number of administrative reforms in the army and established an élite field army (the so-called imperial *tagmata*) at Constantinople in the 760s. Political stability internally, the beginnings of economic recovery in the later eighth century, and dissension among their enemies, enabled the Byzantines to re-establish a certain equilibrium by the year 800. In spite of occasional major defeats (for example, the annihilation of a Byzantine force following a Bulgar surprise attack in 811, and the death in battle of the Emperor Nikephoros I [802–811]), and an often unfavourable international political situation, the Byzantines were able to begin a more offensive policy with regard to the Islamic power to the east and the Bulgars in the north – in the latter case, combining diplomacy and missionary activity with military threats. From the early ninth century imperial authority was re-established over much of the southern Balkans and the Illyrian coastal regions; while successive Byzantine victories in central Asia Minor from the 860s on (and in spite of occasional setbacks, such as the Arab sack of the important fortress town of Amorion in 842) stabilised a new frontier and pushed the Caliphate onto the defensive. By

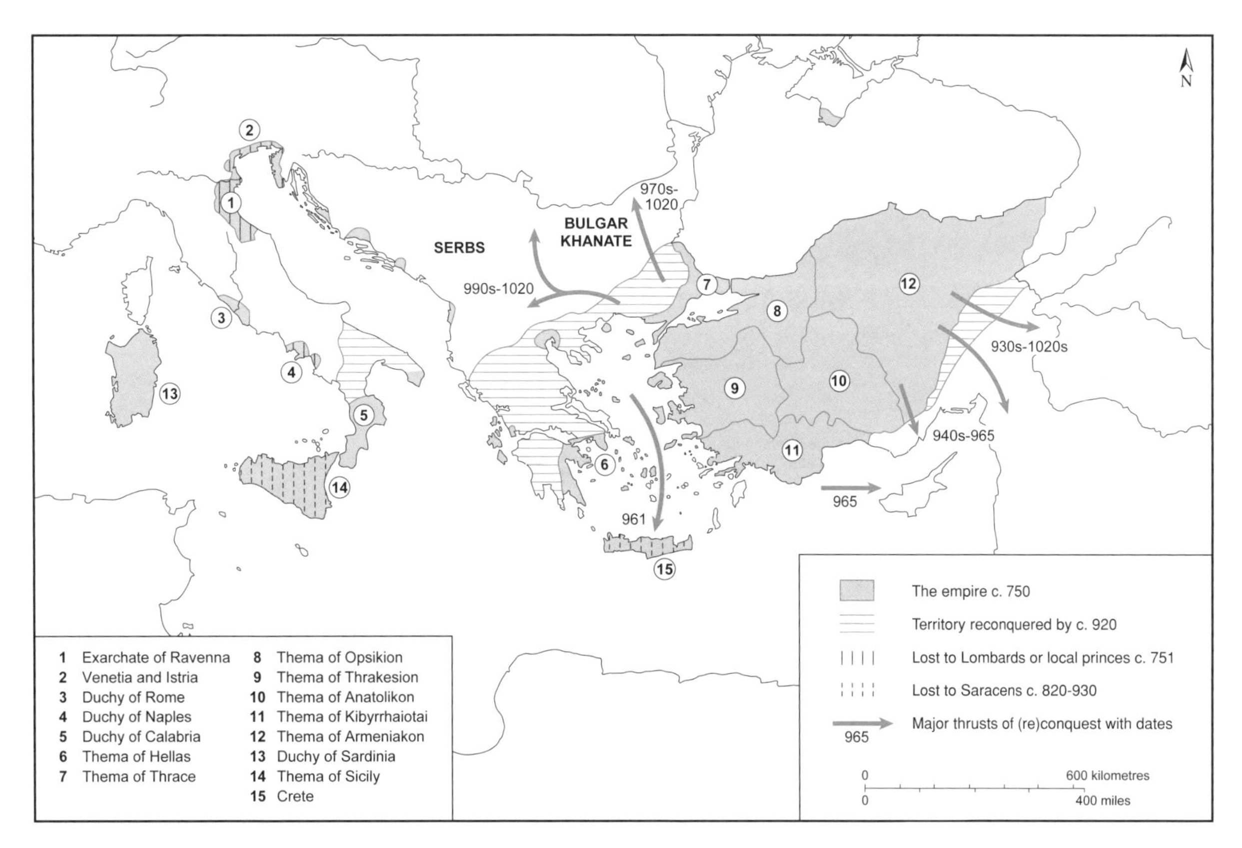

Map 5.2 The east Roman empire c. 632–1050: transformation and recovery.

the early tenth century, and as the Caliphate was weakened by internal strife, the Byzantines were beginning to establish a certain advantage; and in spite of the fierce and sometimes successful opposition of local Muslim warlords (such as the emirs of Aleppo in the 940s and 950s), there followed a series of brilliant reconquests of huge swathes of territory in north Syria and Iraq, the annihilation of the Second Bulgarian Empire, and the beginnings of the reconquest of Sicily and southern Italy.

During the last years of the ninth century and into the first two decades of the tenth, the recently-Christianised Bulgar state posed a serious threat to the empire – Constantinople was briefly besieged – but peaceful relations (followed by an increasing Byzantine influence on Bulgar culture and society) lasted for much of the tenth century. Resurgent Bulgar hostility resulted in a long and costly series of wars, culminating in the eventual destruction of an independent Bulgar Tsardom after 1014 and its absorption into the empire. By the time of the death of the soldier-emperor Basil II 'the Bulgar-slayer' (1025) the empire was once again the paramount political and military power in the eastern Mediterranean basin and south-east Europe, rivalled only by the Fatimid Caliphate in Egypt and Syria.

The offensive warfare that developed from the middle of the ninth century reacted, in its turn, upon the administration and organisation of the state's fiscal and military administrative structures. The provincial militias became less and less suited to the requirements of such campaigning, tied as they had become to their localities, to what was in effect a type of guerrilla strategy, and to the seasonal campaigning dictated by Arab or Bulgar raiders. Instead, regular field armies with a more complex tactical structure, specialised fighting skills and weapons, and more offensive *élan* began to develop, partly under the auspices of a new social élite of military commanders who were also great landowners, partly encouraged and financed by the state. Mercenary troops played an increasingly important role as the state began to commute military service in the provincial armies for cash with which to hire professionals: by the middle of the eleventh century, a large portion of the imperial armies was made up of indigenously recruited mercenary units together with Norman, Russian, Turkic and Frankish mercenaries, mostly cavalry, but including infantry troops (such as the famous Varangian guard).

The expansionism of the period c. 940–1030 also had negative outcomes. Increasing state demands clashed with greater aristocratic resistance to tax-paying; political factionalism at court, reflecting in turn the development of new social tensions within society as a whole, and in the context of weak and opportunistic imperial government, led to policy failures, the over-estimation of imperial military strength, and neglect of defensive structures. Pecheneg raids in the Balkans, the appearance of Seljuk raiders in the Armenian highlands, and the appearance of Norman mercenaries in Italy were all harbingers of change to come. Yet in 1050 the empire was at the height of its territorial power, its international position appeared unassailable, and its capital city was one of the most populous, commercially vibrant and cosmopolitan in the western Eurasian world.

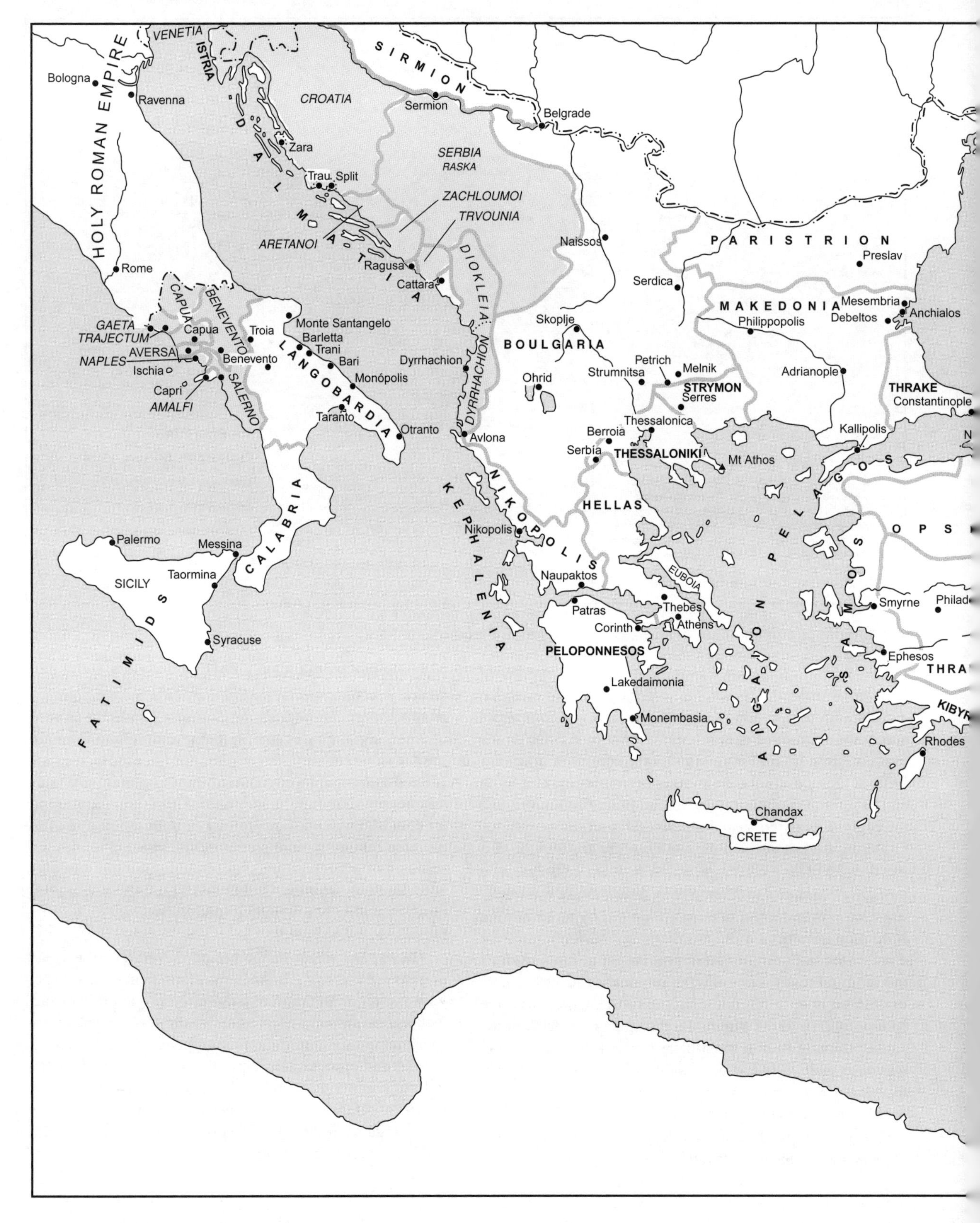

Map 5.3 Territorial losses and gains: the empire c. 1040.

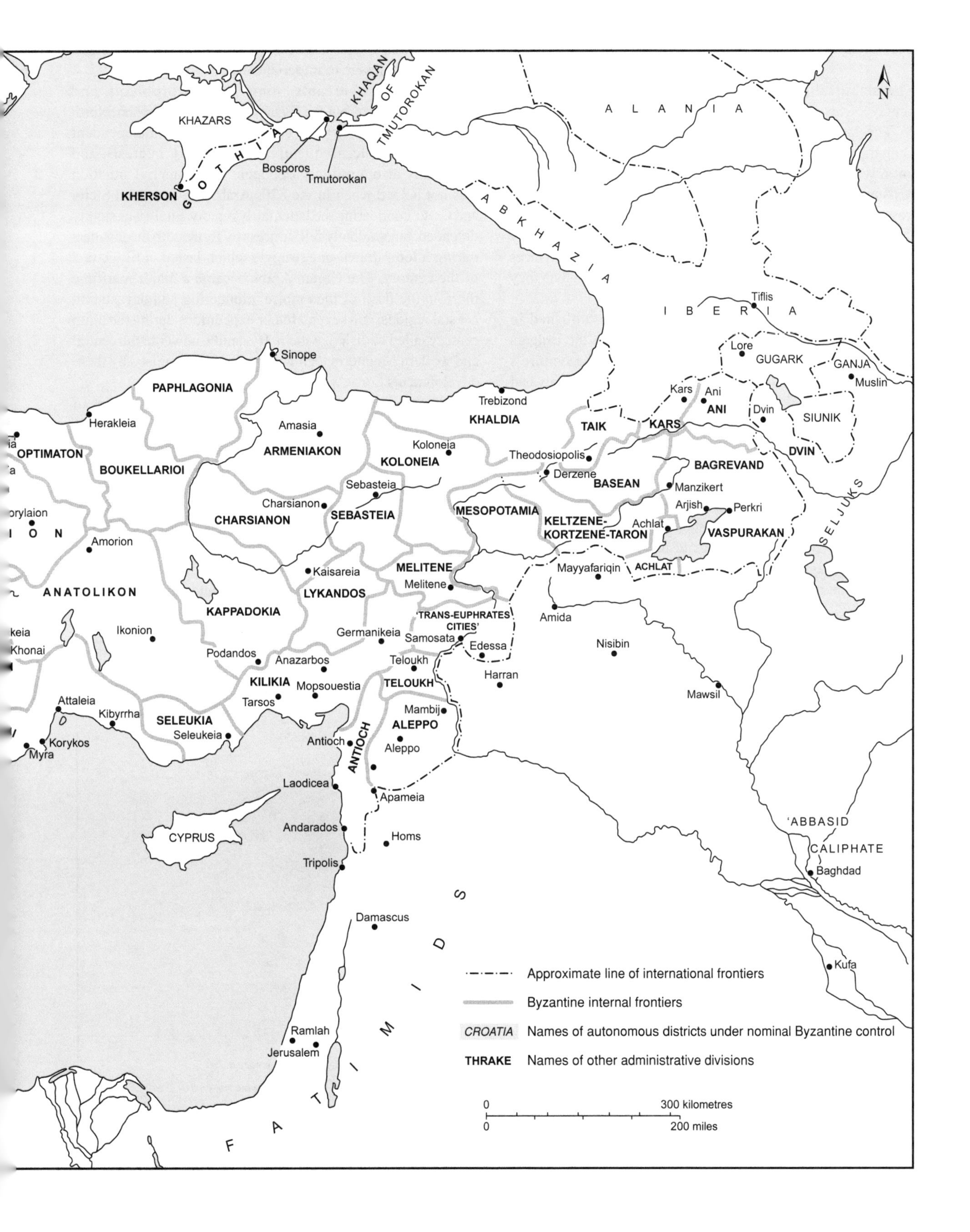
KHAZARS
KHAQAN OF TMUTOROKAN
GOTHIA
Bosporos
Tmutorokan
KHERSON
ALANIA
ABKHAZIA
IBERIA
Tiflis
Lore
GUGARK
GANJA
Muslin
Kars
Ani
ANI
Dvin
SIUNIK
DVIN
Sinope
PAPHLAGONIA
Trebizond
KHALDIA
TAIK
KARS
Herakleia
Amasia
ARMENIAKON
Koloneia
KOLONEIA
Theodosiopolis
BAGREVAND
OPTIMATON
BOUKELLARIOI
Derzene
BASEAN
Manzikert
Sebasteia
Arjish
Perkri
Charsianon
SEBASTEIA
MESOPOTAMIA
KELTZENE-KORTZENE-TARON
Achlat
VASPURAKAN
CHARSIANON
SELJUKS
Amorion
MELITENE
Mayyafariqin
ACHLAT
Kaisareia
ANATOLIKON
LYKANDOS
Melitene
KAPPADOKIA
'TRANS-EUPHRATES CITIES'
Amida
Ikonion
Germanikeia
Samosata
Nisibin
Khonai
Podandos
Anazarbos
Edessa
Teloukh
Harran
KILIKIA
Mopsouestia
TELOUKH
Mawsil
Attaleia
Tarsos
Mambij
Kibyrrha
SELEUKIA
ALEPPO
Korykos
Seleukeia
Antioch
ANTIOCH
Aleppo
Myra
Laodicea
Apameia
Andarados
Homs
'ABBASID CALIPHATE
CYPRUS
Baghdad
Tripolis
Damascus
FATIMIDS
Kufa
Ramlah
Jerusalem
Approximate line of international frontiers
Byzantine internal frontiers
CROATIA Names of autonomous districts under nominal Byzantine control
THRAKE Names of other administrative divisions
0 300 kilometres
0 200 miles
N

Territorial Losses and Gains

The stabilisation of the frontiers was one of the great achievements of the Emperors Leo III and his son Constantine V. The latter's frequent campaigns into the heartland of Bulgarian territory came near to destroying the Bulgar khanate entirely, although the Bulgars offered a tenacious and fierce resistance. In the east, he campaigned against a number of key Arab fortresses, re-establishing military parity between the Roman and Muslim armies, and thus providing the stability economically and politically to permit the devastated provinces to recover from the century and a half of warfare to which they had been subjected.

But although some certain stability was established in the east, and although Constantine's efforts kept the Bulgars quiescent until the end of the eighth century, the empire's political presence in the central Mediterranean and in Italy had markedly worsened. Ravenna, the last outpost of the Exarchate of Italy, fell to the Lombards in 751, and they in their turn soon came under Frankish domination. The papacy had for decades been effectively autonomous and independent, since imperial military support was minimal, and from the 750s, exacerbated in part by the tensions caused by the imperial espousal of iconoclasm, the alienation had increased. The popes forged an alliance with the kings of the Franks, Pepin I and then Charlemagne, who now replaced the eastern emperor as the dominant power in Italy (excluding Sicily); and in 800 the pope crowned Charlemagne emperor, an act seen in the east as a direct challenge to imperial claims.

Diplomacy overcame some of the problems and misunderstandings, but the Byzantine emperors had henceforth to reckon with a 'revived' empire in the west, independent of Constantinople, frequently with contrary interests, and potentially also a military opponent. The imperial situation was not helped when in the 820s Arab forces invaded Sicily and Crete, conquering the latter fairly rapidly. Sicily was stoutly defended, but gradually fell, fortress by fortress, to the invaders during a long-drawn-out struggle which lasted until the end of the century. The Cretan Arabs became a major maritime thorn in the flesh of the empire, plundering and devastating coastal regions, and several major expeditions during the ninth century failed to dislodge them. Byzantine power in the central and western Mediterranean was fatally compromised by these developments.

Although the Bulgar Khan Krum had inflicted a series of heavy defeats on the Byzantines in the period 811–813, the empire was able to recover and establish a peaceful relationship with his successors. The situation in the Balkans improved further under Basil I with the conversion to Christianity of the Bulgar Khan Boris, who took the Christian name Michael (852–889) and the title of Tsar (Caesar); a strong Christian, pro-Byzantine party developed at the Bulgar court. But during the reign of the Tsar Symeon (893–927), who was brought up in the imperial court at Constantinople and who had evolved

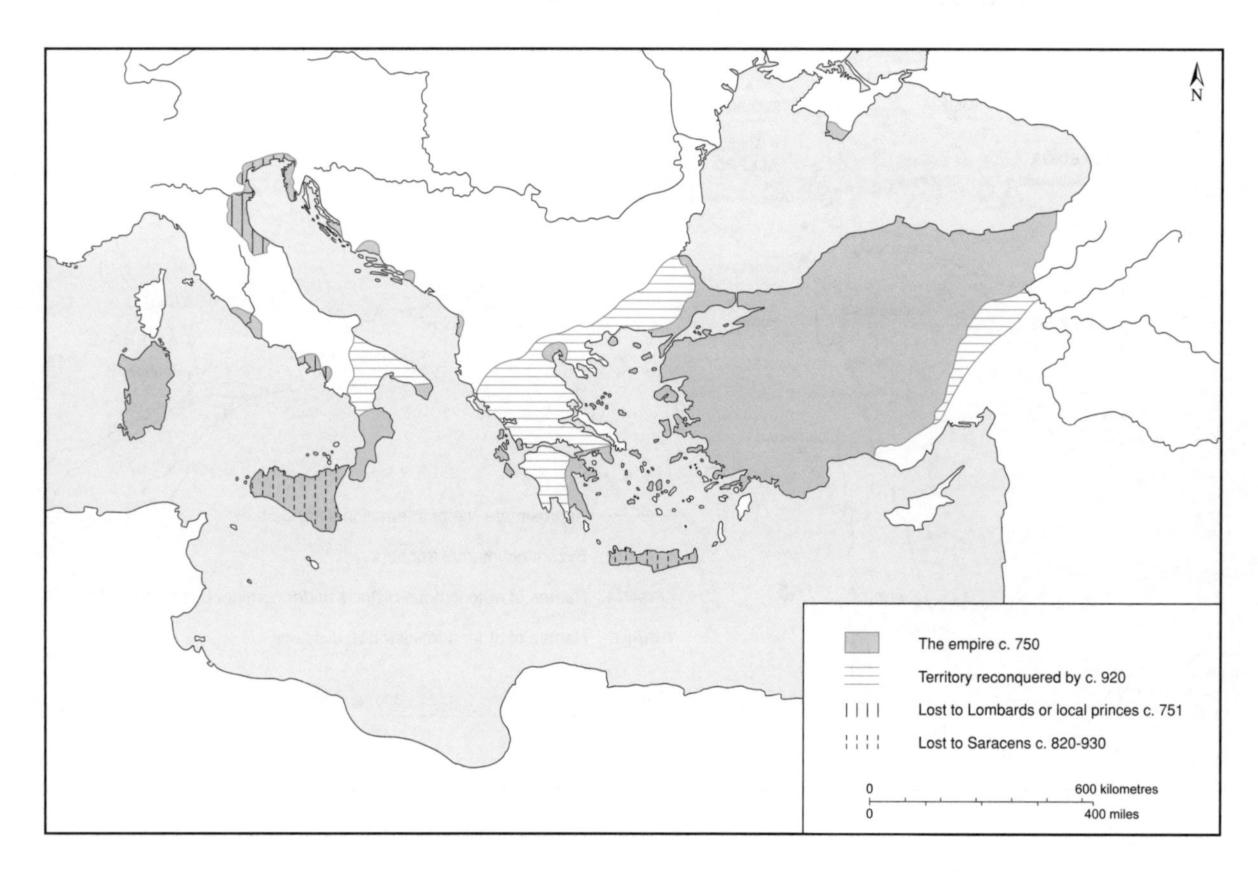

Map 5.4 Territorial losses and gains: 7th–10th centuries.

his own imperial pretensions, war broke out once more, a war which, with pauses, lasted until the 920s, and which at one point saw Constantinople besieged by a powerful Bulgar army. When peace was restored, it was through the efforts of the Emperor Romanos I, previously a commander of the imperial fleet, who had seized power and who, along with his sons, shared the imperial position with the legitimate heir Constantine VII. The peace lasted into the late 960s under a succession of pro-Byzantine Bulgarian rulers.

One other zone was of importance to Byzantine rulers, as it had been to the emperors of the sixth century and before. The steppe region stretching from the plain of Hungary eastwards through south Russia and north of the Caspian was the home of many nomadic peoples, mostly of Turkic stock, and it was a fundamental tenet of Byzantine international diplomacy to keep the rulers of these various peoples favourably disposed towards the empire, achieved predominantly through substantial gifts of gold coin, fine silks, and imperial titles and honours. After the collapse of the Avar empire in the 630s, Constantinople had been able to establish good relations with the Khazars whose Khans, although converting to Judaism, remained a faithful ally of most Byzantine emperors, duly exploiting the Byzantine invitation to attack the Bulgars from the north, for example, when war broke out in this region, but serving also to keep the imperial court informed of developments further east. The Khazar empire began to contract during the later ninth century, chiefly under pressure from the various peoples to the east who were set in motion by the expansion of the Pechenegs and allied groups. The Magyars (Hungarians) were established to the north-west and west of the Khazars by the middle of the ninth century, whence they established themselves in what is now Hungary, destroying local Slav kingdoms in the process, by the early tenth century. Both Khazars and Magyars served as mercenaries in Byzantine armies, particularly against the Bulgars, although the establishment by the later tenth century of a Christianised Hungarian kingdom on the central Danube posed a potential challenge to Byzantine power in the region, which became especially acute during the twelfth century. The Khazars remained important players in steppe diplomacy until the middle of the tenth century, when the growing power of the Kiev Rus' finally brought about their destruction and replaced them in Byzantine diplomacy. The appearance of the Turkic Pechenegs (Patzinaks) during the late ninth century complicated this arrangement: the newcomers clashed with both the Khazars and the Magyars, establishing themselves in the steppe region between the Danube and Don. Their value to the empire as a check on both the Rus' and the Magyars was obvious, particularly in the wars of the later tenth century, but they were dangerous and frequently unreliable allies. During the middle years of the eleventh century groups of Pechenegs began

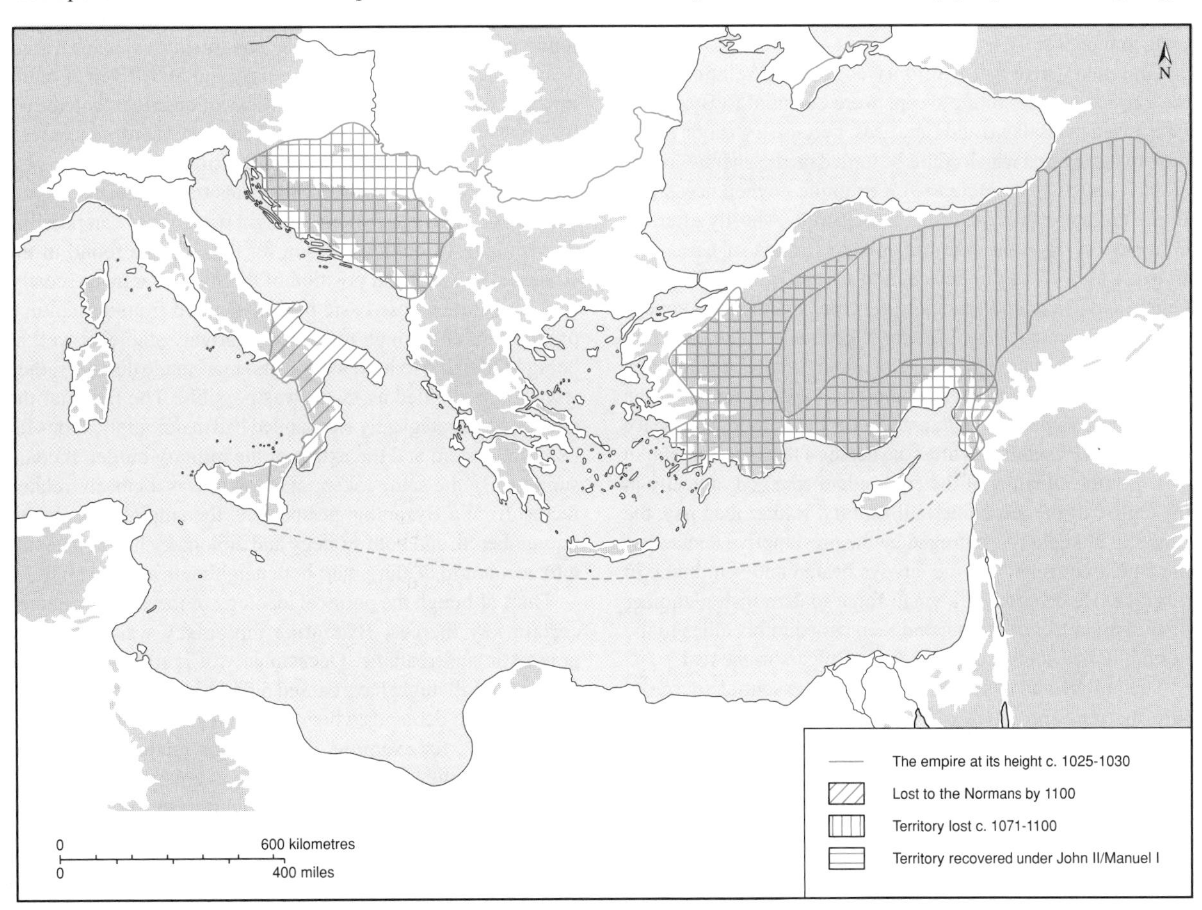

Map 5.5 Territorial losses and gains: 11th–12th centuries.

to move into the Balkans, where they clashed with imperial troops. Until the period of civil wars after Manzikert in 1071, however, they were kept more-or-less in check; thereafter they ravaged and pillaged with little opposition until Alexios I was finally able to crush them in 1091.

Diplomacy and Embassies

Diplomacy was a crucial aspect of the east Roman struggle for survival or for regional political supremacy. But the emphasis placed by Byzantine writers and governments on effective and intelligent diplomacy was not just a question of cultural preference informed by a Christian distaste for the shedding of blood: to the contrary, the continued existence of the state depended upon the deployment of a sophisticated diplomatic arsenal. The whole history of Byzantine foreign relations reflects this, both in the few explicit statements of political theory which survive, most obviously in the tenth-century *De administrando imperio* ('On governing the empire'), as well as in the theory and practice of Byzantine diplomacy. As the Emperor Constantine VII states in the introduction to this treatise, a ruler must study what is known of the nearer and more distant peoples around the Roman state, so that he can understand 'the difference between each of these nations, and how either to treat with and conciliate them, or to make war upon and oppose'.

Diplomacy also had a military edge: good relations with the various peoples of the steppe were essential to Byzantine interests in the Balkans and Caucasus, because a weapon might thereby be created which could be turned on the enemies of the empire – such as the Bulgars, for example – when necessary. In the autumn of the year 965, for example, shortly after the conquest by Byzantine armies of the islands of Crete and Cyprus, as well as the destruction of the Islamic power in Cilicia and its incorporation into the empire, Bulgarian envoys arrived at the court of the Emperor Nikephoros II Phokas. Their purpose was to request the payment of the 'tribute' (or 'subsidy' from the imperial perspective) paid by Constantinople to the Bulgar Tsar as part of the guarantee for the long-lasting peace that had been established after the death of the Tsar Symeon in 927. But the situation of the empire had changed radically in the course of the preceding half century. Rather than pay, the Emperor Nikephoros, outraged by the presumptive demand of the Bulgarian ruler, had the envoys beaten and sent home in disgrace. He despatched a small force to demolish a number of Bulgarian frontier posts, and then called in his allies to the north, the Kievan Rus', to attack the Bulgars in the rear.

Such allies and contacts were also an essential source of information, and much effort was expended in gathering information which might be relevant to the empire's defence. Many people were involved – diplomatic contacts, embassies, as well as spies, merchants and other travellers, and not excluding churchmen. Military treatises devote considerable attention to information-gathering, which became even more important from the later seventh century when, following 50 years of warfare, both sides began to establish a sort of 'no-man's-land' in Asia Minor, across which information travelled only with difficulty through the usual channels of social and commercial intercourse.

The history of Byzantine embassies to the empire's eastern and western neighbours is complex and full of shifts in emphasis and motive as the empire's political and strategic situation changed over the centuries. The routes used also changed as access was made possible or not according to the particular political situation in specific regions through which travellers had to pass. Several key motifs in imperial diplomacy remained constant, however. In the first place, the emperors needed to be able to persuade their neighbours not to attack them, and the offer of subsidies, the threat of an attack from another imperial ally from the rear, or of a direct imperial military response, were all part of the diplomat's arsenal. Challenging aggression on the basis of a shared faith was also a useful tool, and was used in the case of the empire's nearer neighbours, in particular the Bulgars. In the second place, the image that the east Romans wished others to have of them was important, and much effort was devoted to impressing visiting rulers and ambassadors of the splendour and thus the power of this God-protected empire. As part of this picture, and as one element in the development and maintenance of a set of protective alliances, the emperors also arranged extensive exchanges of gifts, offers of military, diplomatic or material support, cultural exchanges, and marriage alliances. Imperial diplomacy was certainly successful across the life of the empire, for its beleaguered strategic position rendered it extremely vulnerable, and its survival owes a great deal to factors other than the purely military. Even if many imperial plans foundered on the rocks of hostile intent abroad (or opposition at home), yet still there were many successes.

Apart from ideological considerations, there were also pragmatic concerns. A thread that runs throughout Byzantine history is a general reluctance to fight wars if they can possibly be avoided. An obvious reason for this is to be found in the strategic-geographical position of the state. Wars were costly, and for a state whose basic income derived from agricultural production, and which remained relatively stable as well as being vulnerable to both natural and man-made disasters, they were to be avoided as far as was possible. The fact that the empire was strategically surrounded had major implications for its fiscal system and the extent of the military burden it could support. By the same token, manpower was a closely related factor: from a Byzantine perspective, the empire was always outnumbered, and both strategy and diplomacy had to take this into account in dealing with both neighbours and enemies.

Thus, although the political ideology of the empire dictated certain key themes, Byzantine diplomacy was extremely pragmatic and realistic. Occasionally, it is true, a particular emperor's will might have caused difficulties, particularly when responding to demands which were taken to be unreasonable. Thus in 572, for example, the Emperor Justin II refused to pay the subsidies demanded by Persian envoys guaranteeing peace in the east. The Persians declared war, and although there followed some minor but successful Roman raids into Persian territory in Arzanene, there was also a major Persian incursion into Syria, with great loss of life and property to the Romans, followed in 574 by the successful Persian siege and capture of Daras, a strategic disaster for the Romans. The difficulties

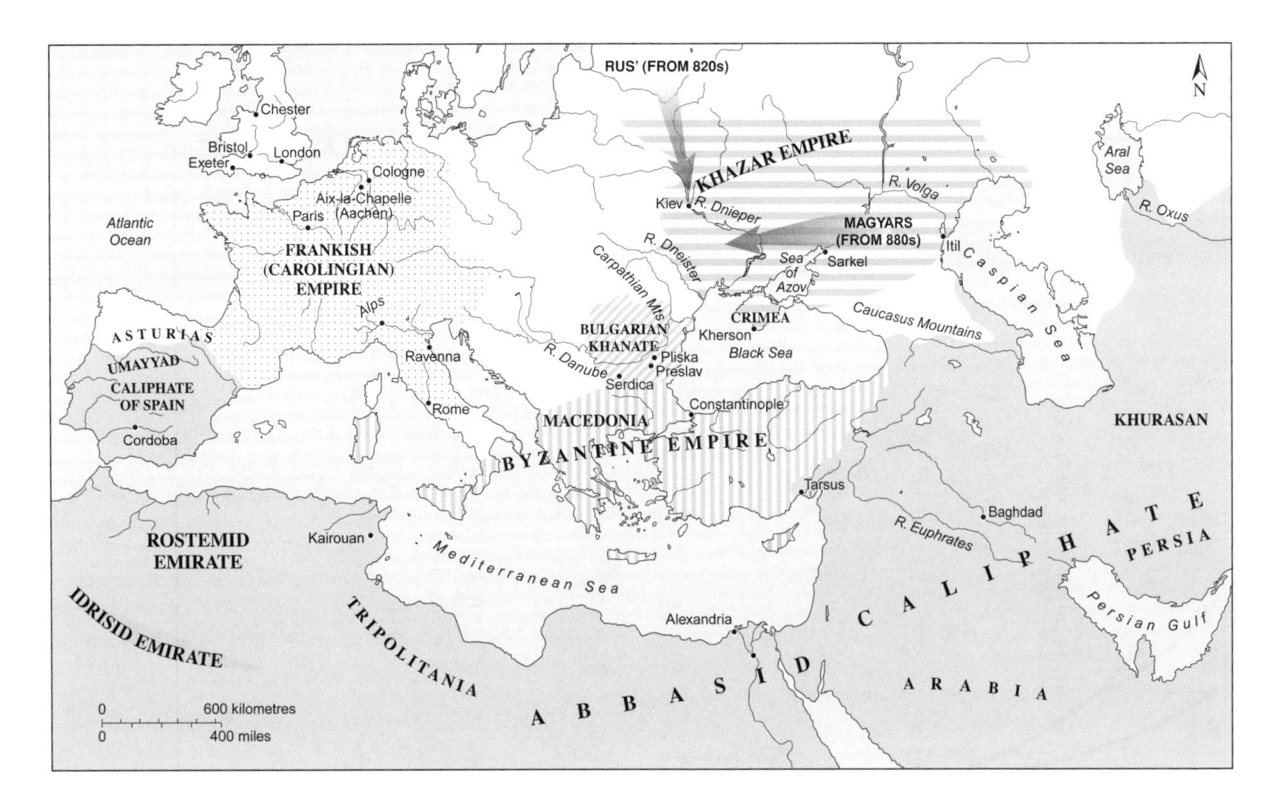

Map 5.6 The diplomatic world of Byzantium c. 840.

caused by the Avars on the Danube, as well as the emperor's illness, meant that the Roman response was ineffective, and a truce on the Mesopotamian front was bought for the years 576–578: Justin's rashness had brought some serious medium-term problems for the Romans.

In spite of such vagaries, however, continuity of purpose and effort is remarkable across many centuries, and again the fact of the empire's survival is a testament to the effectiveness, responsiveness and flexibility of the imperial government when faced by a vast range of issues.

Church Politics: Heresy, Schism and Expansion c. 641–1060

By the later sixth century, the disaffection brought about by Constantinopolitan persecution of the Monophysites rendered a compromise formula essential for the re-incorporation of the territories which had been lost to the Persians. Under Heraclius two possible solutions were proposed. The first was known as 'monoenergism', whereby a single energy was postulated in which both divine and human aspects were unified. At this point, the arrival of Islam on the historical stage made the need for a compromise even more pressing. When monoenergism was rejected, an alternative doctrine – of a single will ('monotheletism') – while initially attracting some support, was eventually also rejected, but survived as an imperial policy, enforced by decree after Heraclius' death in 641. By this time the Monophysite lands had been lost to the Arabs and the purpose of the compromise was lost. The government, which ruled in the name of the young emperor Constans II, was obliged to maintain the policy, and Constans himself fiercely imposed a ban on further discussion. Only in 680, some 12 years after the assassination of Constans in Sicily, was his son and successor Constantine IV able to summon a general council of the church and restore ecclesiastical unity by quietly abandoning official monotheletism.

After the middle of the seventh century, Christological issues faded into the background. Only the iconoclast controversy of the eighth and ninth centuries (which did, however, at a later stage have a Christological element) stimulated further major internal rifts, and it is by no means clear that these commanded the interest or commitment of more than a handful on either side, at least until the iconophiles rewrote the history of the period during the ninth and tenth centuries. Traditionally it has been assumed that the sources describing the mass persecution, harassment and death of many iconophiles, as well as the destruction of icons themselves, were more-or-less accurate accounts, and that the Emperor Leo III was to blame. In fact, it seems that much of the story consists of later legend and exaggeration. Leo III may have been a mild critic of the use of images (although there is no reliable evidence that he issued an edict condemning them). His son, Constantine V, while theologically more involved, only adopted a strongly iconoclastic policy after the first eight or so years of his reign. Constantine's concern was for images to be removed from those positions in churches where they would be the object of mistaken veneration. In the mid-780s it became convenient for

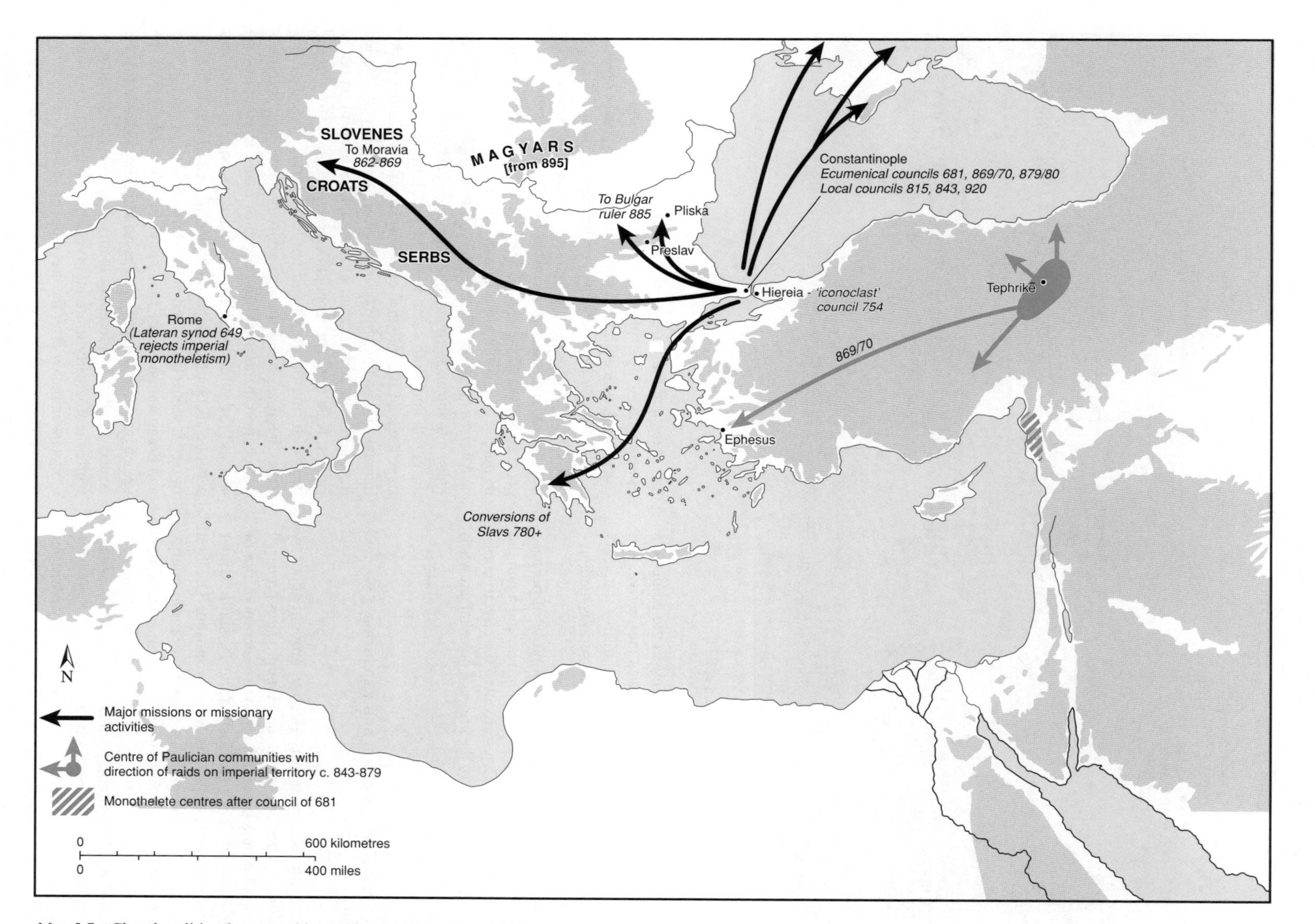

Map 5.7 Church politics: heresy, schism and expansion c. 641–1060.

the Empress Eirene, acting for her young son Constantine VI, to shift her allegiance. The Ecumenical Council of 787 'restored' images, but it is also clear that it was only from this time that a formal theology of images, so important for later orthodox doctrine, was first elaborated.

After a period during which the 'restored' cult of images flourished, the general Leo V (813–820) re-introduced imperial iconoclasm, seen as the ideological force behind the victories of Constantine V 50 years earlier. From then until 843 the iconoclastic controversy once more divided church and state, being resolved only after the death of the Emperor Theophilos in 842. Under the influence of a leading court official, the eunuch Theoktistos, the empress and regent (for the young Michael III, 842–867) agreed to the restoration of sacred images and their public display at a series of small private meetings in 842–843. The change was made public by a triumphal procession through Constantinople, an event still celebrated in the orthodox liturgical calendar.

Heresy and heterodoxy were two of the constant issues which the church, and the emperors, had to confront. The geographical and cultural variety of the Byzantine world meant that in many regions traditional, pre- or non-Christian practices could linger on unobserved for centuries, albeit in isolated and relatively limited groups. By the same token, heterodox beliefs could evolve that might, and did in some cases, evolve into major challenges to the imperial authority. The local and ecumenical councils tried to grapple with some of the causes for heresy, namely the lack of clerical discipline or supervision in far-flung regions, the ignorance of some of the lower clergy as well as of the ordinary populace, or the arrival of immigrant population groups with different views or different understanding of the basic elements of Christianity.

Occasionally state and church had to confront a major heretical movement. Such was the case, for example, with Paulicianism in eastern Asia Minor in the middle of the ninth century, named probably after one of its early exponents, Paul of Samosata. By the ninth century a mixture of dualist and neo-Manichaean elements, it became powerful in eastern Asia Minor. State persecution led to military mobilisation of the Paulicians under a series of very able commanders, an alliance with the Caliphate in the 870s, and a full scale war, waged by the Emperor Basil I, which led eventually to its destruction. The transfer of populations by the government from eastern Asia Minor to the Balkans brought also Paulicianism or beliefs influenced by it, and led directly to the development of a heretical tendency, primarily among the Slavic populations of Bulgaria and the western Balkans, known as Bogomilism. Although fiercely persecuted by the emperors, especially by Alexios I, and eradicated from Constantinople, it spread throughout the Balkans and represented a major strand in the religious culture of the region.

By the time Leo III came to the throne in 717 an increasing alienation between Constantinople and Rome was apparent. Chiefly at issue were matters of ecclesiastical jurisdiction and imperial taxation policy in Italy (although in the period after 754 iconoclasm may also have contributed). Strained relations with the papacy erupted into full-scale conflict in the middle of the ninth century in the so-called 'Photian schism', which followed the forced resignation of the Patriarch Ignatios in 858. The appointment (preceded by a rapid ordination) as his successor of the learned layman Photios, permitted the former patriarch and his supporters to enlist the support of Pope Nicholas I, who was able to use the situation to intervene in eastern church politics and justify the papacy's claim to a superior status within Christendom. When, on his accession in 867, the Emperor Basil I removed Photios and restored Ignatios, however, things did not improve, since Ignatios was equally hostile to papal claims. Reconciliation finally took place at the Council of Constantinople in 879.

A second, more serious break took place in the 1050s, and involved both political and doctrinal issues. At the beginning of the sixth century the word '*filioque*' was added to the Chalcedonian creed in the Frankish lands, in an attempt to clarify the fact that the Holy Spirit proceeded from both Father and Son. Frankish churchmen used it during their efforts to convert the Bulgarians in the ninth century, and the Patriarch Photios later wrote a detailed treatise condemning it. At the Council of Constantinople in 879 the Roman legates accepted its redundancy and it was withdrawn. But by the early eleventh century it had been reintroduced, and formed the basis of disagreement. In 1054 the stubbornness of the pope's legate, Cardinal Humbert, and of the Patriarch, Michael Keroularios, led to mutual anathemas – formal pronouncements of condemnation – being proclaimed. The two churches remained estranged thereafter, in spite of a gradual lessening of the tension during the reign of Alexios I.

Nevertheless, after the loss of the three eastern patriarchates of Antioch, Alexandria and Jerusalem to Islamic domination, the Byzantine Church was able to expand, as the conversion of, first, Bulgaria (from the 860s) and, later, the Kievan Rus' (from the 990s) to the Byzantine form of orthodoxy led to the creation of what has been termed the 'Byzantine commonwealth', and a permanent feature of the cultural history of those lands (see pages 160–161).

6 Economy, Administration and Defence

Strategic Change: The East Roman Response

At the end of the Persian wars (c. 628/629) Heraclius had re-established the pre-existing arrangements familiar from the end of the sixth century (see pages 24–25). Two changes seem to have been the merging of the two praesental field armies into one, and the disappearance of the army of Illyricum as the area was overrun by Slav and other invaders or immigrants. There was also a partial re-establishment of Arab allies along the eastern frontier, along with the restoration of the system of at least some *limitanei* posts and garrisons. The regional command structure was restored to the situation before Heraclius. The system of defence in Italy and Africa had been unaffected by the Persian wars and remained unchanged.

The Arab Islamic conquests radically altered the strategic and political geography of the whole east Mediterranean region. Following the disastrous defeat of 636, the field armies were withdrawn first to north Syria and Mesopotamia, and shortly thereafter back to the line of the Taurus and Anti-Taurus ranges. The regions across which they were based were determined by the ability of these districts to provide for the soldiers in terms of supplies and other requirements. The imperial field army was pulled back to its original bases in north-west Asia Minor and Thrace, where it becomes known as the *Opsikion* division. That of the *magister militum per Orientem* (or 'Master of Soldiers of the east'), occupied southern central Asia Minor, and became known as the *Anatolikon* army; and that of the Master of Soldiers of Armenia, now known as the *Armeniakon*, occupied the eastern and northern districts of Asia Minor. The army of the Master of Soldiers of Thrace, which had apparently been transferred to the eastern theatre in the mid-630s, and had been employed unsuccessfully to defend Egypt, was established in the rich provinces of central western Anatolia, and known thenceforth as the *Thrakesion* army. By the last decades of the seventh century, the districts across which these armies were garrisoned were known collectively by the name of the army based there. While the distribution of the various units of the field armies across the provinces in this way was certainly connected with logistical demands, it had obvious strategic implications, since it meant that Roman counter-attacks were relatively slowly to organise, and that defence was fragmented and organised locally on a somewhat piecemeal basis.

The provinces which had belonged to the *quaestura exercitus* established by Justinian did not survive the Slav and Avar invasions of the Balkan provinces (although the empire still controlled much of the Danube itself, through isolated fortresses on the Danube delta and along the coast of the Black Sea); but the Aegean regions continued to function as a source for men, ships and resources, and a maritime corps, known in the later seventh century as the 'ship troops', or *karabisianoi*, seems to have been based to begin with on Rhodes. In the light of the considerably increased threat posed to the empire's exposed coastline and its hinterlands, brought about by the rapid development of Arab seapower from the 660s, these 'ship troops' were to develop into the core of middle Byzantine provincial naval power. In addition to these naval units, the imperial fleet at Constantinople (equipped from the 670s with 'liquid fire' projectors) was complemented by squadrons from the *thema* of Hellas. The armies of the *magistri militum* or exarchs of Italy, and Africa (which included Sardinia) continued to function, although the latter disappeared with the completion of the Arab conquest of North Africa in the 690s, the army of Italy surviving, on an ever more localised basis, until the demise of the Exarchate of Ravenna in the middle of the eighth century.

The *themata* or themes were at first merely groupings of provinces across which different armies were based. By 730 or thereabouts they had acquired a clear geographical identity; and by the later eighth century some elements of fiscal as well as military administration were set up on a thematic basis, although the late Roman provinces continued to subsist. The number of *themata* expanded as the empire's economic and political situation improved, partly through the original large military divisions being split up into different 'provincial' armies (a process begun under Leo III and continued by his successors), and partly through the recovery in the last years of the eighth century and the re-imposition of imperial authority over lands once held in the southern Balkans (begun under Eirene and Constantine VI). The first large *thema* to be thus subdivided was the *Opsikion*, which was broken up into three corps, the names of which reveal their late Roman origins – the *Boukellarioi*, the *Optimatoi* and the *Opsikion*. The *themata* were complemented along the eastern frontier by a series of special militarised districts intended to control key passes and roads into and out of the empire, known as *kleisourai*. As the empire went back onto the offensive in the later ninth century and after, these were converted into themes in their own right.

The localisation of recruitment and military identities that resulted from these arrangements led to a distinction between the regular elements – full-time soldiers – and the less competent or well-supplied militia-like elements in each theme region. In the 760s a small élite force, known as the *tagmata* ('the regiments') was established under Constantine V (741–775), which quickly evolved into the élite field division for campaign purposes. It had better pay and discipline than both the regular and the part-time provincial units, and this was the first step in a tendency to recruit mercenary forces, both foreign and indigenous, to form special units and to serve for the duration of a particular campaign or group of campaigns. As the empire reasserted its military strength in the east in the ninth and tenth centuries, the role and the proportion of such full-time units became ever more important.

Strategic Change: From Defence to Offence

In spite of some significant defeats, a more offensive policy in both east and north, combining diplomacy and missionary

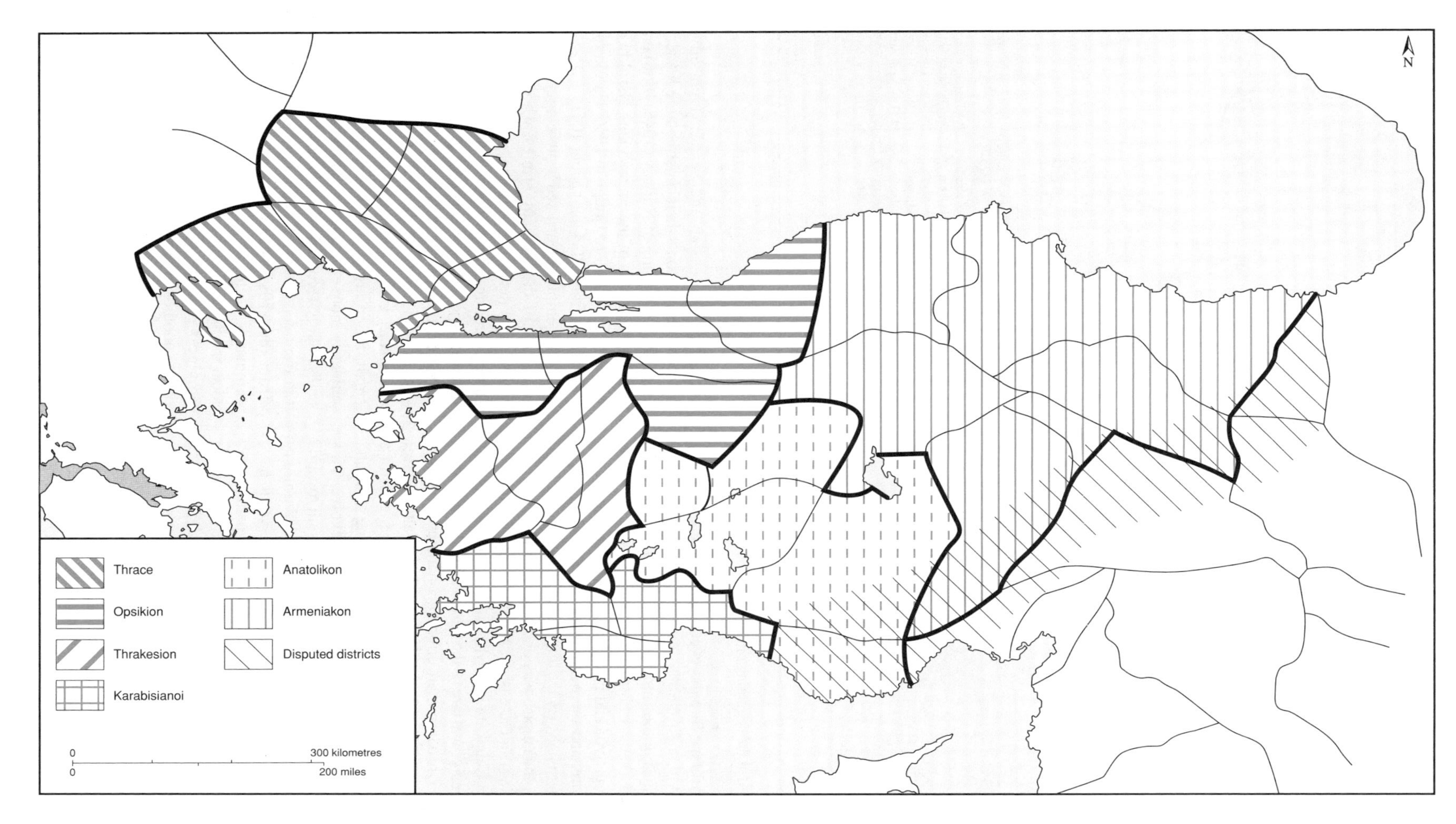

Map 6.1 Schematic map of the first *themata* (▬) and the late Roman provinces (——) c. 660–740.

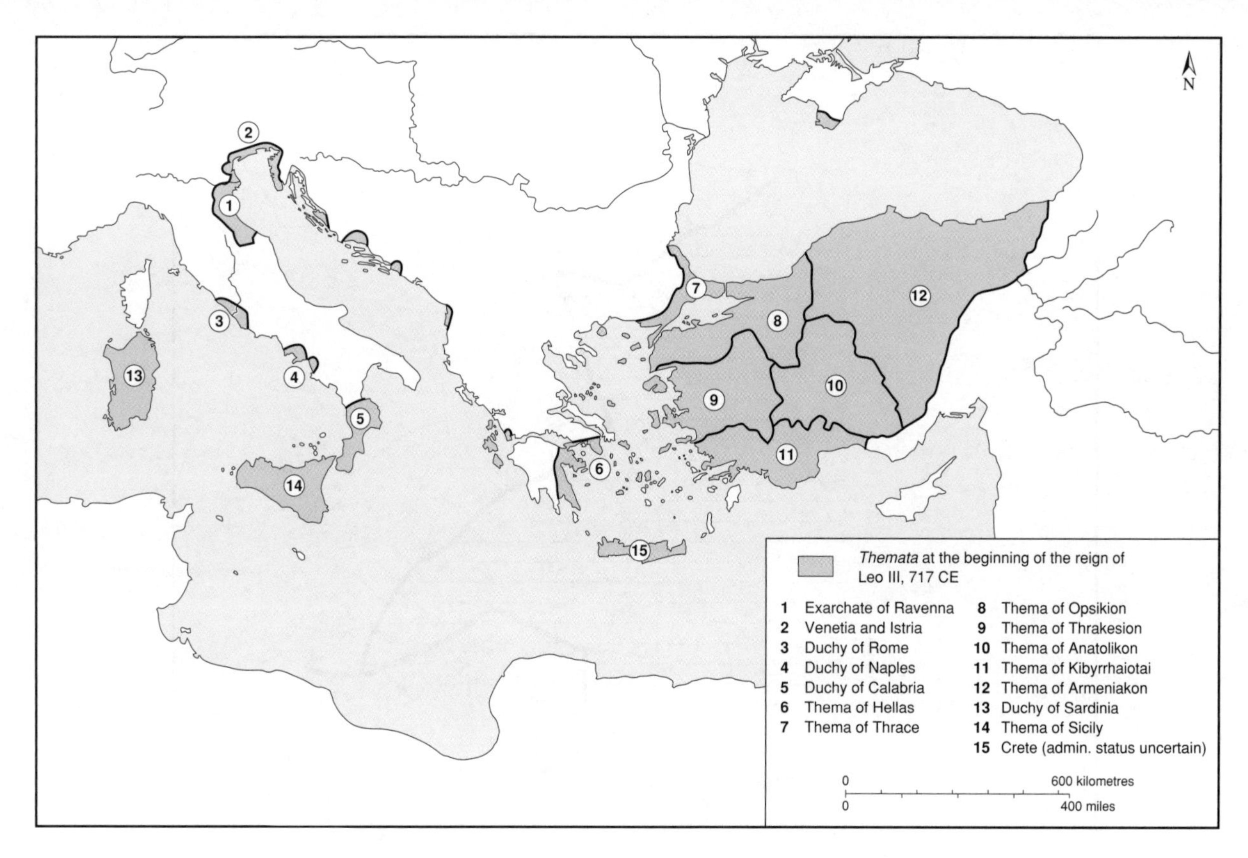

Map 6.2 *Themata* at the beginning of the reign of Leo III, 717 CE.

activity with military threats, eventually helped to confer an advantage. Although facing fierce opposition from local Muslim emirs during the first half of the tenth century, the empire achieved a series of brilliant reconquests of large tracts of territory in north Syria and Iraq, the destruction of the Second Bulgarian Empire, and the beginnings of the reconquest of Sicily and southern Italy. By the time the soldier-emperor Basil II died in 1025, the empire was once more a major political and military power in the eastern Mediterranean basin, with only the Fatimid Caliphate in Egypt and Syria to challenge its power. The offensive warfare of the period from the mid-ninth century, however, had important effects upon imperial military organisation. As the thematic militias became less able to meet the needs of aggressive warfare, regular field armies with a more complex tactical structure, specialised fighting skills and weapons, and a more offensive spirit began to evolve. This process was aided by and was also one of the stimuli towards the growth of a social élite in the provinces, landed military officers whose dominance of the regional military system gave them both expertise and political weight at the imperial level. Full-time professional units played a growing role as the state began to commute thematic military service for cash payments, which were then used to hire mercenaries. The result was a colourful and international army – remarked on by outside observers – consisting of both indigenous mercenary units as well as Russians, Normans, Turks and Franks, both infantry and cavalry. Perhaps best-known among these are the famous Varangian guard, first recruited during the reign of Basil II, consisting of Russian and Scandinavian adventurers and mercenaries. Among their most notable leaders was Harald Hardrada, later King of Norway (1046–1066) until he met his death at the hands of the English king Harold Godwinsson at the battle of Stamford Bridge in 1066. Harald fought with the Varangians from 1034 until about 1041.

As the empire prioritised a more aggressive strategy, the relevance of the thematic armies, whose primary function had become defensive in nature, meant that new tactical and strategic command structures evolved. New military districts under independent commanders evolved, beginning with the conversion of former *kleisourai* – small frontier commands – to *themata* along with the incorporation of conquered regions as *themata*. Unlike the older *themata* these were usually quite small, based around a key strongpoint. As ever larger and militarily more effective detachments of the imperial *tagmata* and similarly-recruited professional units were established along the frontiers so this system grew in extent and significance. From the 970s, these divisions were grouped into larger commands, each under a *doux* or *katepano*, independent of the local thematic administration. They formed a screen of buffer provinces protecting the old *themata*, tactically independent of one another in terms of their available manpower. Similar arrangements were established in the Balkan and western provinces. Such forces, whether on the frontiers or within the provinces, consisted increasingly of mercenary, professional troops or of forces sent by the dependent rulers of the various smaller states bordering the empire.

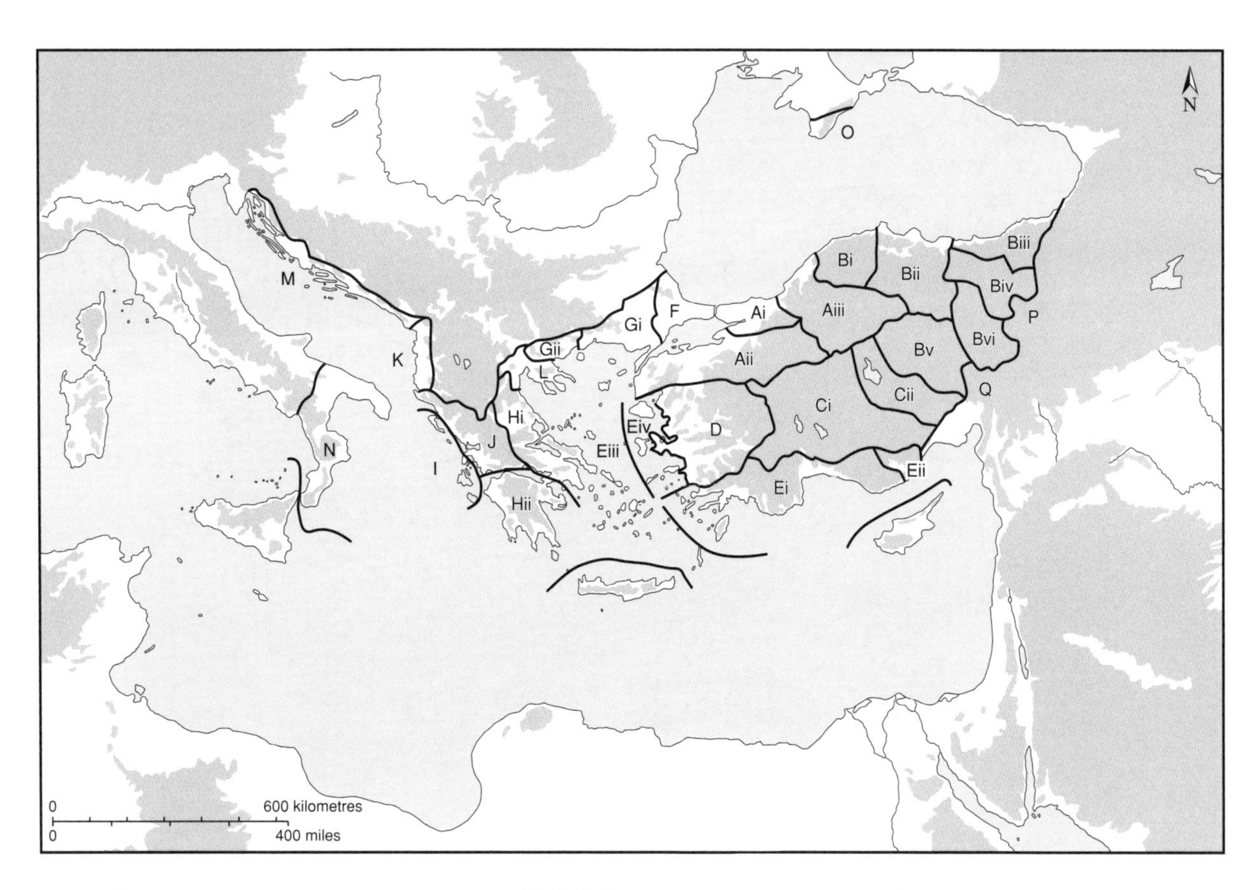

	New *themata*	**Original *themata***		**New *themata***
Ai	Optimaton	Opsikion	I	Kephallênia
Aii	Opsikion			
Aiii	Boukellarion		J	Nikopolis
Bi	Paphlagonia	Armeniakon	K	Dyrrhachion
Bii	Armeniakon			
Biii	Chaldia		L	Thessaloniki
Biv	Koloneia			
Bv	Charsianon		M	Dalmatia
Bvi	Sebasteia			
			N	Laggobardia
Ci	Anatolikon	Anatolikon		
Cii	Kappadokia		O	Kherson
D	Thrakêsion	Thrakêsion	P	Mesopotamia
Ei	Kibyrrhaiotai	Karabisianoi/Kibyrrhaiotai	Q	Lykandos
Eii	Seleukeia			
Eiii	Aegaios Pelagos (Aegean Sea)			
Eiv	Samos			
F	Thrakê	Thrakë		
Gi	Makedonia	Makedonia		
Gii	Strymôn			
Hi	Hellas	Hellas		
Hii	Peloponnêsos			

Not shown here are short-lived commands such as *Leontokômê*, created in the region around Tephrikê after Basil I's armies destroyed the town c. 879, originally a *kleisoura*, then renamed and established as a *thema* by Leo VI

(After Haldon, *Warfare, State and Society*)

Map 6.3 Themata c. 920.

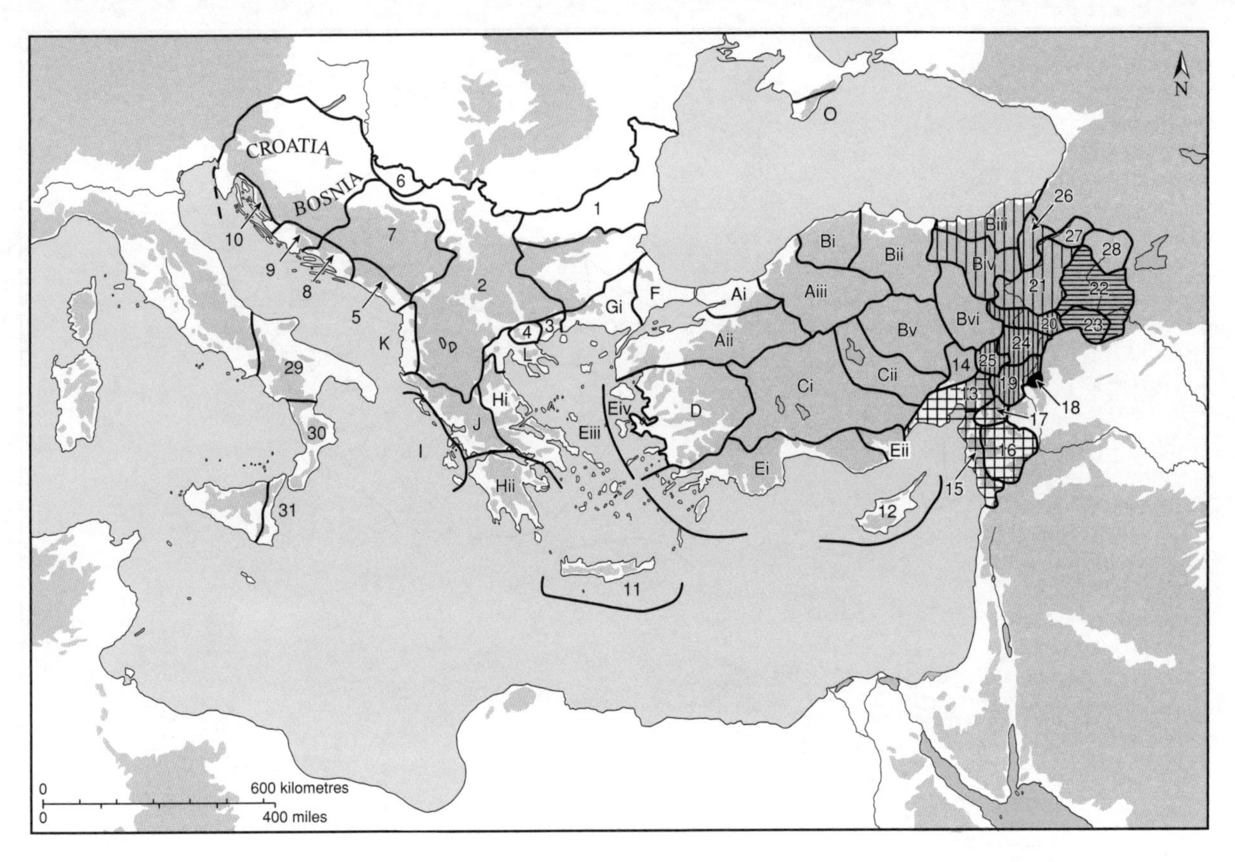

1 Paristrion
2 Boulgaria
3 Strymôn
4 Neos Strymôn
5 Diokleia
6 Sirmion
7 Serbia
8 Terbounia
9 Zachloumoi (autonomous)
10 Arentanoi (autonomous)
11 Crete
12 Cyprus
13 Kilikia
14 Lykandos
15 Antiocheia
16 Aleppo (autonomous)
17 Dolichê (Teloukh)
18 Edessa
19 Trans-Euphrates cities
20 Keltzinê-Chortzinê
21 Derzênê/Phasianê (Basean)
22 Vaspurakan
23 Taron
24 Mesopotamia
25 Melitênê
26 Iberia
27 Kars
28 Shirak/Ani
29 Laggobardia
30 Kalabria
31 Sikelia (1038–1042)

A–L as in Map 6.3.

Districts under *Doux* of Antioch

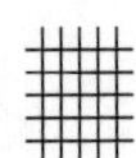

Districts under *Doux* of Edessa

Districts under *Doux* of Chaldia

Districts under *Doux* of Vaspurakan

Districts under *Doux* of Mesopotamia

(After Haldon, *Warfare, State and Society*)

Map 6.4 Themata and ducates c. 1050.

Over the same period the empire's naval arrangements expanded from the single provincial fleet and the Constantinopolitan imperial squadrons of the seventh century. By the 830s there were three main naval *themata*, of the Aegean, of Samos and the Kibyrrhaiotai, in addition to the imperial fleet, and the much smaller provincial fleets of Hellas and the Peloponnese. The maritime front was thus covered in the east, and while continued raiding and piracy was not stopped, it was at least checked and occasionally thrown back. In the west a different situation prevailed. The definitive loss of Carthage and the remaining North African provinces by the late 690s had deprived the empire of its naval bases there, although Sicily probably continued to support imperial flotillas, while there is some slight evidence for imperial naval activity in the Balearics. Sardinia remained an imperial possession. But by the early ninth century the empire seems to have lost interest in the western Mediterranean. Adequate naval support at the time when Sicily and then Crete were invaded in the 820s was not forthcoming, a costly strategic error, since the latter in particular became the source of disruptive raiding activity against the empire's coastal lands. From the late 840s the Balearics too were providing shelter for Muslim pirates and raiders.

The appearance and rank of the new frontier commands illustrates the imperial strategy of expansion and conquest on both the eastern and western frontiers. A new array of such commands – the ducates of Chaldia, Mesopotamia and Antioch – covered the eastern frontier by the 970s, expanded to include ducates of Iberia, Vaspurakan, Edessa and Ani in the period from 1000–1045. A similar process can be observed in the west, in the establishment in the 970s and 980s of a ducate of 'Mesopotamia in the west', of Adrianople and of Thessalonica; and after the destruction of the Bulgar empire by Basil II (by 1015), the commands of Sirmion, Paristrion and Bulgaria. Similar commands appear a little later in Byzantine southern Italy – partly associated with the aggressive activities of the Normans in that region – and in the southern Balkans.

Administration and Taxation

The changes which occurred in the administration and structure of the departments of the sacred largesses (*sacrae largitiones*) and the private finance department (*res privata*) during the course of the sixth century prefigured changes throughout the whole apparatus of fiscal and civil administration which followed the drastic shrinkage of the empire in the middle of the seventh century. By the middle of the eighth century, a logothete for the general finance office (*genikon logothesion*) was responsible for the land-tax and associated revenues; similarly a department for military finance (*stratiotikon logothesion*) dealt with recruitment, muster-rolls and military pay; while another department, the *idikon*, or special *logothesion*, dealt with armaments, imperial workshops and a host of related miscellaneous requirements. The various departments which were once part of the *res privata* became similarly entirely independent and placed under their own officials. The public post, previously under the *magister officiorum*, the master of offices, became independent under its own logothete. Other departments that had originally been part of the imperial household, such as the sacred bedchamber, evolved into specialised treasuries and storehouses for particular state needs, while the bedchamber itself, known as the *koiton*, evolved its own personal imperial treasury for household expenditures.

While the *themata*, or military garrison districts and their armies, had achieved a clear territorial identity by the early eighth century, the substructures of the older provincial administration survived until the early ninth century. The dioceses disappeared, replaced in effect by the themes; but within the themes the old provincial names continue to be used for fiscal districts. Each such district was supervised in terms of tax assessment and collection by a 'director' or 'manager' – *doiketes* – with a staff of officials for the province and for the central *sekreton* or bureau at the capital. During the later seventh century supervisors 'of all the provinces' or 'of the provinces' of a particular theme appear, but after the early eighth century only individual provincial supervisors are known. By the 830s and 840s the late Roman provinces had been eclipsed by a more up-to-date structure, headed in each *thema* by a *protonotarios* or chief notary, responsible to his chief in Constantinople for running the thematic fiscal administration. Each theme had also a judge or *krites* responsible for civil administration and justice; and a *chartoularios*, responsible to the military finance department at Constantinople for the maintenance of military registers and related issues. They were all under the nominal authority of the *stratêgos*, the general, successor to the older *magistri militum*, but retained a degree of autonomy. This structure developed quite slowly: the old idea that the thematic general was a military supreme who was in charge of the whole thematic administration from the beginning is clearly incorrect – this was the case probably only from the time of Theophilos (829–842). These arrangements remained in place until the late eleventh century.

Late Roman and Byzantine taxation aimed at maximising revenues. Up to the middle and later seventh century this was achieved by attributing land registered for taxation, but not cultivated, to neighbouring landlords for assessment (known as *adiectio sterilium*). Tax was assessed by a formula tying land (determined by area, quality and type of crop) to labour (the *capitatio-iugatio* system). Unexploited land was not taxed directly. Tax was reassessed at intervals, originally in cycles of five, then of 15 years, although in practice it took place far more irregularly. From the seventh or eighth centuries a number of changes were introduced. Each tax unit was expected to produce a fixed revenue, distributed across the tax-payers, who were as a body responsible for deficits, which they shared. The tax-unit (in effect, the community) was jointly responsible for the payments due from lands that belonged to their tax unit but were not farmed, for whatever reason. Remissions of tax could be requested or bestowed to compensate for such burdens, but if the community took over and farmed the land for which they had been responsible, they had also to pay the deficits incurred by the remission. Since during the same period the cities appear to have lost their role as intermediaries in the levying of taxation, this now devolved upon imperial officials in each province and upon the village community.

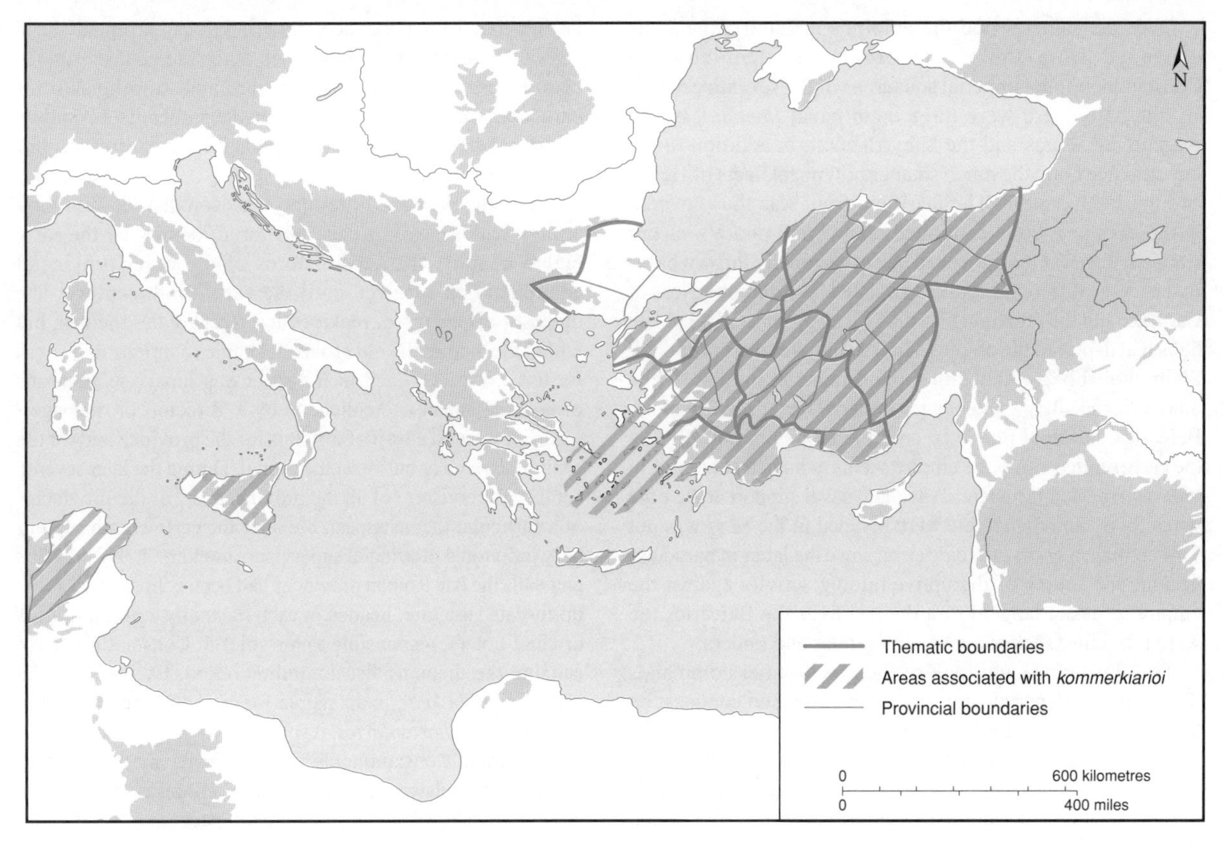

Map 6.5 Provinces associated on lead seals with general *kommerkiarioi* and their warehouses, c. 660–732.

The most important change which took place after the seventh century seems to have been the introduction of a distributive tax assessment, whereby the annual assessment was based on the capacity of the producers to pay, rather than on a flat rate determined by the demands of the state budget. This involved accurate records and statements of property, and the Byzantine empire evolved one of the most advanced land-registration and fiscal-assessment systems of the medieval world. These changes had been completed by the middle of the ninth century.

The regular taxation of land was supplemented by a wide range of extraordinary taxes and corvées, including obligations to provide hospitality for soldiers and officials, maintain roads, bridges, fortifications, and to deliver and/or produce a wide range of requirements such as charcoal or wood. Originating in Roman times, they continued into the middle and later Byzantine periods, their Latin names largely being replaced by Greek or Hellenised equivalents. Some types of landed property were always exempt from many of these extra taxes, in particular the land owned or held by soldiers, and that held by persons registered in the service of the public post, partly a reflection of administrative tradition, partly because they depended to a degree on their property for the carrying out of their duties. But the extra demands placed upon less powerful taxpayers complicated the system, which became immensely ramified. During the second half of the eleventh century, depreciation of the precious-metal coinage combined with bureaucratic corruption led to the near-collapse of the system.

Crisis Management, Military Supplies and Customs

The problems faced by the government after the loss of the eastern provinces to Islam in the 630s and 640s is reflected in the crisis measures it adopted to deal with them, and in particular by the transformation in the role of officials called *kommerkiarioi*, the earlier *comites commerciorum*. These officials had been originally under the authority of the *comes sacrarum largitionum*, although during the sixth century they had come under the praetorian prefecture. Their chief role lay in supervising the production and sale of silk, which was a state monopoly, and in functioning as customs officials dealing with imports and exports of precious goods. During the middle years of the seventh century they were also made responsible for supplying troops with equipment and provisions, and the levying and storing of fiscal income in kind. The high-ranking *kommerkiarioi* and the warehouses (*apothekai*) which they supervised and administered seem to have filled a gap created by the new situation, with which the administration of the prefecture could not cope. The essential task of supplying the army fell to them because of the suitability of their administrative competence and the network of state warehouses they managed.

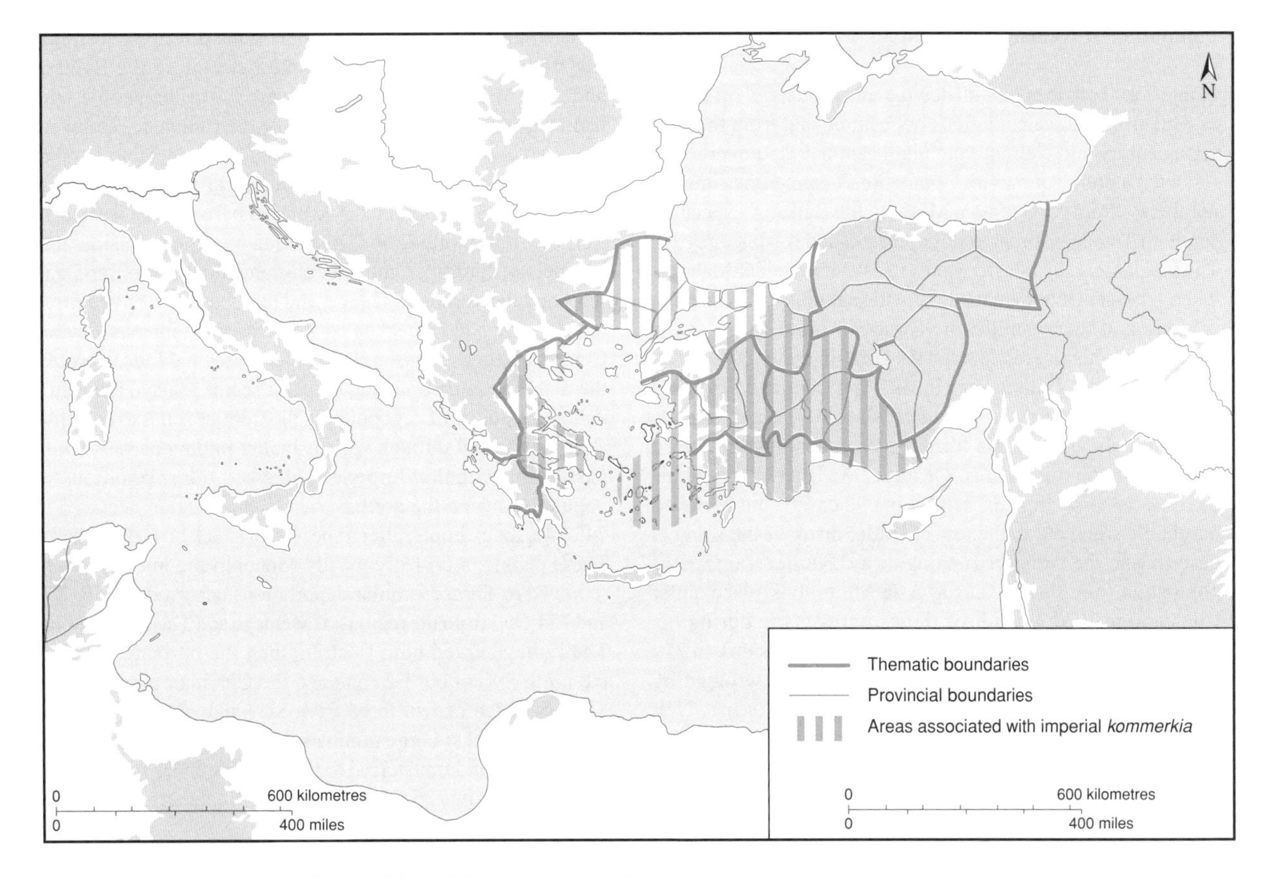

Map 6.6 Provinces/ports associated with imperial *kommerkia* from 730.

This arrangement operated until certain reforms and changes were introduced c. 730 by Leo III (717–741).

From about 730/731 there seems to have taken place a gradual reduction in the importance of individual *kommerkiarioi*: instead of high-ranking general *kommerkiarioi* associated with warehouses there appear instead *kommerkiarioi* associated with no specific region and with no warehouse. Institutions called imperial *kommerkia* appear at this time, and seem to have fulfilled a related but more limited function until the first decades of the ninth century, when the establishment of the thematic *protonotarioi* and the system of supplying the armies which they administered from their *themata* made them redundant. The *kommerkiarioi* themselves appear thereafter in association with specific military provinces (*themata*) or, more usually, specific ports or entrepôts, underlining their reversion to the role of customs officials controlling trade and exchange activities with regions outside the empire. From the later eighth century a duty on trade was levied, the *kommerkion*, and *kommerkiarioi* were associated with its collection.

The *kommerkiarioi* often worked in partnerships, sharing responsibility for their allotted tasks, and frequently managed several different warehouses at the same time. Since these were not necessarily geographically contiguous, nor constituted, apparently, on an annual or five-yearly basis, and since the same combinations were not repeated with any degree of regularity or frequency, it seems that they were not associated with regular taxation, which took place on an annual, or at least a regular and repeated basis. On the other hand, some evidence suggests that the coincidence of groups of warehouses or provinces under a single or several *kommerkiarioi* with certain military events must reflect a relationship between the two. On this evidence (although the issue is still debated), the system of warehouses administered by the *kommerkiarioi* was associated with supplying and equipping, and probably also feeding, expeditionary or field forces assembled for particular campaigns.

After the early ninth century the more regular system managed by the thematic *prôtonotarioi* was made permanent (it had been developing probably since the middle of the eighth century) and the activities of *kommerkiarioi* appear to be wholly connected with trade and customs dues. But with the expansion of the empire and the offensive warfare which predominated in the later ninth century onwards this system too began to change. The marginalisation of the thematic militias, as they had become, meant that the partially 'self-supporting' theme armies were more and more replaced by professional mercenaries, who were maintained both by the collection and delivery of supplies as before, but in addition were often quartered on the provincial populations, whom they were permitted to exploit in terms of accommodation, food and other necessities, thus placing an increasingly heavy burden on the tax-payers. During the second half of the eleventh century this placed increasing strains on the taxation system and on the producing population.

Population Movement and Population Transfer

In its efforts both to cope with demographic and fiscal problems, as well as to eradicate religious opposition, and from the later sixth century and well into the ninth century, the government followed a policy of moving populations – sometimes in huge numbers – from parts of Asia Minor to the Balkans, especially southern Thrace, in order to re-populate that devastated region. The first large-scale transfer seems to have been under Maurice, when arrangements were made to settle a number of Armenian soldiers and their families in Thrace in order to boost the number of potential recruits for the army in the impoverished and devastated Balkan provinces. Apparently on a much grander scale, the Emperor Constans II moved large numbers of captive Slavs into Asia Minor, a precedent also followed by the Emperor Justinian II. Of these, the latter recruited his own special army from the settlers in 690 or 691, but it proved unreliable when put to the test of battle, many of the soldiers deserting to the Arabs or running away. Further transfers of population from the Balkans to Asia Minor took place under Constantine V as a result of the captures made during his campaigns into Bulgaria, and on at least two occasions, in 759 and in 762. These efforts reflect the destruction wrought by the constant incursions of the Islamic raiders, and the need felt by the central government to stabilise the situation. The Slav populations of Bithynia in north-western Asia Minor retained an identity as such for several centuries thereafter.

The movement was not all in the same direction. Justinian transferred the Mardaites from the Lebanon to the Balkans and Asia Minor. The Mardaites were a warlike people who had caused particular difficulties for the Caliphate, and were removed under treaty in the 680s; and a considerable number of people were removed from the region around Germanikeia in northern Syria under the same emperor, to be settled in Thrace. In the mid-750s Constantine V settled considerable numbers of forcibly removed emigrants from north Syria and the Anatolian region in Thrace, and this appears to have caused the Bulgar leadership some concern, no doubt increased when Constantine constructed a chain of fortresses and forts to protect them. At the same time, the emperor seems to have pursued a deliberate policy of depopulating the frontier zones to establish a no-man's land through which smaller raiding parties would pass with difficulty. Imperially-decreed transplantations of populations from the north Syrian frontier region occurred in 745/746, for example, after a successful attack on Germanikeia. These people, reportedly mostly Monophysite in belief, were removed to Thrace; similar deportations occurred in 750–751 and 754–755 from the regions of Mélitene and Theodosioupolis. The policy served both to strengthen the no-man's-land of the north Syrian border zone and the Christian population in Thrace, which had suffered from Slav and especially Bulgar raids and attacks. Large numbers from the same regions were again seized and transferred to the Balkans under Leo IV after an expedition in 776, and Constantine VI deported mutinous

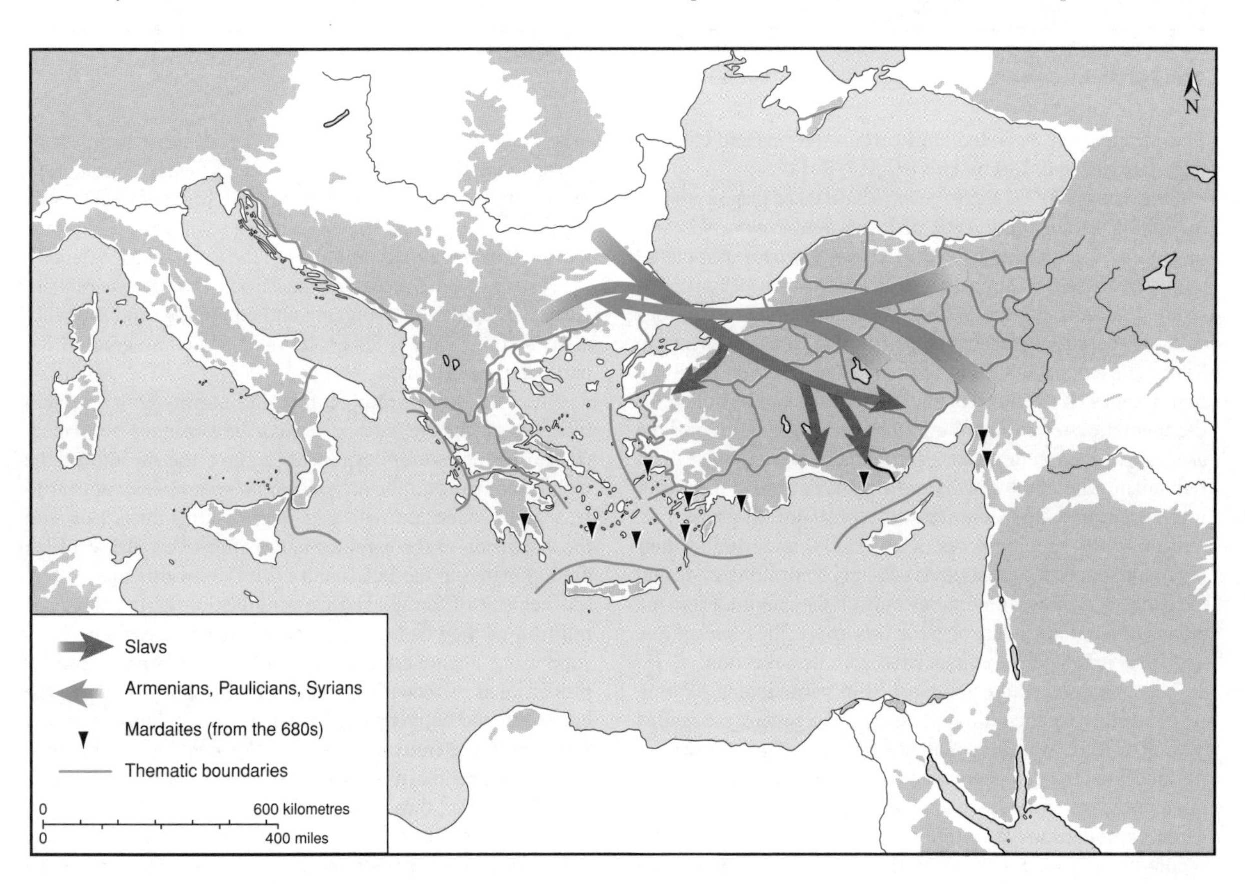

Map 6.7 Population movement c. 660–880.

soldiers from the Armeniakon thema to the western provinces in 793. Michael I (811–813) deported heretical members of the sect of the Athinganoi to western provinces also.

The transfers affected not just captives, however: under Nikephoros I many soldiers from Asia Minor registered on the military rolls of their region were forced to move to Thrace, with their families, for similar reasons. Under Basil I again the defeated Paulicians were removed in considerable numbers from eastern Anatolia to the Balkan provinces, to which they brought their own version of a dualist heretical system which, on new ground, seems to have prospered and eventually given rise in the Bulgarian territories during the tenth century to Bogomilism. During the eleventh century and after there was a large-scale migration of Syrians and especially Armenians into south-western Asia Minor, partly a result of imperial expansion eastwards in the tenth century, partly a result of the Seljuk threat in the middle of the eleventh century.

This policy, carried on intermittently over some four centuries, had the effect of introducing significant changes into the demographic and cultural composition of many provinces (and also the placenames of some regions), and it emphasised the multi-ethnic character of the orthodox Byzantine state. It may also have played no small part in the continued flexibility of the empire when faced with substantial economic and military challenges from its enemies over the same long period. But the changing demographic structure of the east Roman state was not all a result of official policy, or at least of compulsory transfer. Considerable numbers of Goths, for example, settled by treaty agreement in north-western Asia Minor during the fifth century, and this group was strengthened by further additions of soldiers and their families in the later sixth century, probably making up the division of the *Optimatoi* in the Opsikion army, represented perhaps also by the so-called *Gotthograikoi* (Goth-Greeks) who appear in this region in the early eighth century and who had their own fiscal administrator. Many soldiers of the army of Illyricum were likewise settled in the same region, to become part of the imperial field army in the first half of the seventh century and later constitute a separate *thema*. The ethnic make-up of Asia Minor was further complicated by the arrival and settlement in the eleventh century of small numbers of Frankish mercenaries and their families and, much more significantly, of the nomadic Türkmen.

The Transformation of Urban Life: *Polis* to *Kastron*

The effects of the warfare of the seventh century – first the Persian invasions, then the devastation of the Arab invasions and raids – proved too much for the majority of provincial cities and their localised economies. The great majority shrank to a fortified and defensible core which could support only a very small population, housing the local rural populace and, where present, a military garrison and an ecclesiastical administration. Byzantine towns became merely walled settlements. Civic buildings were for the most part non-existent; the state and the church built, for their own use (churches, granaries, walls, arms-depots), but the cities had no resources of their own, no lands, no revenue, no corporate civic juridical personality. Wealthy local landowners invested in building, but there is very little evidence until the eleventh century. Most invested whatever social and cultural capital they had in Constantinople and in the imperial system which became after the loss of the eastern provinces focused almost exclusively on the capital, for this 'Constantinople factor' was important for the contours of middle Byzantine society.

The cumulative result of these changes was that the 'city' effectively disappeared, and in its place there evolved the middle Byzantine *kastron* or fortress-town. The government and its military establishment contributed to these changes since they had the needs of local administration and the army firmly in the foreground. Distinguished by its limited extent and strongly defensive character, the *kastron* becomes the middle Byzantine urban settlement par excellence. In many contemporary sources the traditional Greek word for city, *polis*, is replaced by the new term, even when, in the later tenth and eleventh centuries, urban life began to flourish once more.

The archaeological and literary evidence bears eloquent testimony to the changes. The major city of Ankyra shrank to a small citadel during the 650s and 660s, the fortress occupying an area of a few hundred metres only. Amastris (mod. Amasra) offers similar evidence, as does Kotyaion (mod. Kütahya); many more formerly major centres underwent a similar transformation. The city of Amorion, defended successfully in 716 by 800 men against an attacking army more than ten times larger, was reduced in effect to the area of the citadel, or *kastron*, with an area of a few hundred metres. Excavations there and at several other sites show that while the very small fortress-citadel continued to be defended and occupied, discreet areas within the late Roman walls, often centred around a church, also continued to be inhabited. In Amorion there were at least two and probably three such areas. Sardis similarly shrank to a small fortified acropolis during the seventh century, but it appears that several separate areas within the circumference of the original late ancient walls remained occupied. At Ephesos, which served as a refuge for the local rural population, as a fortress and military administrative centre, but also retained its role as a market town, survey and excavation suggest that it was divided into three small, distinct and separate occupied areas, including the citadel. Miletos was reduced to a quarter of its original area, and divided into two defended complexes. Didyma, close by Miletos, was reduced to a small defended structure based around a converted pagan temple and an associated but unfortified settlement nearby. Other evidence for Euchaita, on the central plateau, may also support this pattern of development – a permanently-occupied settlement or settlements, perhaps concentrated around key features such as a church within the original late Roman circuit, with the citadel or fortress as the site of military and administrative personnel and the centre of resistance to attack. Archaeological survey and excavation show that the same holds also for the formerly thriving city of Sagalassos in Pisidia, and is probably true of many similar medium-sized towns.

The occupied medieval areas of most cities appear to have been similar in nature. It seems thus often to have been the case that separate areas within the late Roman walls of many cities continued to be inhabited, functioning effectively as distinct communities whose inhabitants regarded themselves (in terms of their domicile quite legitimately) as 'citizens' of the city within whose walls their settlement was located, and that the *kastron*, which retained the name of the ancient *polis*, provided a refuge in case of attack (although not necessarily permanently occupied or garrisoned).

As well as these 'urban' centres, there was also a large number of small forts, outposts and refuges, sometimes associated with nearby villages or towns, and generally sited on rocky outcrops and prominences (and often the sites of pre-Roman fortresses). Together with the provincial *kastra*, these characterised the provinces well into the eleventh century and beyond. And many larger or more important sites in Byzantine Asia Minor also fit the pattern. Apart from some already mentioned, such as Amaseia and Amastris, the fortresses at (Pontic) Koloneia, Herakleia Pontike/Kybistra, Charsianon, Ikonion, Akroinon, Dazimon, Sebasteia in the central and eastern regions, Priene, Herakleia in Caria, and several others along the western coastal provinces, provide good examples, defended by natural features, adequately supplied with water, positioned to control the region around it together with the main routes, or means of access and egress serving the district, but often with a lower town located within the late Roman walls which remained occupied during times of relative peace. As long as the defences of the lower town were kept in reasonable repair, they might also serve as an appropriate refuge for the surrounding rural population during hostile raids, since small raiding parties rarely had the time or the strength to concern themselves with a siege, logistically demanding, very time-consuming and potentially very costly in manpower. This is a pattern typical also of Byzantine southern Italy and, with a different topographical context but a similar structural relationship to the surrounding territory, the Balkans.

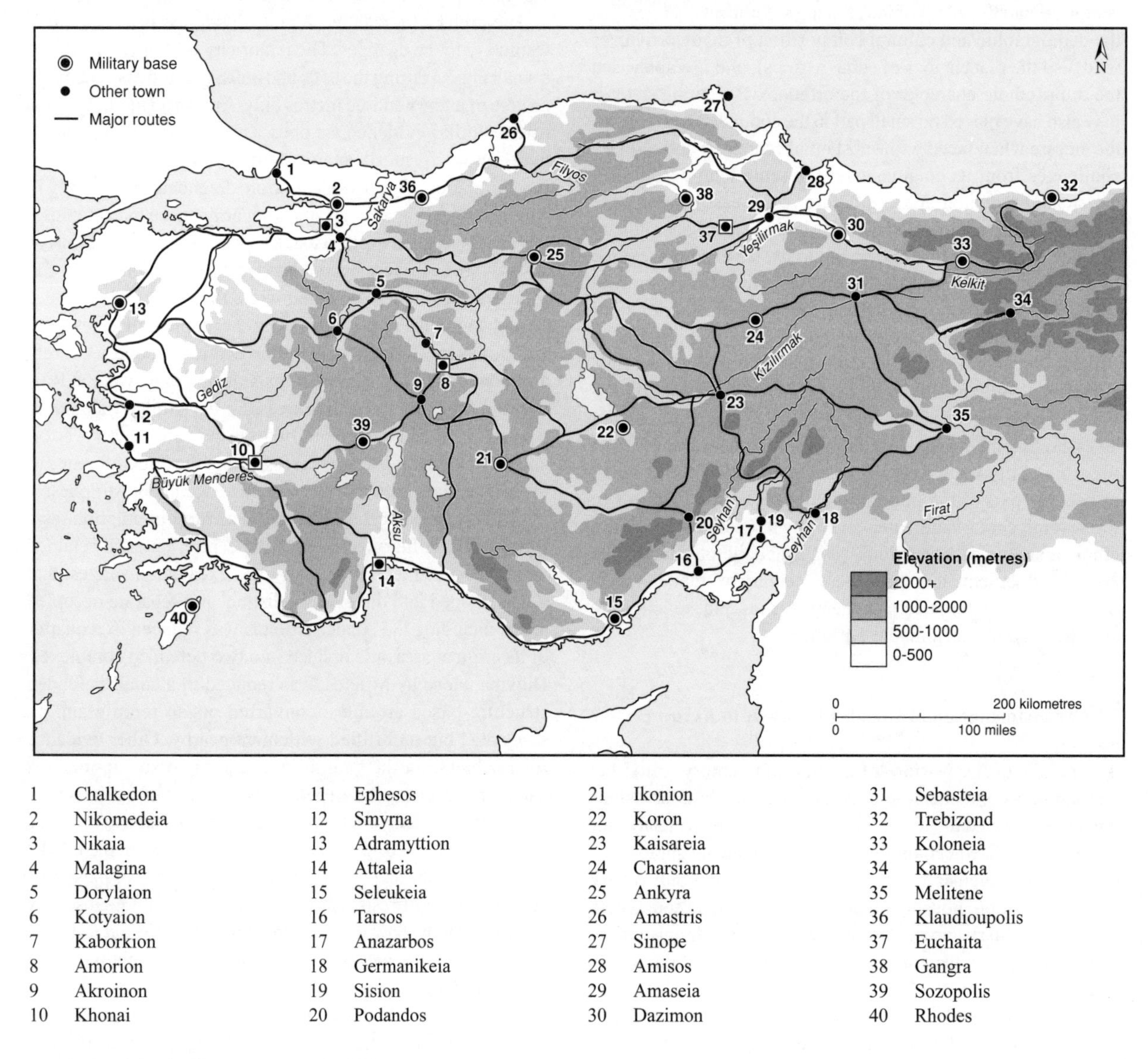

Map 6.8 Major fortified centres c. 700–1000.

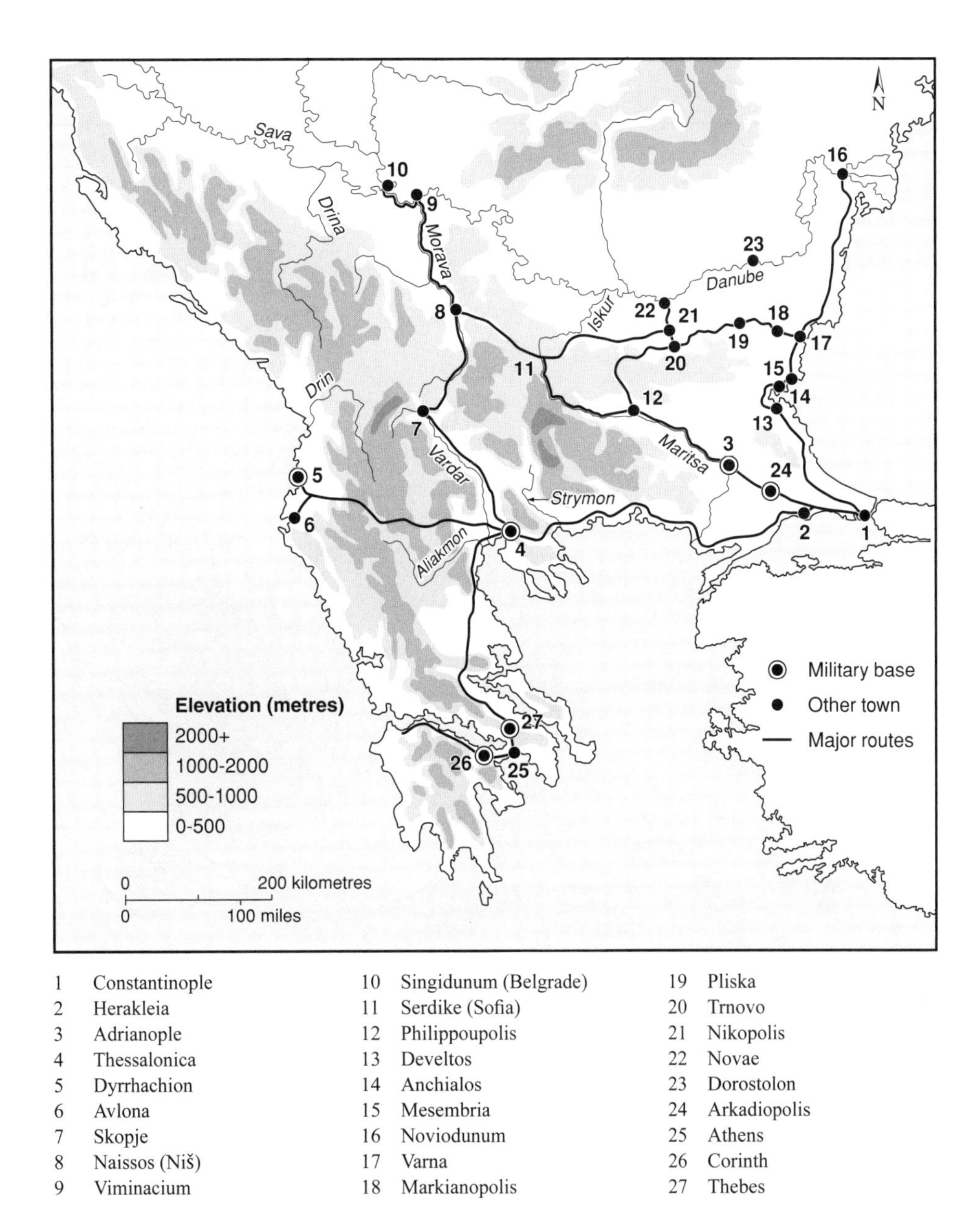

Map 6.9 The Balkans: military bases.

These developments also impacted upon the demography and settlement pattern of the empire. Survey evidence suggests that villages around many cities attracted some of the urban population, becoming larger, sometimes defended or re-sited to defensible locations, and more nucleated. From the fifth century onwards there is evidence of a downward demographic trend, which is paralleled but somewhat preceded by the beginnings of a colder climate, and which lasts into the later eighth century. The arrival of plague (bubonic and pneumonic, although there is still some disagreement) in the eastern empire in the 540s had a significant negative impact on the population as a whole. Its reoccurrence thereafter into the middle of the eighth century, while it affected different regions in different degrees, nevertheless continued to affect population adversely. Colder winters, lower agricultural outputs, a reduction in the amount of land farmed and the encroachment on marginal areas of habitation and cultivation of forest and woodland, all played a role. Byzantium was not alone, of course, since these changes affected the whole Eurasian zone. But the political and economic effects of warfare and conflict exacerbated the consequences for established patterns of settlement and land-use in the Byzantine context. The changing appearance and function of towns is a part of this broader picture. From the ninth century, in contrast, an amelioration of climatic conditions seems to have contributed to the revived fortunes of urban settlement as well as the demographic and economic improvement of the empire's position, culminating in the twelfth century.

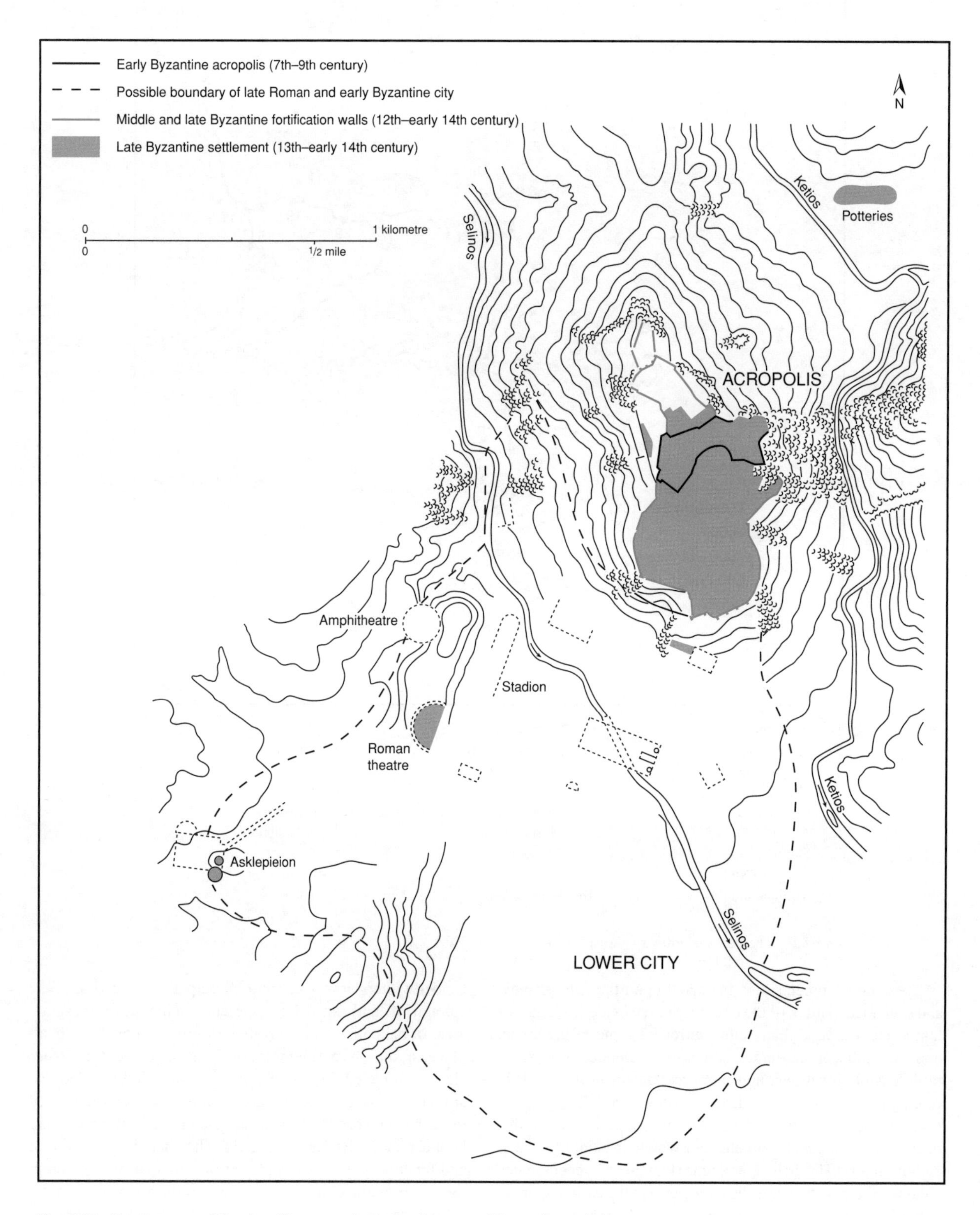

Map 6.10 Development of the city of Pergamon in the late Roman and Byzantine period.

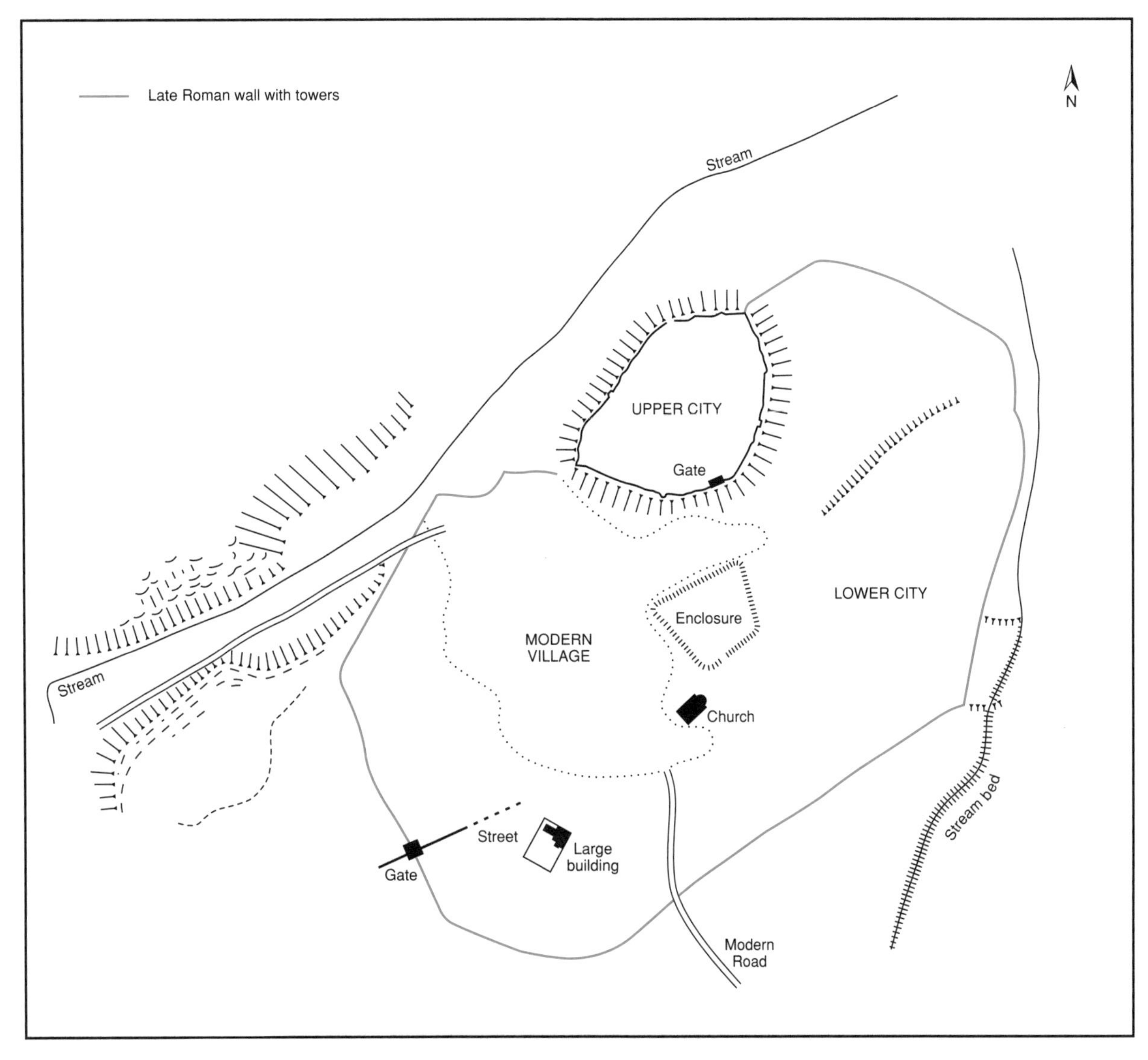

Map 6.11 Late Roman and Byzantine Amorion in the 6th–9th centuries. (After Lightfoot, 'The public and domestic architecture of a thematic capital'.)

Resources, Industry and Trade

There was always a tension between the fiscal interests of the east Roman or Byzantine state and the private sector of merchants, bankers, shippers and so on. The state represented a set of ways and means of regulating the extraction, distribution and consumption of resources, determined by the need to balance consumption of wealth and agricultural and other forms of production. Three key factors determined the export of finished goods, the flow of internal commerce between provincial centres, or between the provinces and Constantinople, and the movement of raw materials and livestock. These were: the needs of the army and treasury for raw and finished materials and provisions; the state's need for cash revenues to support mercenary forces and the imperial court; and the demands of the imperial capital itself, which dominated regional trade in the western Black Sea and north-western Asia Minor, north Aegean and south Balkans.

Evidence from a wide range of sources shows a decline in inter-regional trade and exchange from the middle of the seventh century, with a nadir in the first quarter of the eighth and a plateau thereafter until a slow recovery – regionally accented and with a number of false starts – setting in from the 750s and 760s. The ceramic evidence provides some information about trade and exchange, and although the archaeological record is still so patchy that it is difficult to generalise, it appears that pottery production had, by the end of the seventh century, become highly localised. The distribution of finds of the various coarse and fine wares produced at Constantinople, in the southern Balkans and eastern Peloponnese, and in the Aegean and Crete provides good evidence for the maintenance of a considerable degree of maritime commerce or exchange which

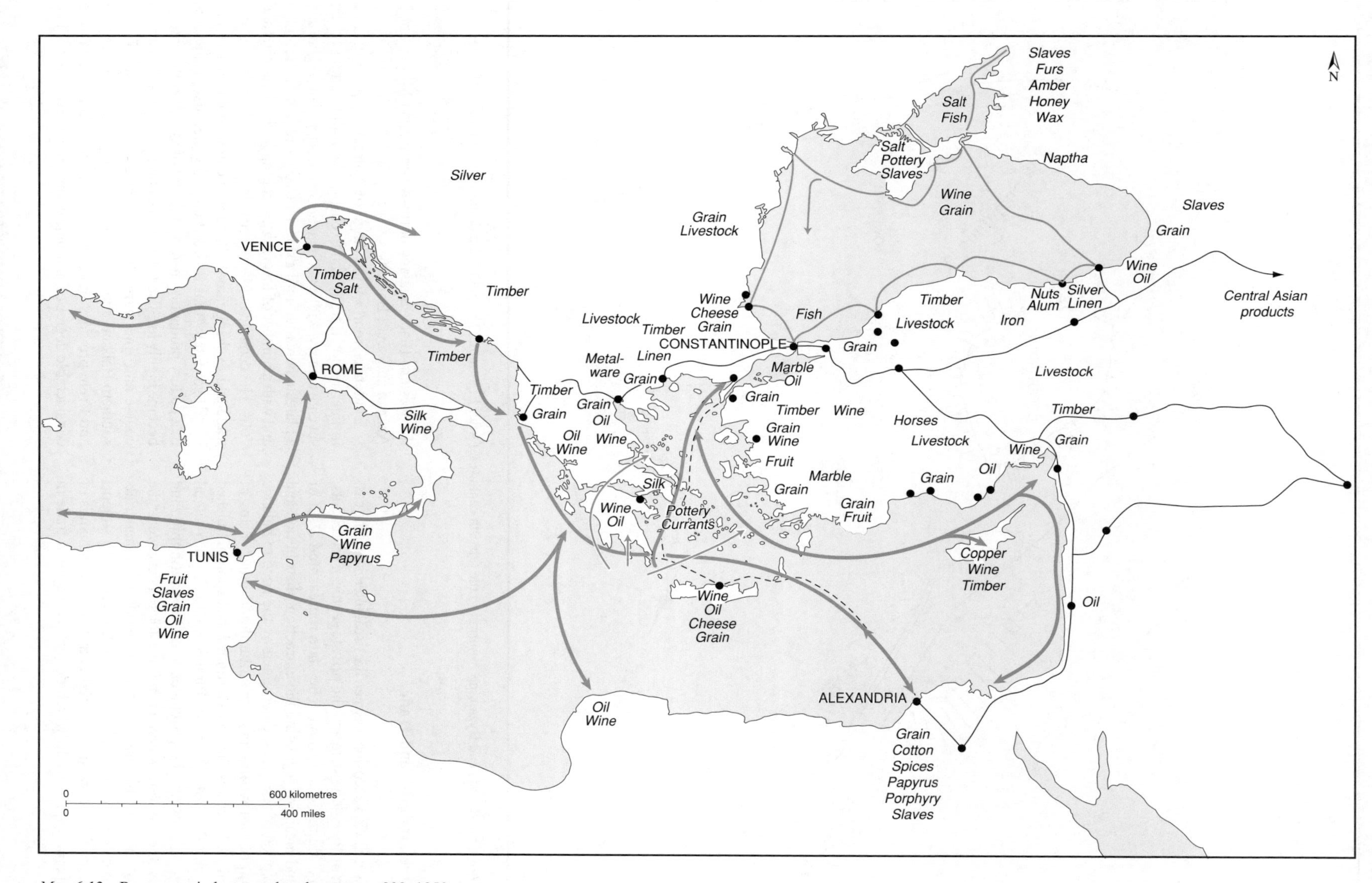

Map 6.12 Resources, industry and trade routes c. 900–1050.

was limited to the territories within the political boundaries of the empire. At the same time, there is little evidence at present for much commercial activity extending far inland. Thus the pattern in the Byzantine world is much the same as the pattern in the rest of the Mediterranean world at this period, with a strong tendency towards localisation of production and regionalisation of patterns of exchange. This evidence tells us little or nothing, however, about either the levels of production within each locality, nor about local patterns of consumption of locally-produced goods, nor again about the relative wealth of the provinces and sub-regions.

As noted in previous sections, the pattern of supply and demand was heavily slanted towards Constantinople. Trade within the empire was largely from the provinces or the empire's neighbours to Constantinople, or between the provinces, although there was always some movement from the empire to adjacent neighbouring territories – trade agreements and evidence for commerce between Byzantium and the Bulgars in the early eighth and ninth centuries, for example, between the Kiev Rus' and the empire in the tenth century, or between the empire and various Muslim lands to the east illustrate this clearly. And after a low-point in the period c. 680–750, both internal and international trade began to be more significant. An expansion of imperial coin production in the 820s and after reflected an increase in exchange activity not just related to state needs, although these certainly played a role. From this time on, the relationship between the government's coinage, with its strongly fiscal emphasis and function, and non-state enterprise and exchange, into which the state coinage was inevitably drawn, becomes increasingly complex, so that government minting policy had necessarily to take some aspects of market demand and commercial use into account.

From the 840s and 850s there were the beginnings of a real and permanent recovery, often with new routes, reflecting very different economic and political circumstances in the west and east from those of the late Roman centuries, dominating the pattern of international exchange. The evidence from the later ninth century suggests that internal exchange and commerce were flourishing, and large numbers of traders and entrepreneurs were associated with them. Apart from the capital, entrepôts such as Corinth, Thebes, Adrianople, Thessalonica, Kherson, Smyrna, Ephesos, Sinope, Trebizond, Mélitene, Attaleia, centres which lay on key crossroads, or possessed good port facilities, or served as centres of local production for goods which would travel (or all three), played an important role in international exchange. State-dominated movement of goods in bulk (grain, for example, to Constantinople) may also have encouraged trade and commerce along the routes most exploited by the state itself, as in the late Roman period, since private entrepreneurial activity could take advantage of state shipping and transportation. In view of the number of trading ports around the Black Sea, from which the Byzantine government deliberately excluded Italian merchant shipping before the Fourth Crusade, long-distance trade by Byzantine merchants before 1204 may well have been substantial. But the real beneficiaries of the opening stage of greater stability and growth in the ninth to eleventh centuries were the state, on the one hand, and private or institutional landowners, such as the church and some monasteries, on the other. Only in the later eleventh century do commercial and external pressures exert sufficient influence to destabilise the imperial monetary system.

The picture which emerges for this period is therefore of an economy – or set of overlapping economic sub-systems – which experienced radical contraction and localisation of production and exchange during the later seventh and eighth centuries but which, partly as a result of the key role of the state, began from the later eighth century to expand both in terms of productive potential and in respect of the extension of exchange networks from the intra-regional to inter-regional and international levels. This is not to say that international exchange died out. Indeed, quite the reverse was the case, for continuous trade in certain luxury commodities with the Indian Ocean and, via the central Asian steppe zone, with China, was maintained without a break, although with fluctuating fortunes. By the end of this period Constantinople was at the centre of an international network which reached westwards to southern and western Europe, northwards via the Black Sea into Russia and Scandinavia; and eastwards into the Islamic world, especially to Egypt and Syria and, beyond them, into the Indian Ocean, on the one hand, and the central Asian steppe, on the other.

The Revival of Urban Life

Following the stabilisation of the political and military situation in both the Balkans and in Asia Minor after the early tenth century, and the beginnings of the demographic recovery that accompanied a period of warmer climatic conditions, many urban centres recovered their fortunes. The most obviously favoured were those that had an obvious economic and market function in their locality. Thebes in Greece provides a good example of such a recovery: by the middle of the eleventh century it had become the centre of a flourishing local silk industry, local merchants and landowners had houses there, attracting artisans, peasant farmers with goods to sell. Landless peasants looking for employment also gravitated to such foci, thus further promoting urban life. This urban regeneration was also connected with the growth of a social élite of office and birth, which had the wealth to invest in urban or rural production.

Towns therefore grow in economic importance during the later tenth and especially in the eleventh and twelfth centuries. This reflects in part the improved conditions within the empire for trade, commerce and town–country exchange-relations to flourish. It also reflects the demands of Constantinople on the cities and towns of its hinterland for the provision of both foodstuffs and other goods. Towns begin to play a central role in political developments – in the period from the later seventh to the mid-eleventh century most military revolts had been based in the countryside and around the headquarters of the local general; during the eleventh century and afterwards political opposition to the central government is almost always rooted in towns, whose populace also appear in the sources as a body of self-aware citizens with specific interests. Unlike in contemporary Italy in the tenth and eleventh centuries, however, communal identity did not go much beyond this –

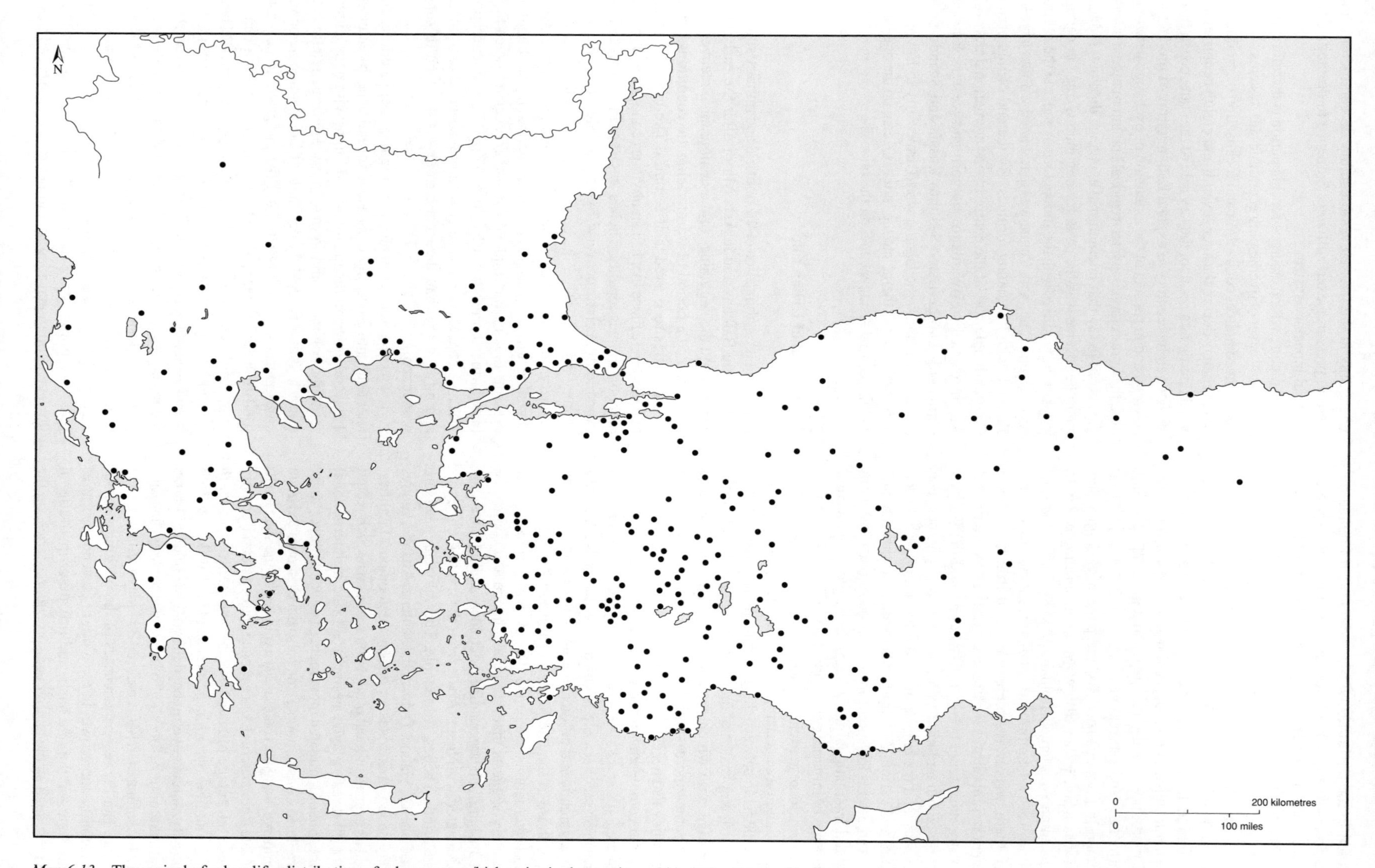

Map 6.13 The revival of urban life: distribution of urban centres/bishoprics in the empire c. 900. (After Hendy, *Studies*.)

local magnates who held both landed wealth as well as imperial titles and offices tended to dominate, and their attention was divided between their own town and locality and the attractions of the imperial capital, the centre of Byzantine society.

Towns were also affected by the military organisation of the empire – from the middle of the tenth century many towns became the seats of local military officers and their soldiers, a reflection of the improved ability of the state to supply and provision its soldiers through cash payments and a reliance upon the existence of local markets. The consequence of all these factors was a reversal of the process of ruralisation of economic and social life that characterises the later seventh and eighth centuries. But however much towns now came to flourish as centres of local economic activity, they still retained crucial functions, and all the appearance of, *kastra*, fortresses, and represented a very different sort of urban culture from that of the late Roman world, which they now replaced.

Equally, a large number of fortress-towns underwent only limited change at this time. In many cases (although lack of evidence makes generalisation dangerous), it is clear that there was little to differentiate between an undefended village settlement and a *kastron*. The inhabitants of many *kastra* were assessed for their taxes on a communal basis, just like any village. Size was certainly not an important feature. A major difference between the typical late Roman 'city' and the medieval town was that public buildings were no longer funded from 'public' sources – the church, monasteries and private individuals were the only sources of wealth, except where the state was involved (in constructing defensive works, for example). At the same time, the medieval Byzantine town was cramped within its defences, with few large public spaces and no planned network of streets. Tradition determined the siting of cemeteries, and the sort of buildings that might be erected near to churches or the houses of the local magnates (*archontes*) or the bishop – whose presence was possibly the only obvious differentiating feature between small town and defended village.

But there were important functional differences between towns and villages. Towns had a greater role as markets, as residences for representatives of the military or other state administrators, as foci for traders and artisans, for an ecclesiastical establishment with economic requirements and effects, a more regular market or fair, and a range of other services and functions not available in a rural village context. The structure of town society was also very different from that of the countryside. Communal, non kinship-based organisations, such as confraternities, specialist 'societies' focused on a particular saint's cult, for example, or the supporters' groups associated with chariot- or horse-racing, did not exist in villages.

State Structures 700–1050

Most of the administrative posts typical of the middle Byzantine period and found in the sources of the eighth and ninth centuries can be traced in some way to a late Roman equivalent, sometimes directly, and involving the continued use of the same title, sometimes indirectly or with a change of title but a continuity of function. The east Romans did not necessarily differentiate by functional category in the way that modern historians tend to do, in order to make sense of structures in our own terms, so any breakdown of the Byzantine administration will necessarily do some injustice to the ways in which Byzantines themselves perceived their system to operate.

One significant difference between the late Roman 'pyramidal' structure is that the emperor has, in theory at least, a direct oversight over the affairs of many departments, rather than having everything channelled through a few high-powered officials such as the master of offices or the praetorian prefects. The Byzantine system is much 'flatter'. A glance at Figure 6.1 will show that the administration can be broken down into several areas of competence: state finance, justice and prisons, transport and the post, the imperial household, provincial military and palatine military. Each set of departments – *sekreta* – had its own staff, some substantial (as with the department of the general treasury – *genikon logothesion* – for example, whose bureau had some eleven different grades, including sub-sections for each *thema* and many other finance-related activities), others very small (as with the *orphanotrophos*, the curator of the great imperial orphanage and its estates, whose department had just three grades and a limited number of sub-departments for the different estates). The figure also illustrates the complex inter-relationships pertaining across many sections, and the overlapping nature of the competences of many departments. The central role of the imperial household needs to be underlined, both because access to the emperor was through one or another household department, and because the distinction between public, palace and private (family) treasuries was never very particularly observed. This meant that state funds often flowed into what were essentially private hands, while the imperial family or the emperors themselves often invested substantial funds drawn from their personal revenues in state-related ventures.

A key aspect of the structure of imperial administration was the system of precedence embedded within it. While this was always fluid, with new titles being introduced at times, with shifts in status between different ranks, and in particular with the relationship of any individual to the emperor being of crucial significance in determining what position they attained and how that was described through the system of titles, a certain regularity in these relationships did exist, and is described in a variety of documents dating from the late Roman period through into the late Byzantine period. By the tenth century this system had settled down and it became possible to draw up lists of precedence by which imperial ceremonial, public meals, processions and so forth could be regulated. The master of ceremonies, the staff of the imperial palace and the prefect of the city all played a key role in the maintenance and observance of tradition, although 'tradition' was itself constantly evolving.

A career in the state administration was attractive because of the potential for illicit as well as regular rewards, and could be achieved through various means. Before the collapse of the middle of the seventh century, study of the law was always a good qualification for court posts as well as provincial positions of authority and responsibility, although a general acquaintance

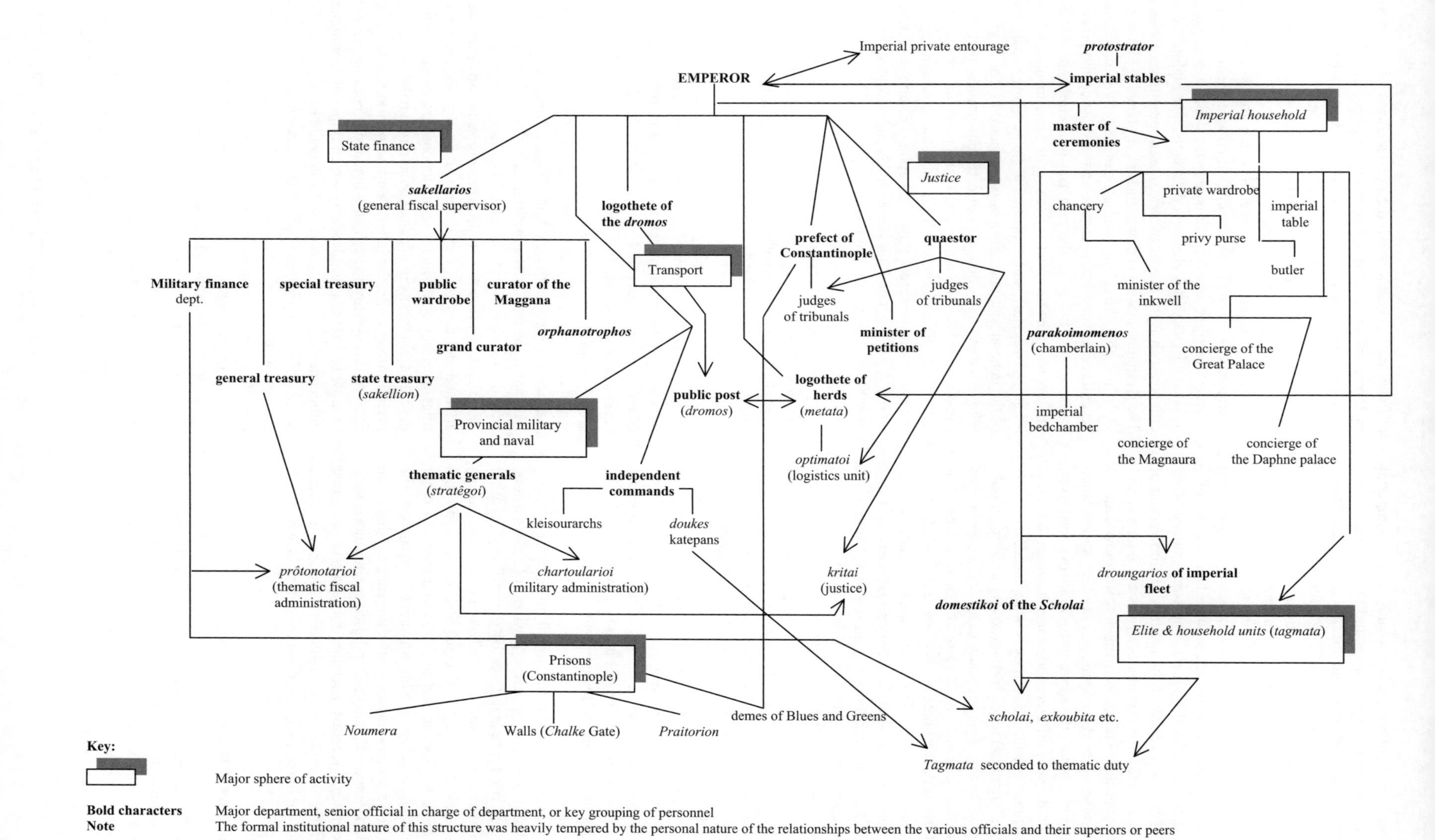

Figure 6.1 The imperial administration c. 700–1050

with traditional classical scholarship was sufficient. During the later seventh and eighth centuries this changed, and it seems that many provincial officials were entirely ignorant of the law and of the administration of justice. But literacy was generally the norm, since this was a literate and record-keeping state administration which depended upon the transmission of vital information in written form, not just between officials, but from one generation to the next. By the tenth and eleventh centuries a knowledge of the law was once again an important part of the education of senior officials. In theory, all posts were open to all persons, but in practice, the system was heavily inflected by the existence of a powerful social élite and the networks of patronage which were a part of any medieval society. The administrative hierarchy was graded according to military and non-military posts, as well as, by the tenth century, ranks normally held by eunuchs and non-eunuchs, although the system was by no means exclusive or rigid.

Officials were inducted into their posts by a formal ceremony at which they received the signs of their office – a ceremonial military girdle and a robe or other garment specific to their department and rank – and during which they swore an oath of loyalty to the emperor and declared their orthodoxy. By the ninth century the great majority of junior posts were conferred by the award of a token of office, so that the emperor did not need to be present. Senior posts, in contrast, which were of greater significance to the emperor and which were often directly chosen by him, were appointed by word of mouth at a ceremony formally conducted by the ruler and during which the official or officer, if the post was military, did formal obeisance to the emperor. Such ceremonies applied to the clergy of the Constantinopolitan churches also, since they, too, were members of this hierarchy of state positions.

Promotion depended upon a regular rhythm of movement within each department – during the eighth and ninth centuries, for example, thematic commanders were rotated fairly frequently, sometimes across to an alternative post, sometimes upwards. Where the move was from one post to another comparable one, however, the incumbent would sometimes receive a higher-ranking title, so that salary and social standing would rise accordingly. If all went well, an individual of reasonable talents could expect to rise to a fairly senior position by the end of his career and, if he came to the attention of the emperor or another powerful senior official, perhaps even become a senior minister or official himself. Salaries rose incrementally with promotion, and upon retirement, since there was no system of pensions as such, officials received an enhanced sum, together with certain judicial rights and sometimes also fiscal exemptions. Some administrative officials, especially in the period from the tenth century, sold their posts in advance of their retirement as a means of putting a sum aside.

Mints and Money

The copper coinage, represented by the *follis*, of which there were 288 to the gold *solidus* (or *nomisma*, the standard gold coin, of which there were 72 to the Roman pound), suffered a number of fluctuations during the seventh century, reflecting the financial problems the empire faced, and was, for example, reduced to less than half its weight under Heraclius. A reform under Constantine IV had only short-term effects, and the reduction in weight and value soon set in once more. There was also a drastic curtailment in production of the *follis* from the end of the reign of Constans II. Under Leo III a reformed silver coin, the *miliaresion*, was introduced, valued at 1/12 of a gold *nomisma*, smaller than its predecessor of the fourth century, and initially struck at a rate of 144 to the pound. But this appears to have had as much a ceremonial as a functional role as a medium of exchange, and it has been argued that its introduction was connected with the introduction shortly before of the new Muslim silver coin, the *dirhem*. The reformed silver coinage affected the gold, however, because the minting of fractional issues (halves – *semisses* – and thirds – *tremisses*) of the *nomisma* declined during the eighth century and after. But apart from relatively minor fluctuations in the weight of the gold coinage, and more significant ones in the relationship of copper to gold, the system as a whole remained unchanged in its essentials until the later tenth century.

During the first half of the ninth century the copper coinage underwent a major transformation., with an increase of issues beginning during the reign of Michael II (821–829), and the establishment of at least one, probably two new mints for copper (Thessalonica and Kherson in the Crimea). There was also an increase in weight of the standard copper coin, the *follis*. An initial limited increase in copper coin production, associated with a slightly larger coin under Michael II in the 820s, was followed by a dramatic sixfold increase in the issue of a fully reformed and still larger coin type. This may suggest a recognition by the government of a market-led demand for copper coin, and the connection between that and the state's fiscal requirements, although most excavated sites demonstrate such an upturn in finds of such coins only from the later years of the ninth century. This numismatic evidence, together with other evidence, seems to imply an economic recovery and the beginnings of growth in the economy, or at least in the non-state sector, especially in the southern Balkans. The increased production of coinage may also reflect an increased demand by the state for taxable resources in cash, which in turn may imply an expansion of the monetised sector of the economy as a whole. Under Nikephoros II (963–969) a new reduced-weight *nomisma* appeared, known as the *tetarteron*, and weighing some 2 carats less than the full-weight coin. Although discussion continues, this was probably intended as a means of rendering the system as a whole more flexible, although it seems also to have acted as a destabilising element in the price structure of the empire for a while.

The gold coinage was not unaffected by the changes of the seventh and eighth centuries, and a gradual reduction in the purity of the gold *nomisma* took place from the time of Justinian II, under whom the fineness of the *nomisma* was reduced from 98% to 96% gold. With very minor incidental fluctuations this then remained constant until the time of Constantine VII (913–959), when a further slight reduction in gold content was made, to 94.4%. Again, another reduction was made under Michael IV (1034–1041), where 90% became standard for a time. Thereafter, as a result of increasing demand and limited revenue and bullion, devaluations occurred more frequently and led to the collapse of the system after 1070.

Calculations of the state budget are difficult and fraught with methodological problems. But recent estimates, based on the numbers of dies employed at different points, changes in the weight of the gold *nomisma*, the standard gold coin, the distribution of finds and a range of other factors, suggest that whereas during the reign of the Emperor Constantine V (742–775) the annual budget amounted to a value of some 1.7 million *nomismata* (based on an estimated minting of approximately 250,000–300,000 *nomismata* annually), by the reign of Basil II (976–1025) it had expanded to a value of over 4 million *nomismata*, reflecting both greatly increased revenues and a much more active economy.

The number of mints actually contracted in the period from the middle of the eighth century (the mint at Carthage had ceased production in 695 when the city fell): that at Ravenna ceased production in 751, when the city finally fell to the Lombards; those at Rome and Naples in 776 and 842; the mint in Sicily, at Syracuse, ceased production when the city fell to the Arabs, and was for a while removed to Reggio in Calabria (until 912). The result was that the main mint for the production of coin for the empire as a whole was now Constantinople, from where coin was delivered to centres for distribution according to demand as assessed by the relevant government officials.

Sources of bullion remained very limited – Armenia and other sources in the Caucasus had been exploited from Roman times on, and the conflicts between Rome and Persia and between Byzantium and the Caliphate in this region are partially explained through competition for this resource. Other sources included the Black Sea coast and river mouths, where panning for gold and other minerals occurred, and the Balkans, although political conditions determined access, as in the Armenian highlands. The Taurus and Anti-Taurus ranges also included sources of various ores, including gold and iron. But the supply of metal was basically very inelastic, and the state developed a remarkably efficient system – through its fiscal apparatus – of retaining as much precious metal as it could through the process of taxation. Even so, regular crises in supply occurred, and given the lack of developed banking facilities (at least until near the end of the empire's history), recourse was had to measures such as seizing or borrowing gold and silver plate from private individuals, the palace itself, or the church – there are several instances of this from the early seventh century onwards. An alternative was to devalue the gold in order to maintain levels of supply or meet demand, but this inevitably led to an inflationary cycle and, in the eleventh century, to a fatal reduction in precious metal content which brought about the collapse of the established late Roman system.

Table 6.1 The money system c. 650–1050

Nomisma (gold)	*miliaresion* (silver)	*keration* (carat: unit of account)	*follis* (copper)
1	12	24	288
*	1	2	24
*	*	1	12
*	*	*	1

72 *nomismata* = 1 Roman pound = 324.72g (although this also evolves: in the later period the Byzantine pound is valued at only 318g)
1 *kentênarion* = 100 Roman pounds or 7,200 *nomismata*

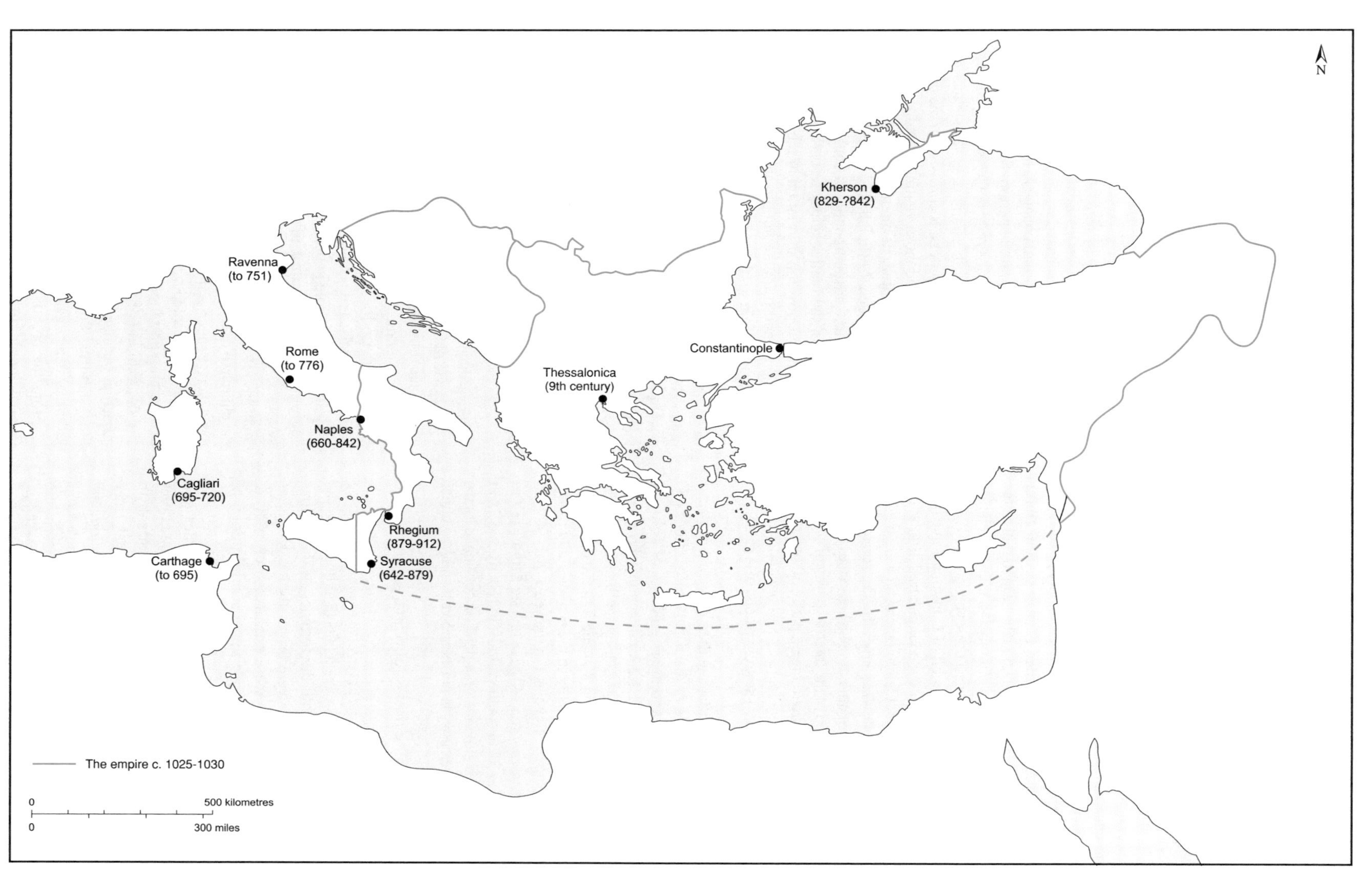

Map 6.14 Imperial mints c. 640–1050 with dates of closure or period of operation.

7 *Church and Monastic Organisation*

Church Administration

Between the 650s and the middle of the eighth century the territory of the see of Constantinople was subject to the same threats and to the same losses as the secular state. The provincial infrastructure of the church in particular was jeopardised in many of these areas, for the constant raids and invasions, and the economic damage which was caused, brought about in many outlying areas the flight of the local clergy to safer regions. This was an issue addressed in the so-called Quinisext Council (or Council in Trullo, because it convened in the domed hall of the imperial palace) held in 692, when a number of matters relating to church discipline and the fate of the exposed provincial dioceses were debated. At the Council of Constantinople held in 680 the total of bishops who attended numbered 174; at the Quinisext in 692 the total was 211, although not all were present at both, so that the total is somewhat larger. By the council held at Nikaia in 787, the number of attending bishops rises to 319, a result partly of the stabilisation of the internal political situation and a more secure travelling environment, partly of the creation of new bishoprics to compensate for lost sees now under enemy authority, especially in the east. The archbishops of Alexandria, Jerusalem and Antioch continued to send representatives, of course, and they continued to manage their own ecclesiastical administration, following the original pattern of dioceses. Continued internal stability, and the beginnings of political and territorial expansion in the later ninth century, brought a new phase of expansion to the Constantinopolitan church. At councils held at Constantinople in 869 and 879 the number of sees has increased again, especially in the Balkans, and by the time of the patriarchate of Nicholas I in the years 901–907, an episcopal list enumerates some 442 sees in Asia Minor, 139 in the Balkans, as well as 34 in southern Italy and Sicily and 22 in the Aegean region. This situation of expansion continued apace with the reconquest of territory in northern Syria and Iraq under the emperors of the late tenth and early eleventh centuries, and was only reversed after the loss of central Anatolia to the Turks in the 1070s.

Unlike the state, the church did not evolve new administrative units. The older diocesan names were retained, although the site of some bishoprics changed with the fortunes of the various towns and cities in which they were originally located, or as new towns and new bishoprics were established or revived. And bishops were important not just to the administration of the church and the pastoral care of the Christian community. They were the spiritual leaders of their communities and representatives of the church, but as managers of sometimes substantial resources in land, their views were important. In times of political turmoil, the role of bishops was crucial, since it was they who might give a lead to a particular faction, and they were in any case expected to judge the rights and wrongs of such matters. But since they were invariably drawn into political events, they could also suffer the consequences if they sided with the wrong faction. The political relevance of senior clergy was well understood by the emperors, who had a vested interest in the selection and appointment to such posts. Thus, during the iconoclastic period, the support of the vast majority of the bishops for the imperial cause was probably a major factor in the stability of the rule of the emperors from Leo III to Leo IV. Senior clergy often acted effectively as imperial officials, also, representing the government or an emperor on foreign missions and embassies.

There was one important change in the political influence of the clergy during this middle period. Since the fourth century there had been a resident synod at Constantinople, chaired by the patriarch, to deal with affairs of ecclesiastical discipline, dogma and liturgical matters. It consisted of the bishops in the metropolitan region and those visiting from more distant sees. But from the ninth century its membership was limited to senior bishops and patriarchal officials and it begins to play a more important role in Constantinopolitan church politics.

Beginning in the tenth century also canon law takes on a more significant position in relation to the (Roman) civil law of the empire, and from the middle and later eleventh century church courts begin to play a greater role in the administration of justice and in the everyday affairs of the ordinary population. Eventually this meant that the influence and the moral and political status and authority of the church in the provinces was thereby considerably enhanced, encouraging greater feelings of local pride and autonomy, as well as the readiness of provincial élites to question the actions and motives of the court or the elements which dominated it.

Monasteries and Centres of Orthodoxy

Whereas the period from the fourth to sixth centuries witnessed a great growth in the importance of individual holy men, hermits and monks (although monasteries were a significant feature of society and culture – see pages 51–54), the focus in the middle Byzantine period is chiefly on monastic communities rather than individuals; and while hermits and holy men continued to play an important role in the ensuing period, as the flourishing hagiographical literature of the ninth and tenth centuries suggests, they came also to be more closely associated with monastic communities. This was partly a result of the efforts of the church to exert a greater degree of control over them, since there were concerns that the spiritual power of unsupervised individuals presented a potential threat to its own authority. Yet at the same time monastic communities remained very fluid, and the tradition of the wandering monk retained a powerful attraction for many, a factor which contributed in large part to the instability of many smaller monasteries and their eventual failure or dependence on larger establishments.

From the eighth century monastic centres flourished in Constantinople, in the Aegean region, in north-western Asia

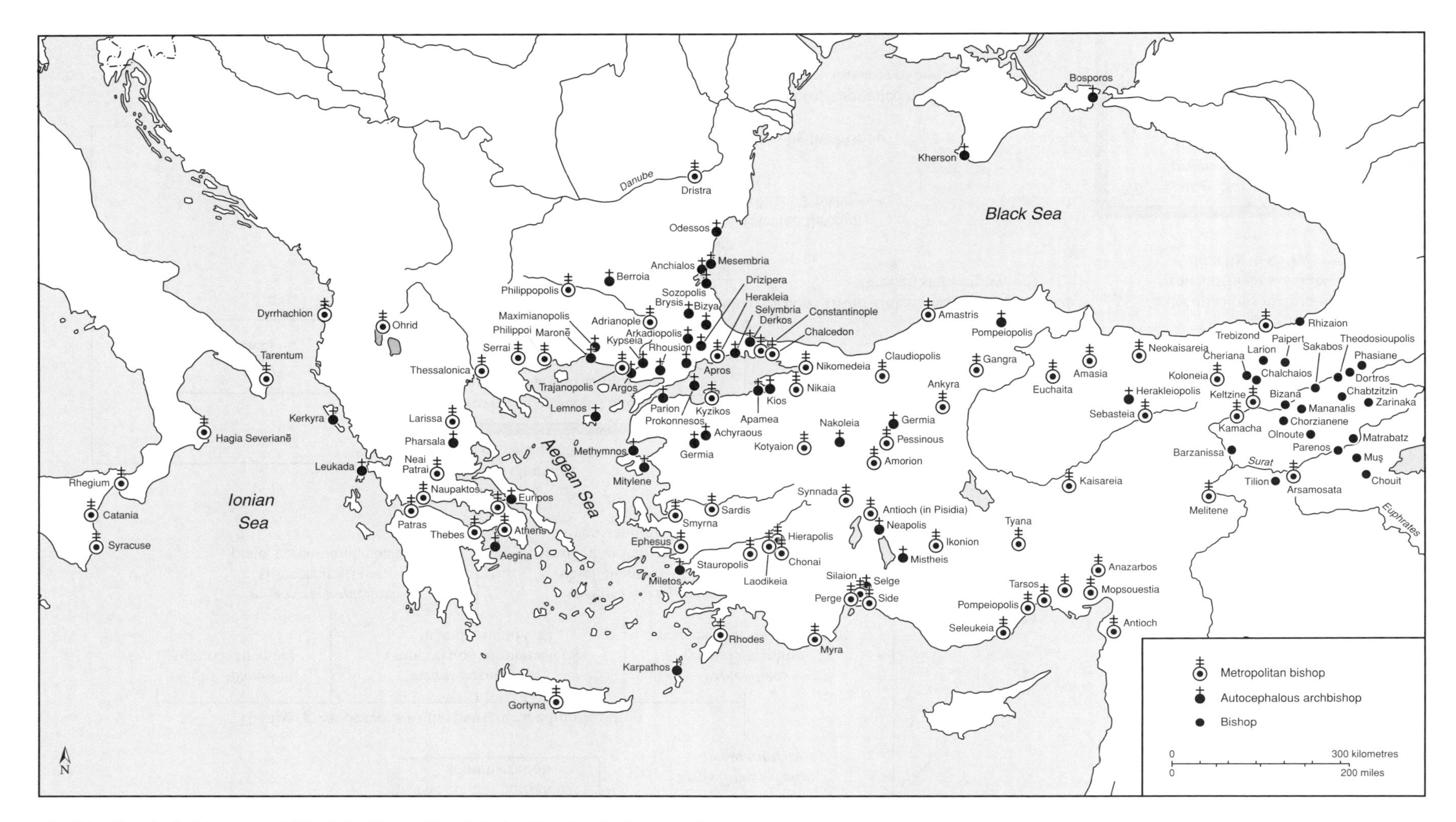

Map 7.1 Church administration c. 1000. (After Hussey, *The Orthodox Church in the Byzantine Empire*.)

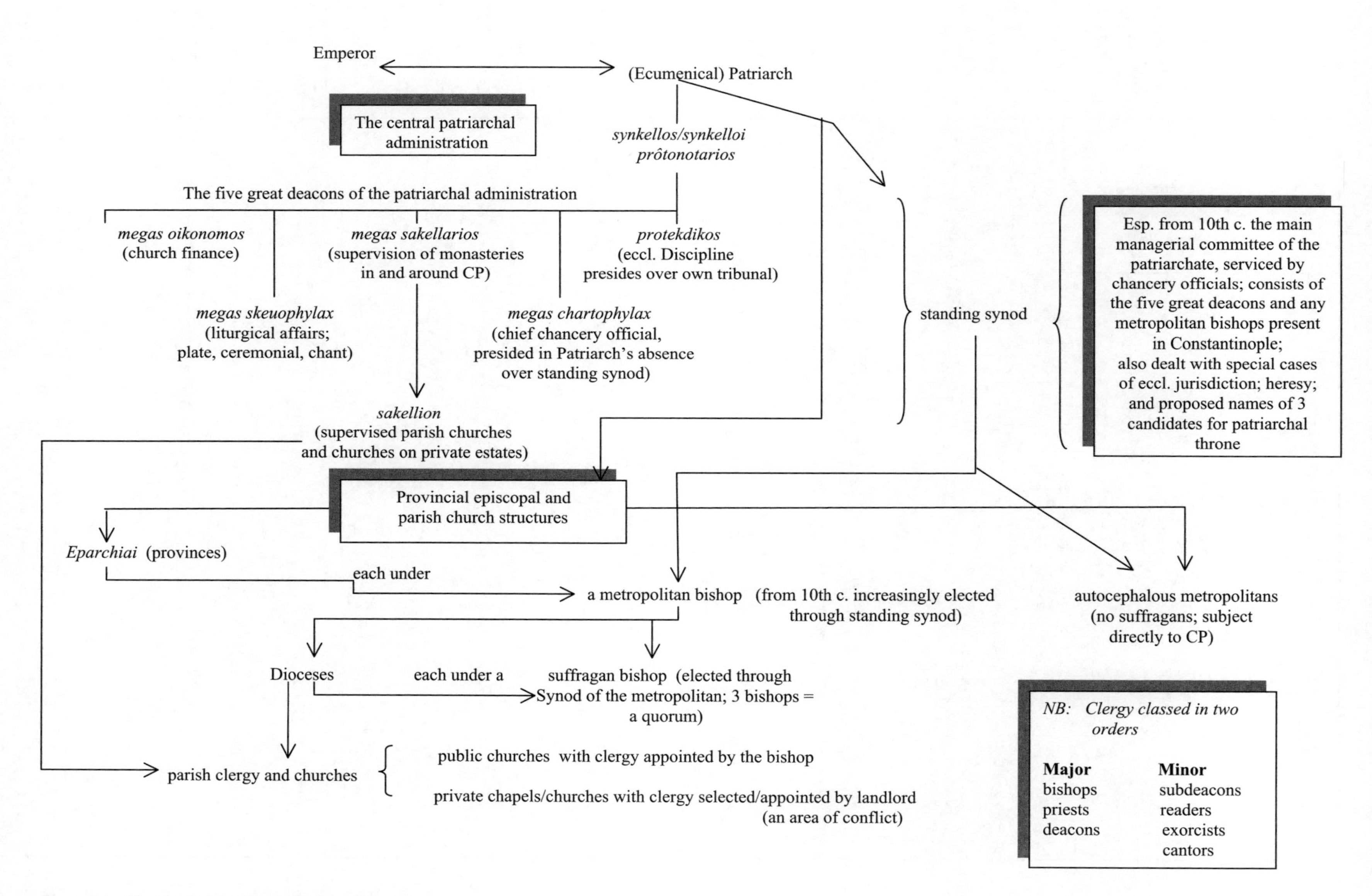

Figure 7.1 Church administration in the later 11th century

Minor, in parts of the Balkans, in Cappadocia and in the Pontus. The most famous, with its origins in the later ninth century, remains to this day a substantial focus of monastic activity, on the Chalkidike peninsula to the east of Thessalonica, at Mt Athos, the 'Holy Mountain'. Before the ninth century some monasteries were granted annuities in cash or produce or both, and occasionally received gifts of land also. Thereafter it became increasingly usual to endow those which could exert sufficient influence at court with lands and rents, and many became substantial landowners in their own right. This brought monastic houses increasingly into the world of commerce and business, since many monasteries put the produce they received as rent on the market. Some of the Athonite monasteries were already engaged in trade by the middle of the eleventh century, possessing their own ships, and cultivating connections at court in order to protect and enhance their interests.

Monasteries also became increasingly important in church politics – and therefore in imperial politics – from the later eighth and ninth centuries, partly because monks in and around the capital were successfully able to present themselves as the heroes of orthodoxy in the struggle against iconoclasm and in the fight to keep the church and the Christian empire as untainted by scandal or ignorance of scripture as possible. Leading monks such as Theodore of Stoudios at the end of the eighth and beginning of the ninth century were especially prominent in public political debates, openly challenging emperors over their interpretation of scripture and canon law, inviting persecution, and building up thereby a powerful reputation as arbiters of the correct interpretation of the church Fathers and the scriptures. Monasteries took great pains to promote their own particular cause, encouraging the writing of lives of saints, especially of the holy men and women with whom they were associated, and emphasising their heroic qualities. One indication of the importance of monasteries is the increasing number of patriarchs of Constantinople who had monastic careers behind them, relatively unusual before the ninth century.

The chief role of monasteries, apart from prayer and contemplation, was primarily philanthropic, and the great wealth accumulated by some establishments was to be employed in this direction, funding orphanages, homes for the elderly, and so forth. Many distributed food and alms of one sort or another to the needy in their locality, and also maintained infirmaries or hospitals that cared for all comers who needed their attention. Yet they were only marginally involved in education, teaching those destined for the monastic life but only rarely offering the sort of education available in western monastic communities. On the other hand, monks made up a very large proportion of all the scribes in the empire, and one estimate suggests that up to 50% of scribes in the tenth and eleventh centuries were monks. Many monks were knowledgeable theologians, however, and

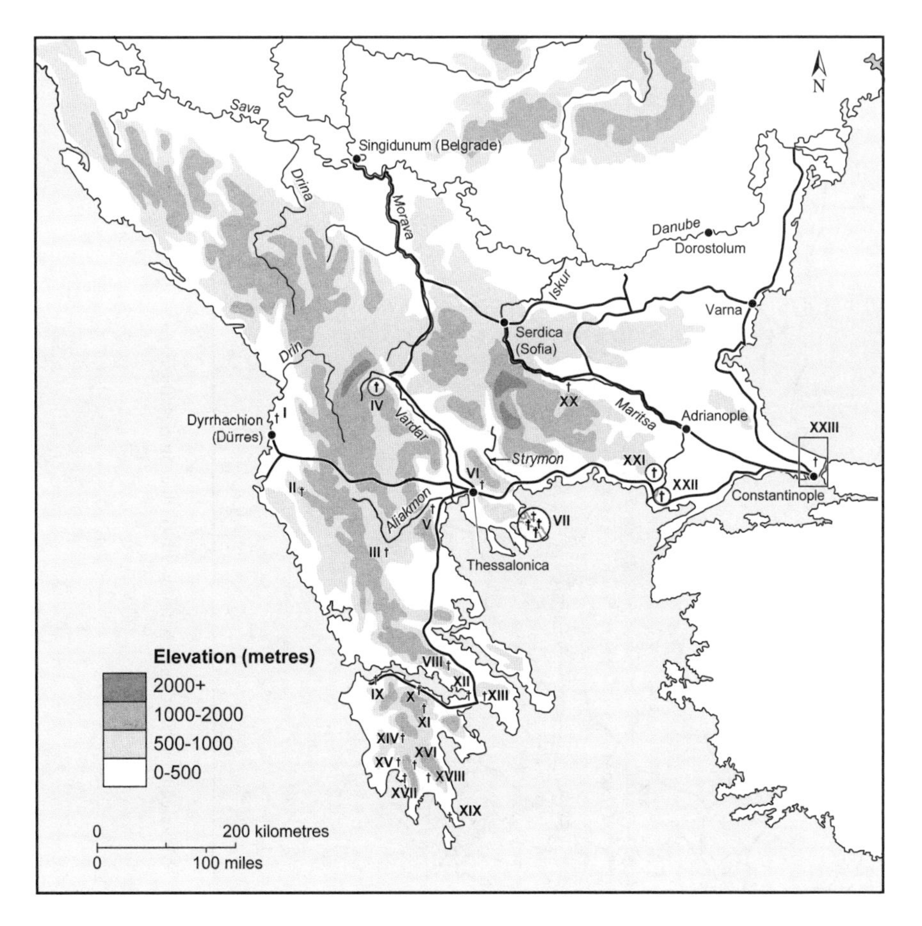

† Monastery
(†) Monastery with imperial connections

Key to monastic names/locations

I	Theotokos
II	Pogonianē
III	Meteora
IV	Panteleēmōn
V	Nea Petra
VI	Thessalonica (6 monasteries)
VII	Mt Athos (many)
VIII	Hosios Loukas
IX	Gerokomeiou
X	Taxiarchōn
XI	Mega Spēlaion
XII	Symbolōn
XIII	Daphni
XIV	Philosophōn
XV	St Theodore
XVI	Mistras (several)
XVII	Pantanassa
XVIII	Brontochion
XIX	H. Martha
XX	Theotokos
XXI	Theotokos Kosmosōteira
XXII	Skalotē
XXIII	Constantinople region (over 20)

Map 7.2 Major centres of monastic activity, 9th–11th centuries: the west.

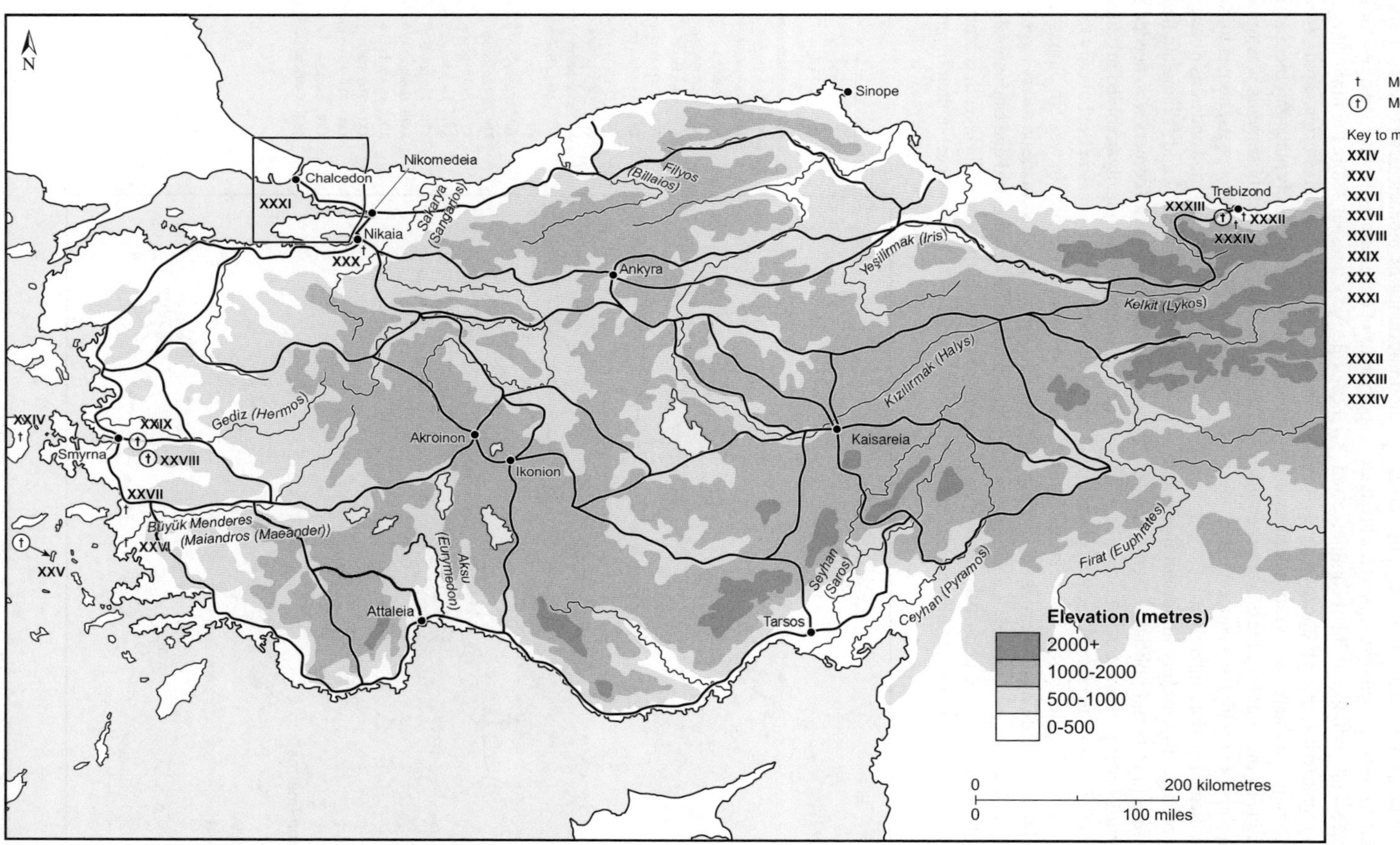

Map 7.3 Major centres of monastic activity, 9th–11th centuries: the east.

monks were actively involved in matters of dogma and church politics from the fifth century onwards, during the christological conflicts and debates, again during iconoclasm (at least according to their own traditions and propaganda, although the accuracy of this material can be challenged), and thereafter.

During the later tenth and eleventh centuries the emperors or their relatives became more involved with founding or supporting monasteries, in return for prayers for the soul of the patron or benefactor on specified occasions. By the middle of the eleventh century the most powerful monasteries were competing also for further privileges from their imperial or aristocratic patrons, in particular exemptions from certain categories of taxation (such as the duty to provide hospitality for soldiers or imperial officials, for example). Monasteries also offered an attractive retirement, and many senior soldiers, for example, retired to a monastery, or founded one (to which they could retire as abbot), partly as a means of assuaging their guilt for the killing for which they had been directly or indirectly responsible, partly as a means of securing their own future.

Unlike the medieval west, no monastic orders developed, each community instead being ruled by its own set of regulations encapsulated in a foundation charter or *typikon*. Monasteries were generally described as belonging to one of four basic types – patriarchal, imperial, episcopal or private – but there were problems in many cases from the under-endowment of some foundations, which did not always attract the manpower to cultivate the lands with which they had been endowed. This had a direct impact on state taxation, of course. Although the Emperor Nikephoros II Phokas attempted to address this issue and related matters by limiting the right of monasteries to acquire more than a certain amount of land relative to their resources, his measures were later revoked and the reform failed. Smaller monasteries often fell into the hands of larger establishments, which thus obtained land and property across a wide area and developed a number of dependent sub-communities or *metochia*.

Constantinople, Rome and the Emperors: Politics, Religion and Spiritual Conflict

The relationship between emperor and patriarch was frequently a tense one. Emperors were regarded as the defenders of Orthodoxy, and in this capacity they invested considerable sums in building and decorating churches and endowing monasteries as symbols of their piety, and many also developed an advanced knowledge of Christian theology. They also intervened directly in matters of strictly theological import, an aspect of their authority inscribed in the definition and assumptions about their role. This frequently led to clashes between emperors and patriarchs, on occasion the deposition of a patriarch, and the polarisation of opinion within the church. In the period from the sixth to the fifteenth century more than a third of all patriarchs were forced to resign or were deposed from office when they clashed with an emperor over some matter or other. The Emperor Justinian had attempted a general definition of this relationship in the sixth century, in which the church and the clergy were defined by their role as pastoral and spiritual guardians of the Christian community, but within which the emperor's position, while not above the law, was nevertheless seen as the embodiment of the law, since he was chosen and appointed by God. The application of these concepts in reality was problematic.

Emperors were more often successful in their conflicts with patriarchs than vice versa: in the ninth and tenth centuries alone four patriarchs – Ignatios, Photios, Nicholas I and Euthymios – were deposed when they refused to accept the imperial line, yet all four are recognised as outstanding churchmen and theologians in their own way. But the ambiguous relationship between emperors and patriarchs, ranging between friendly support to open opposition, is summed up in the act of penance which an emperor might on occasion volunteer, or be required, to perform, to atone for his sins and other transgressions. And although emperors usually won the day when looked at from the short-term point of view, there were plenty of occasions when patriarchs were able to mobilise sufficient and effective opposition to prevent an emperor having his own way, as several of the examples already cited show.

From the ninth century the orthodox world expanded as missionary activity, political conversion and a range of other factors brought extensive territories in the Balkans and in Russia into the church. While the patriarchate of Constantinople did not administer these lands directly, it always retained an ideological authority. The result was a certain imbalance in the relationship between emperor and patriarch, since the latter came to exercise authority over a much wider world, a more 'ecumenical' world than that of the political empire of east Rome. In this context it is understandable that patriarchs often felt they had the authority, and indeed the duty, to pronounce in political as well as moral matters, since the whole orthodox world looked to them, as much as to – and often rather than – the emperor for spiritual and moral guidance.

While the patriarchates of Alexandria, Antioch and Jerusalem were under Muslim domination, a real conflict of interest evolved between Rome and Constantinople. The chief cause of this was simple: Rome was the first see, founded by St Peter, and as such claimed primacy over all others – including the right to intervene in affairs of dogma, liturgical custom and matters affecting members of the Christian community in general. On the other hand, while Constantinople had an apostolic tradition (St Andrew is supposed to have visited the city), it was by no means as strong as Rome's and was, indeed, emphasised only from the later fourth century as the city grew into its role as imperial capital. It was this latter point which caused problems, since after the disappearance of the western empire the eastern ruler was effectively sole emperor, his residence was Constantinople, and the archbishopric of Constantinople could claim equal status with Rome as a result of this imperial position. The problem was that the relationship between emperor and patriarch frequently resulted in frustrated or vexed patriarchs (or their clergy or representatives) who could appeal to the pope at Rome as an independent arbiter of disagreements, but who would see this as a clear recognition of his superior status within the church. Imperial interference in Italian politics, whether secular or religious, did not help.

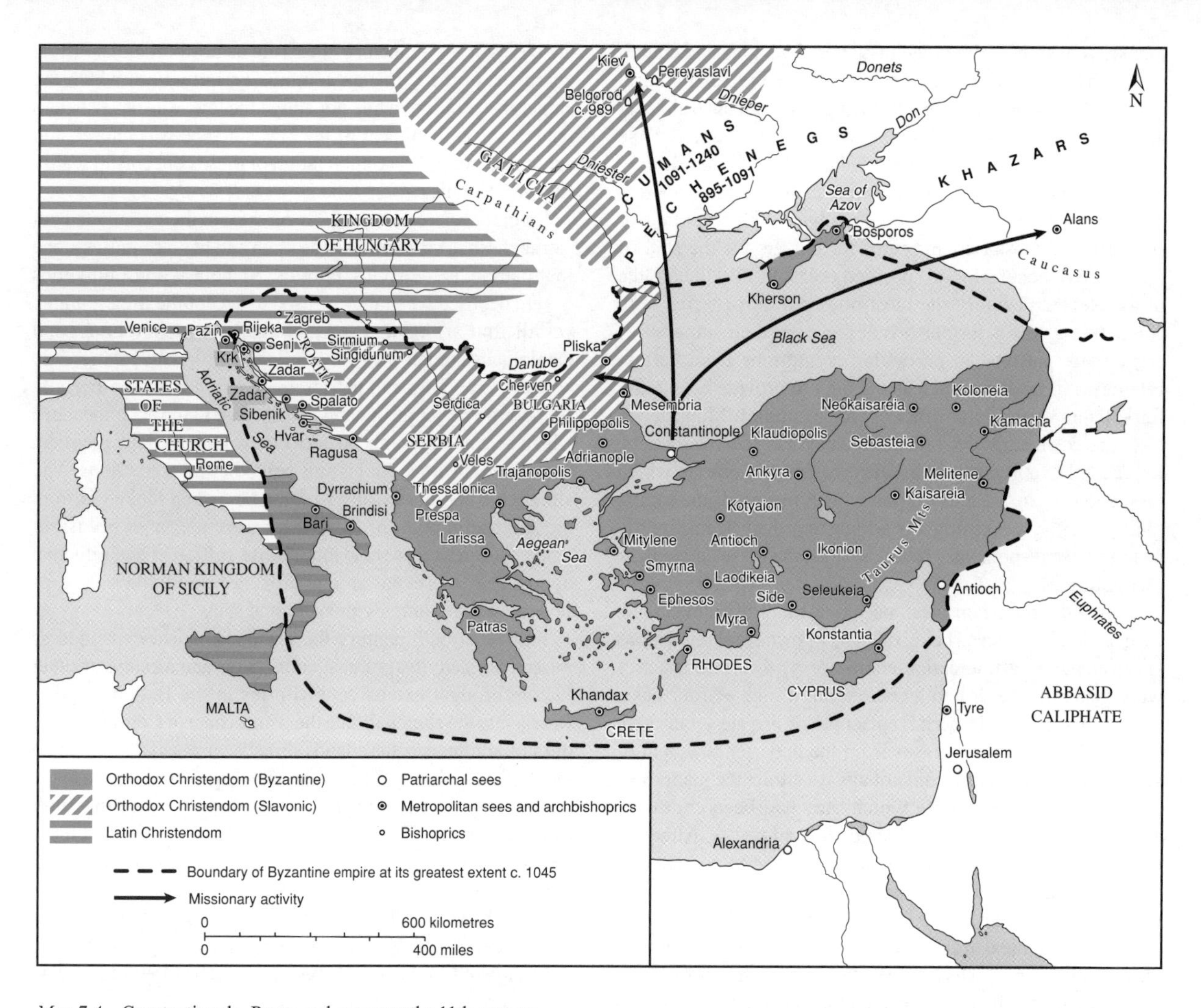

Map 7.4 Constantinople, Rome and emperor: the 11th century.

Disagreement over the line followed by the eastern church was thus one cause for poor relations between the two patriarchates. The tendency in Rome to a more independent position than Constantinople wanted was another. And from the later eighth century the Roman decision to look to the Frankish kings for political support further heightened tensions. The coronation of Charles, King of the Franks, as Roman emperor in the west was the final blow to Constantinopolitan efforts to maintain its position in Italy and the west. The issues came out clearly in a correspondence between Rome and Constantinople in the 860s, in which the Byzantines proudly proclaimed their greater claim to being the 'real' Romans, only to be roundly challenged in a sharply-worded reply from Pope Nicholas I, which pointed out in the clearest terms east Roman inadequacies in these and other areas.

In the end, these tensions reflected much longer-term and much more deeply-rooted differences between the Latin and Greek parts of the former Roman empire. They frequently came to the surface in disagreements which had their origins often in relatively trivial differences of practice or interpretation. The so-called Photian schism, and the schism of 1054 (see page 67), were mere forerunners of the attitudes which were to evolve on both sides following the First Crusade, and which contributed ultimately to the sack of Constantinople by the Fourth Crusade in 1204. But after 1054 the division of the church encouraged ever greater papal intervention in the affairs of the various Balkan powers, further heightening already-existing tensions.

8 The Empire in its International Context

The Wider World: Byzantium in its Cultural Setting

While many aspects of Byzantine culture are seen as very specific to the empire and to eastern orthodox imperial Christianity, Byzantium also shared a great deal with neighbouring and sometimes more distant cultures. Building techniques and traditions in the eastern provinces of the late Roman empire evolved to suit new forms of architecture, but basic methods and materials remained unchanged, so that from Islamic Iraq to the Byzantine provinces of the south Balkans there were many shared elements – techniques of stonemasonry, the construction of the dome, the use of mosaic decoration, for example, are found across the cultural divide between Islam and eastern Christianity. Less obviously, perhaps, but just as importantly, military technology provided an area for cultural exchange which involved the movement of ideas and techniques across many thousands of miles. The introduction of lamellar armour from the steppe in late Roman times, followed by the stirrup, probably via the Avars by the end of the sixth century; the introduction of the curved, single-edged cavalry sabre from the Khazars (probable, but not provable) in the eighth or ninth century; the introduction likewise, via the Avars but ultimately from China, of the traction lever stone-thrower (a precursor of the counterweight trebuchet developed by the twelfth century), and styles of fighting – the heavy cavalry charge developed in the tenth and eleventh centuries owes something to the Franks and Normans – are but a few of the important influences exercised by cultures far from the Byzantine homeland.

Warfare was, of course, only one of many aspects, and by no means the most important, of this cultural openness. In clothing, the world of the steppes and the world of Islam played an important role, with items of personal wear often described by the adjective of the place or people of origin. In food preparation and cooking traditions also, especially in metropolitan contexts, where foreign merchants and traders both settled in their own districts, as in Constantinople, or among the indigenous population, their habits and customs aroused both suspicion and surprise, on the one hand, as well as emulation, on the other, depending upon the context. Indeed trade and commerce were the vehicles most likely to expose Byzantines to foreign ideas and ways, and the degree of external influences seems to increase dramatically as towns revive, as international trade recovers from the nadir of the later seventh and eighth centuries, and as the wealthy élites of the empire look more widely for ways of expressing their social status and cultural sophistication.

The church was another key transmitter, but generally from Byzantium to neighbouring cultures. Most significantly, perhaps, the conversion of the khan – thenceforth the tsar – of the Bulgars to Byzantine Christianity in the 860s and, eventually, the Bulgar élite and the mass of the ordinary population, marks the first phase of a major expansion of east Roman orthodoxy into the Balkans. Just over a century later, the acceptance by Vladimir of Kiev of Christianity in the last years of the tenth century brought a potentially much more massive extension, a corresponding increase in the influence and power of the patriarchate of Constantinople (although the new churches in these distant lands were given autonomy and their own archbishops), and at least for a while a corresponding increase in the international standing of the empire and the imperial court at Constantinople.

Byzantines were aware of the function and purpose of the monuments they produced and embellished: the end result was either to increase the standing of the patron or creator of a work, or in addition, in the case of a religious artefact such as a church or an icon, to glorify God. When an emperor built a new church, he was not simply glorifying God but also increasing his own prestige and affirming his orthodoxy. This applied as much to secular as it did to religious buildings. Art in the Byzantine world was at the same time both original and conservative: older or ancient items were prized for their antiquity and their value as exemplars. The notion of authority was fundamental – a representation, and especially one associated with religion, had to be authentic and conform to the canons of Orthodoxy. The result was a growth in the dependence of many images on earlier works and, from the later eighth century (council of 787), the growth of a wide range of conventions about pose, dress and other features of a representation, as well as particular architectural structures which form the frame and background to many images. Model-books and iconographical guides were compiled, and thus promoted effectively standardised representations. This 'forward-looking conservatism' became typical also of the other regions that adopted Byzantine orthodox Christianity. By the fourteenth and fifteenth centuries there had developed a very clear regionalisation of style, as the provinces of the empire became increasingly independent of Constantinople in social and cultural terms. At the same time, the increasing prevalence of western styles and themes reflected the political, economic and cultural dynamism of the western states. The result was a real efflorescence in the variety of styles and themes in artistic production during the last two centuries of Byzantium.

International Context: the North and West

Although initially able to maintain an effective independence, the Lombard kingdom was extinguished in 773–774 – only 20 years or so after the capture of Ravenna and the extinction of the Byzantine exarchate – when Charles the Great, King of the Franks, seized Pavia and incorporated the Lombard realm into that of the Franks (although the Lombard duchy of Benevento remained autonomous). Between this time and 887 Italy was dominated by the Carolingians until in this year, upon the death of Charles III (the Fat), Berengar, margrave of Friuli was crowned King of Italy, and ruled – with gaps when

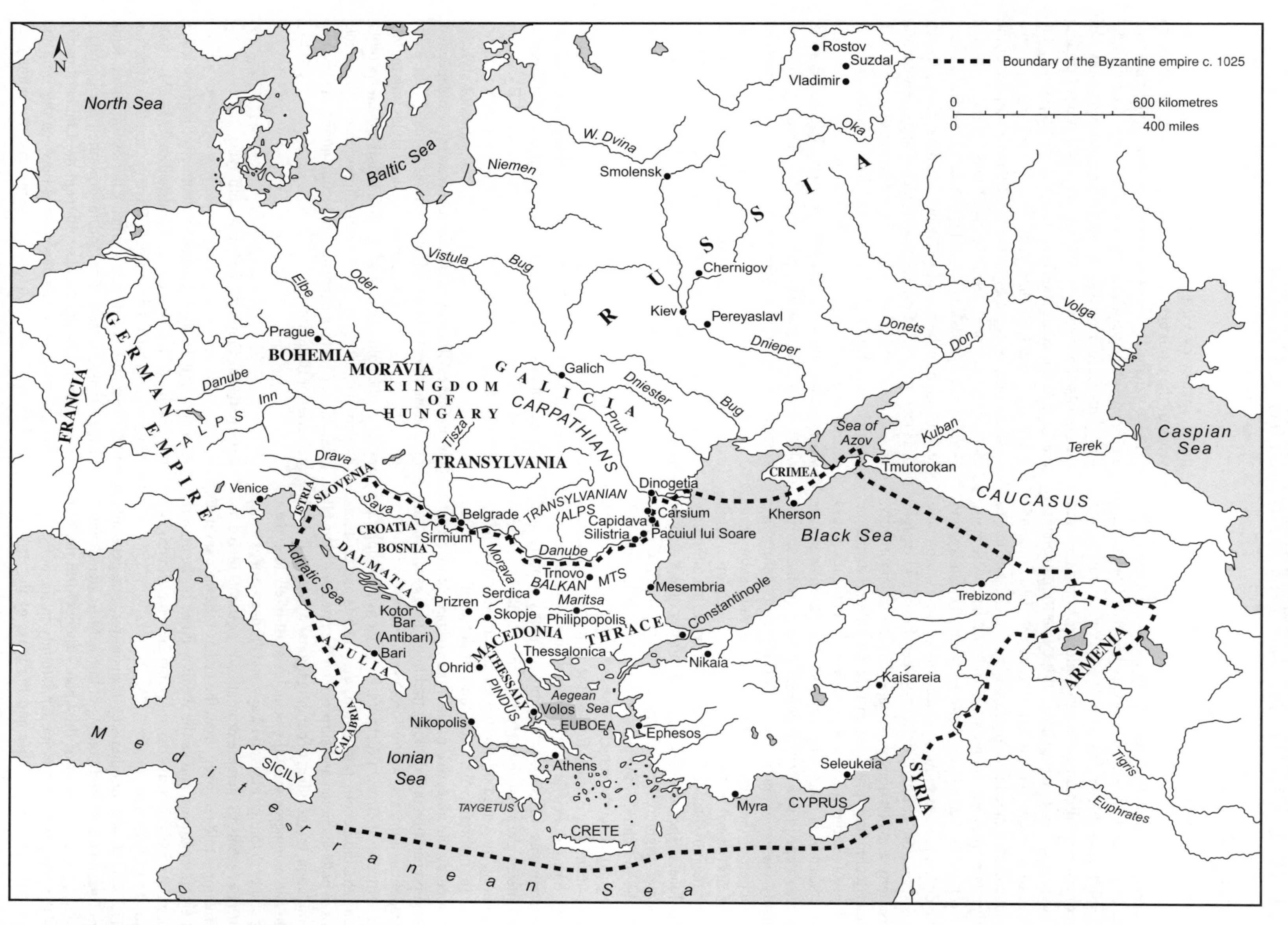

Map 8.1 Byzantium in its wider cultural setting c. 1025.

he was temporarily deposed in the period 891–898 – until 922. Factional division made Italy the subject of constant squabbling among the various claimants to the throne and the (western) empire, although it remained largely under German domination from the middle of the tenth century. After the loss of the exarchate and its territory, Byzantine territory was confined to the south – Calabria and Bruttium – and Sicily, although from the 820s the latter was lost, over a period of some 80 years, to Saracen invaders. In theory cities such as Venice and Naples remained under Byzantine authority but were in all practical respects quite independent.

The Frankish kingdom had fallen into three territorial zones from the middle of the sixth century, Neustria, Austrasia and Burgundy, and these reasserted themselves time after time in the internecine squabbles of the Merovingian dynasty. The last time an effectively unified kingdom was achieved was under King Dagobert I (629–639), and thereafter division and inter-factional strife became the norm. Only with the rise to power of the family of Pepin I of Heristal, *major domo* (mayor of the palace) of the kings of Austrasia, was a degree of stability gradually re-established. Pepin was able to defeat the rival mayor of the palace of the kings of Neustria and Burgundy in 687, thereby bringing the two kingdoms together once more and establishing the basis for a consolidation of his own power and the administrative unity of the realm. Under his (illegitimate) son and successor Charles 'Martel' (who had to fight to retain the unity won by his father) the Alemanni and Thuringians were subjugated, Bavaria was reduced to dependency, and the struggle with the Saxons began in earnest. From 737 he ruled without a Merovingian king, the latter dynasty having been

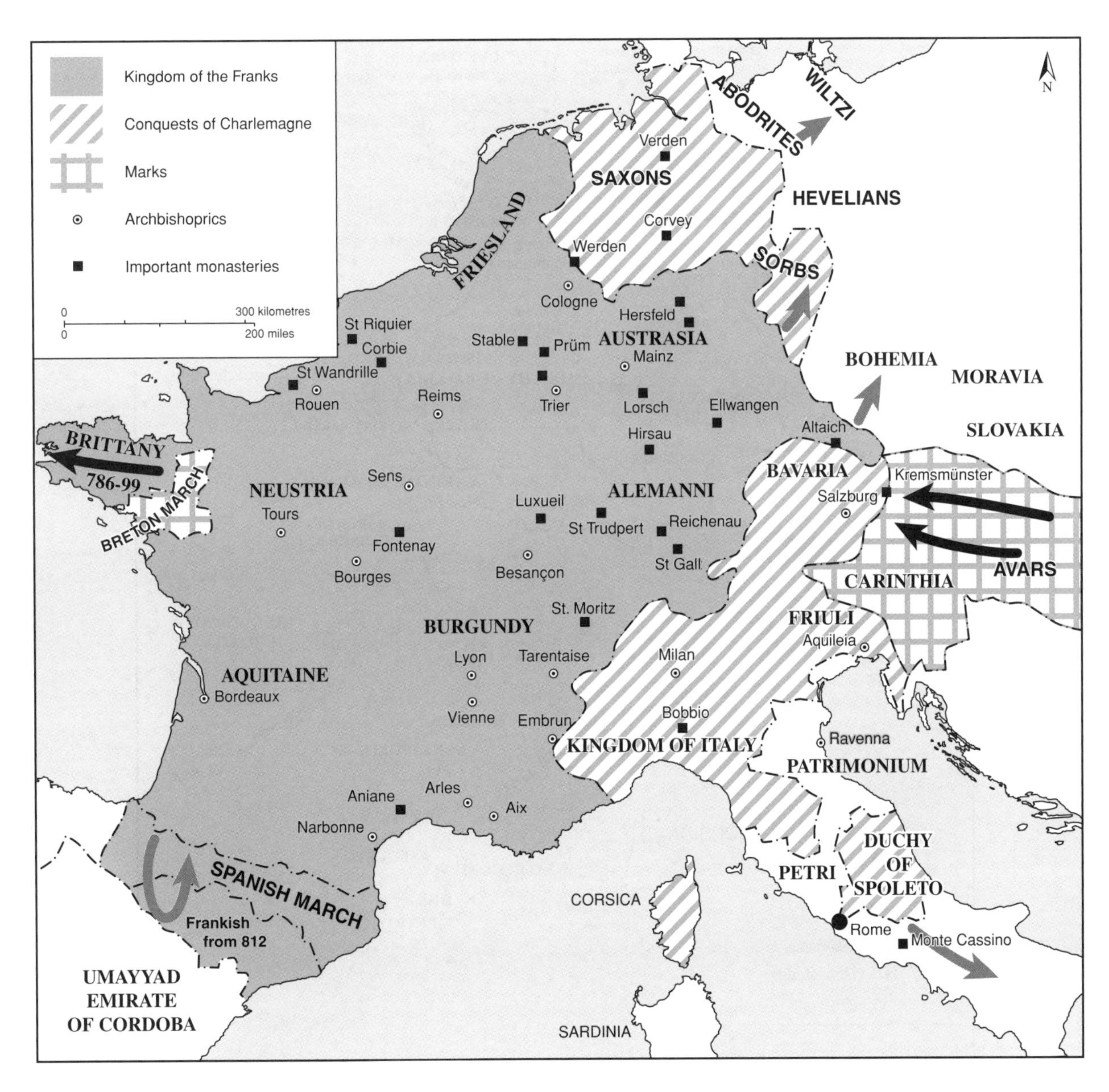

Map 8.2 The empire of Charlemagne c. 814.

reduced to the status of 'shadow kings' (the last of the line, Childeric III, was formally removed, with papal sanction, in 743). Charles died in 741, but divided the kingdom – following Frankish tradition – between his sons Carloman and Pepin, although Carloman retired to a monastery and Pepin became sole ruler in 751. Pepin again divided the kingdom on his own death in 768, but of the two sons Charles and Carloman the latter died soon afterwards and Charles inherited the whole kingdom. Charles the Great, as he is known, turned the Frankish kingdom into an empire. Between 772 and 804 he defeated and subjugated the Saxons, whose leaders converted to Christianity; he conquered and incorporated the Lombard kingdom in Italy; Bavarian independence was effectively ended and the duchy incorporated into the Frankish kingdom; the remaining Avar strongholds were attacked and the Avars reduced to a tenuous dependency; and on 25 December in the year 800 Charles was crowned by the pope as emperor and 'ruler of the territory of the Romans'. It took the Byzantines 12 years to recognise the western emperor, however: only in 812 by the Treaty of Aachen did the Emperor Michael I recognise Charles' title (but only in return for territory in northern Italy and the Adriatic).

Map 8.3 Ottonian central Europe c. 911–1030.

Charles' empire lasted until 842–843, in which year his grandsons divided the empire in three, the eastern, middle and western kingdoms. Within a generation, factional conflict and mismanagement had resulted in the complete fragmentation of the erstwhile empire, with the re-emergence of a host of 'tribal' duchies. Thereafter the political division between eastern (German) and western (French) territories became permanent, with Burgundy (with or without the kingdom of Italy, representing Carolingian conquests from the Lombards) caught between the two.

In the east, the emperors of the Saxon dynasty, which came to power in 919 in the person of Henry I, were able progressively to extend the frontiers of their kingdom to the east and maintain their power in Italy, putting constant pressure also on the Byzantine territories in the south. In the east the campaigns against the various Slav peoples were accompanied by successful missionary activity; the defeat of the Magyars in 955 halted their incursions into royal territory and, in turn, stimulated the development of a settled Hungarian kingdom. Although facing frequent and constant challenges from the dukes and independent lords upon whose power their own authority in part depended, papal support and the Carolingian imperial tradition lent the German empire of the Ottonian and Franconian dynasties a degree of continuity which made it the

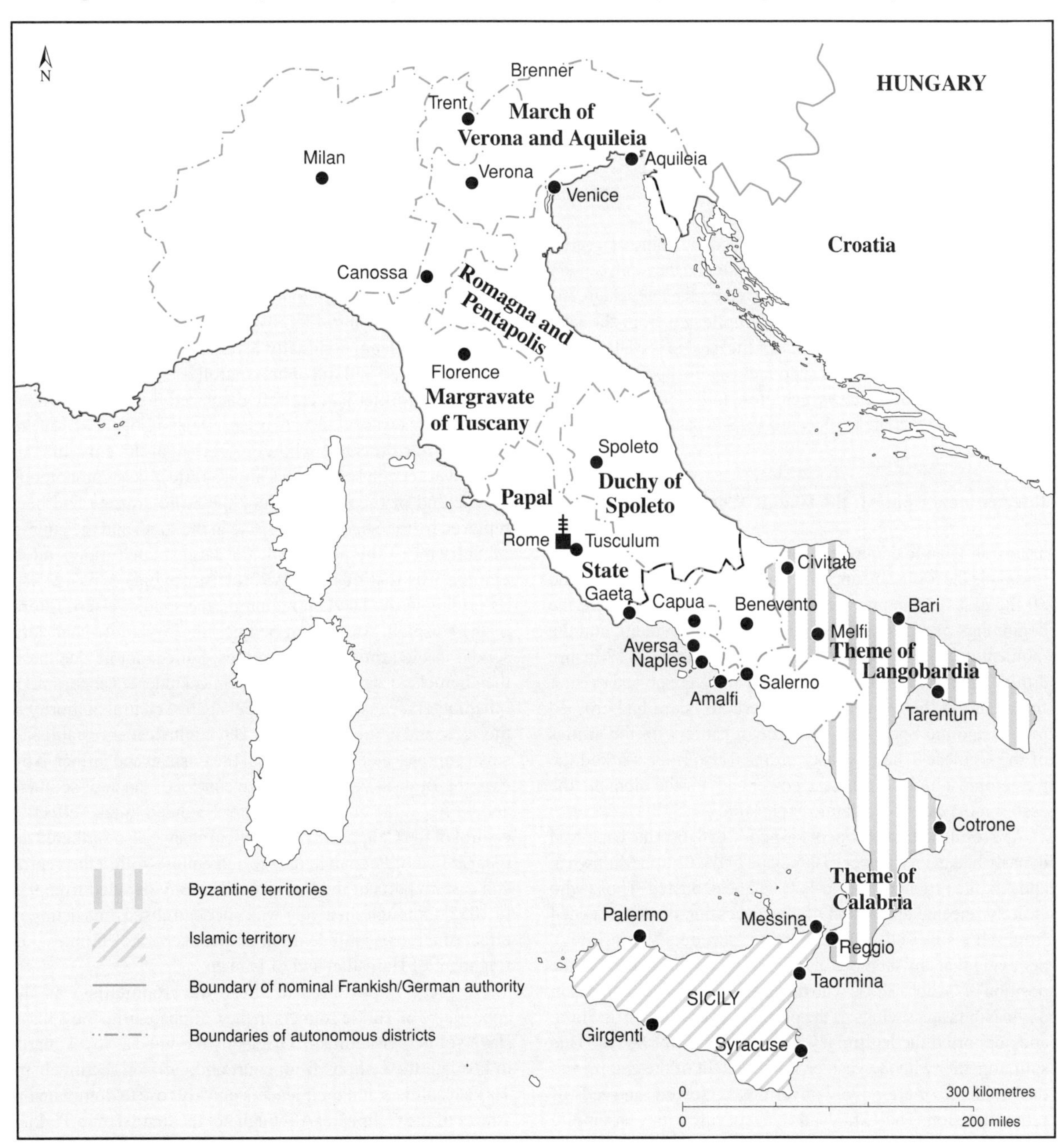

Map 8.4 Italy in the later 9th and 10th centuries.

paramount power in central Europe, able to treat as an equal with the Byzantines and to play a key role in the international politics of the period. The existence of the German empire radically transformed the political and diplomatic world in which the Byzantine empire existed. These relationships were then further complicated during the eleventh century by the arrival of Norman adventurers in Italy, where some served the empire as mercenary cavalry.

Sandwiched between the Franks, Slavic peoples such as the Croats, the Byzantines and the Hungarian plain, the short-lived Moravian kingdom (830–906) turned to Byzantium for support against Frankish pressure. Having adopted Christianity, the Moravian Prince Svyatopluk was able to extend his territory into Slovakia, Silesia and Bohemia, but the arrival of the nomadic Magyars in the late ninth century put an end to the existence of an independent Moravia, which was finally destroyed in 906. The Magyars themselves initially occupied Pannonia and the Hungarian plain, and after an initial 50 years of raiding, established a territorial state which by the 980s was beginning to adopt Christianity and which under King Stephen I became an important Christian power in the region.

Further east the steppe region remained the home of a series of nomadic peoples: from the later seventh into the ninth century the Khazars, then the Magyars, followed by the Pechenegs and Cumans. The picture was further complicated from the later ninth century with the appearance of the Rus' in the Black Sea, Viking traders and warriors who had established themselves in particular around Kiev, which grew to become a powerful principality by the late tenth century. (See page 111.)

International Context: the Islamic World

The Arab Islamic empire expanded extremely rapidly. By the mid-640s the Sasanian kingdom had been destroyed, Egypt and all the east Roman provinces south of Asia Minor taken, the beginnings of a push into North Africa were evident, and the momentum of conquest was still strong. By about 711 Muslim armies had reached Sistan and beyond and were poised to enter India, and in the west Berber converts to Islam had crossed into Visigothic Spain. In 751 a drawn battle with the armies of the Chinese T'ang dynasty on the Talas river marked the westernmost limits of Chinese power and, for the moment, the easternmost extent of Islamic expansion.

As a result of the civil war of the period 658–660 the Umayyad dynasty had gained power in the shape of the Caliph Mu'awiya, and Ali, the prophet's son-in-law, had been ousted. Those who entirely rejected arbitration of the succession dispute seceded from Ali's side and were known as 'seceders' (Kharijites), believing that any member of the faithful could succeed to the position of Caliph. Those who disputed the Sunna (the tradition of the habits and sayings of the prophet), were known as Shi'a, and supported the legitimacy of Ali against Mu'āwiya. This split, together with the vast territorial extent of the empire and the problems of effectively governing it, sowed the seeds of future division. The Umayyad dynasty ruled its vast empire from Damascus, but its removal from power in 750 during the Abbasid revolution brought with it a move to Iran and a new capital created in the 760s, Baghdad. In spite of the Abbasid victory, the Islamic world was thereafter permanently divided. The last of the Umayyads fled to Spain, there to establish the emirate of Cordoba (755–1031, which became the Caliphate of Cordoba from 929 under the emir Abd ar-Rahman) and a flourishing culture.

From the later eighth century the Abbasid empire was rent by factional strife. Several independent dynasties appeared in North Africa, including the powerful Idrisid emirate in Morocco, which spurned the religious authority of Baghdād in 789 and established a Shi'ite caliphate, and the Aghlabids in Tunisia. In the 880s Egypt was effectively independent under the Tulunid emirs. While central Abbasid power in Egypt was restored briefly in the early years of the tenth century, the rebellion of the Shi'ite Qaramati (Qarmatians) in the Hejaz from 899 and in other parts of the Caliphate thereafter caused further political disruption; and the Shi'ite Fatimids replaced the Aghlabids of Tunisia at the same time and began to their rule westwards. To the east they contributed to the collapse of the Idrisid Caliphate, whose lands fell largely to the Spanish Umayyads; and by 972 they were masters of Egypt. From here the Fatimid Caliphate dominated until the twelfth century (and became the major Islamic power in the region during the eleventh century).

From the 820s eastern Iran was effectively independent under the emirs of the Tahirid family; but by the 880s the Samanids ruled Transoxiana, nominally loyal to the Caliph at Baghdad, while the Shi'ite Saffarid emirs controlled eastern Iran, although by 900 the former had entirely displaced the latter. Northern Mesopotamia was ruled by two separate dynasties at Mosul and Aleppo after the 890s; while in Azerbaijan the emir Ibn Abi al-Saj was recognised by the Caliph as effectively autonomous. By the end of the tenth century the Sajid emirate had been replaced by the Shadaddid emirate in the south and the emirate of Shirwan to the north. But the largest such independent emirate was that of the Buyid (or Buwayhid) dynasty, with its origins in the Daylam region on the south-western littoral of the Caspian. An Iranian people, the Daylamites had long served in the armies of the Caliphs; but economic decline in their homeland and a tradition of political independence, among other factors, caused them to rebel against central authority in the 920s, and while a number of Daylamite leaders established small emirates along the shores of the Caspian and farther to the east, the Buyids were able to take control of much of southern and western Iran and, in 945, enter Baghdad to take effective control of the Caliphate. This political change also furthered the renaissance of Persian language and culture within the central and eastern parts of the Caliphate. The Buyids ruled from 932 to 1055, although their rule was not centralised, consisting in effect of a loose confederation of three separate emirates – of Baghdad, of Hamadan and of Isfahan.

Of great importance to these developments was the appearance of Turkic soldiers in the Caliphate from the 830s as slave-soldiers (Mamluks). Originally recruited as loyal guards to insulate the Caliphs from court and garrison factionalism, Turkish soldiers and their leaders quickly rose to dominate the armies of the Caliphate. Although not the first Islamic Turkish dynasty to establish itself (that honour goes to the Karakhanid khanate in Transoxiana in the 980s and early 990s), the emir

Mahmud of Ghazna (998–1030), renowned for his attacks into northern India, seized power from the Samanid emir and established his rule across Afghanistan and eastern Persia. His dynasty ruled in their homeland of Afghanistan until the later twelfth century. But their power in Iran and the central lands of the Caliphate fell prey to the Seljuks, who replaced the Buyids from 1055 and who, having extended their power across the Iranian and middle eastern territories of the Caliphate, began to look to Byzantium and the Fatimid lands as possible targets for further expansion.

Armenia, Georgia and Transcaucasia 550–1000

Armenia, Georgia and the eastern reaches of the Anti-Taurus mountains had always been of strategic and political importance to Constantinople. Yet the fragmented mountainous geography of the region stamped its character on political and social structures, which were dominated by numerous competing clans who, in spite of their fiercely independent outlook, readily called upon outsiders to help in their inter-factional rivalries. Both Persians and Romans thus played a key role, and since the Armenians and Georgians had converted to Christianity during the fourth century, the Christian Roman emperors could also claim to intervene in the interests of their Caucasian neighbours. From the later fourth century Armenia, Albania and Lazica had been under Persian rule, through a series of petty client kings and princes. The Roman parts of Armenia, and Georgia were separated by a border stretching from Theodosioupolis (mod. Erzerum, Arm. Karin) in the north to Dara on the upper Euphrates, and were likewise governed through locally-appointed princes with a variety of Roman titles. They were in practice entirely independent. But as a result of east Roman military successes against the Persians in the period from the 560s Constantinople extended its control over a greater area, including Lazica, and the Emperor Heraclius appointed members of the dominant *naxarar* (noble) clans as *išxans* or 'princes' of Armenia to govern as the emperor's representative. At the same time the Armenian and Georgian churches, which had in the early sixth century rejected the creed of Chalcedon, were reunited with the imperial church (with the exception of those still under Persian authority in the south and east of the country). While the Georgian church remained in communion with the imperial church, however, the Armenian church fluctuated in its adherence according to circumstances.

The Arab conquests ended this period of Roman hegemony. By the 650s the Constantinople-approved 'prince' of Armenia had made a pact with the caliph Mu'āwiya; and although the Romans made several attempts to reimpose their authority, and although many local *naxarars* rebelled against Arab domination, this was firmly established by the end of the seventh century. The Transcaucasus region now became a bulwark of the Islamic world against the steppe nomads, traditionally allies of the Romans and enemies of the Persians, and now of the Arabs. Only western Georgia, in the form of the kingdom of Abasgia, remained relatively independent. Islamic rule was continued through the local princes, however, until a great revolt against new taxation policies in the 770s broke out. This was brutally crushed and many leading noble families were more or less wiped out in the aftermath, although at the same time the Abasgians occupied Lazica, and in the late 780s an independent kingdom of Abasgia was established, under Khazar protection.

The results were twofold. First, the two remaining chief clans, the Bagratuni in the north and the Artsruni in Vaspurakan, were gradually able to establish a complete pre-eminence in their own respective areas; and secondly, the caliphs began to settle large numbers of Arabs and others from outside Transcaucasia in the towns and fortresses of the region, arabising and Islamising much of the countryside and many towns in the process. Local princes were no longer trusted to rule, and a number of small emirates sprang up in their place. But although conflict with the caliphate continued (several leading Bagratunis were in contact with Constantinople, for example, and received imperial court titles), and a major revolt was put down in the 850s, the local dominance exercised by these two families and their numerous subordinate branches and kin, each exercising power in their own locality, was such that many of the smaller emirates were extinguished, while the caliphs of the middle and later ninth century recognised the leading princes of the Bagratid clan as 'prince of princes' ruling on their behalf. The most important of these, Ašot Bagratuni, adroitly exploited the internal troubles of the caliphate to establish the power of his family – through a number of relatives – over Armenia, Albania and the eastern districts of Georgia, and was finally recognised by the caliph al-Mu'tamid in 884 as King of Armenia.

As these events unfolded the Byzantines were also taking advantage of the decline of Abbasid power. As first the Paulicians were crushed in the 870s and then Byzantine armies began to push back into the south-eastern regions of Asia Minor, so diplomatic and political relations between Constantinople and Armenian and Georgian princes became more regular. Basil I and Leo VI recognised the Bagratuni position also, but internal rivalries between various *naxarar* families and resentment at the rise of the Bagratids provoked a Muslim reaction led by the Sajid emir of Armenia and Azerbaijan, resulting in the brief period of Muslim recognition of a member of the Artsrunis, Gagik, as king of part of Armenia and Vaspurakan in 908. Ašot II, the son of the former king Smbat (who had been executed by the emir), ruled the remaining part. But by 915 Byzantine military intervention requested by the *naxarars* and the Armenian church, partly also to counter the danger posed by the Muslim governor's dynastic aspirations and his power-base in Azerbaijan, had restored the Bagratid position. In the course of the Byzantine wars of expansion in the later tenth and early eleventh century much of the Muslim-dominated Transcaucasus fell into Byzantine hands, or at least became politically dependent on the empire. At the same time internal factionalism continued, so that Albania and Siounia were drawn more closely to the Byzantine camp, while Vaspurakan split into several separate principalities. The Armenian nobility were nevertheless able to unite under their king Ašot III in the mid-970s to ward off the approaching imperial armies under John I Tzimiskes, and a treaty of alliance was concluded between empire and kingdom.

Map 8.5 The Islamic world c. 900–920. (After Kennedy, *Historical Atlas of Islam*.)

Yet the Armenian kingdom was already beginning to fragment as different claimants to royal authority set themselves up as kings in their own lands. The Georgian prince David of Tayk' (of the Bagratids) offered military aid to the Emperor Basil II in the war against the rebel Bardas Skleros (976–979), an act which lent the Abasgian kingdom renewed prominence. As Georgian power increased, so Armenian factionalism and strife increased, until Georgia became the dominant force in the region.

The Eastern Frontier c. 700–950

The Arab-Byzantine frontier settled down into almost routine warfare from the late eighth century until the line of the frontier was overrun by the Byzantine advance in the 960s and afterwards. From the 770s and 780s the Islamic side of the frontier was structured in two broad zones, the province of *al-'Awāsim*, a belt of fortified cities stretching from Antioch eastwards along the border provinces, intended to provide both supplies and manpower for the defence of the Caliphate; and the frontier, or *al-Thughūr*, a line of heavily-fortified strongholds intended to deny access to invaders and to serve as advanced warning posts for enemy incursions or as forward bases for Islamic raids into Byzantine lands. Whereas *al-'Awāsim* covered heavily-populated agricultural regions, well able to support themselves, however, the *Thughūr* were often in the 'no-man's-land' zone which both sides seem to have deliberately cultivated from the later seventh or early eighth century onwards. The pattern of warfare this structure reflects dominated the fighting and the cultural relations of both sides for over two centuries and, not surprisingly, encouraged the

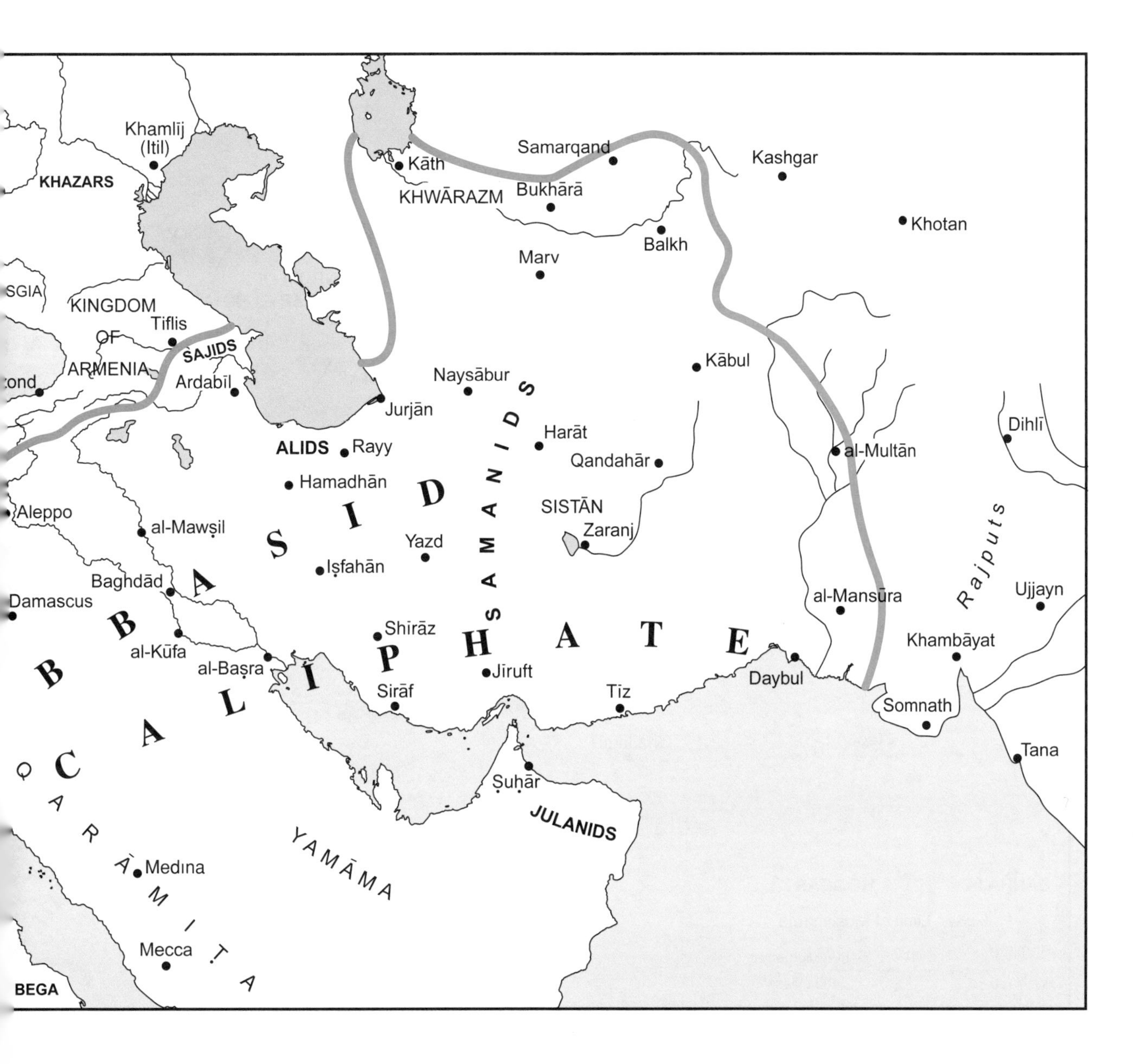

development of a particular frontier culture very different from the metropolitan worlds of Constantinople or Baghdad.

The fundamental principles of Byzantine strategy in the east, as it evolved out of the disasters of the early Arab conquests and raids into Asia Minor, were twofold: where possible, raiding forces should be held and turned back at the passes, before they could do any harm. Where this policy of meeting and repulsing hostile attacks at the frontier did not work, the local forces should harass and dog the invading forces, making sure to follow their every movement so that the location of each party or group was known. A key aspect of this strategy was the garrisoning of numerous small forts and fortresses along the major routes, on crossroads and locations where supplies might be stored, and above and behind the frontier passes through which enemy forces had to pass to gain access to the Byzantine hinterland. As long as these were held, they served to hinder any longer-term Arab presence on Byzantine soil, since they posed a constant threat to the invaders' communications, to the smaller raiding or foraging parties they might send out, and to their logistical arrangements in general. They were a constant threat to any invading force; yet to stop and lay siege to them was more trouble than it was worth for most raiding parties. Although both small and large fortified places frequently changed hands, the Byzantines clearly understood the importance of maintaining their control as a means of preventing efforts at permanent settlement and of minimising the extent and effect of the raids.

From the later eighth and early ninth century the *themata* were complemented by a series of special frontier districts which constituted independent commands. These were known as kleisourarchies (*kleisourarchiai*), created from sub-divisions of the *themata* from which they were detached, which seem to

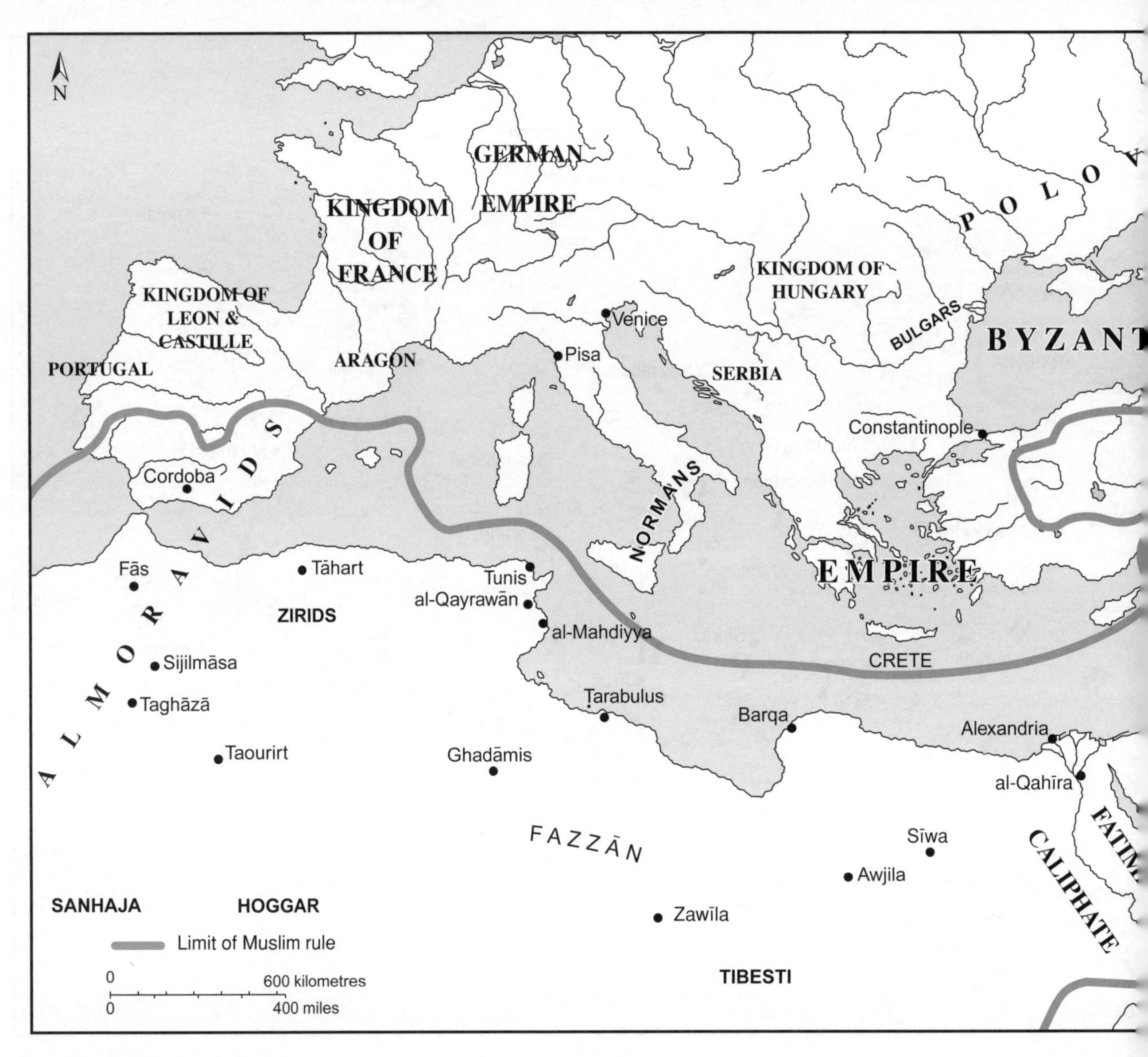

Map 8.6 The Islamic world c. 1071–1100. (After Kennedy, *Historical Atlas of Islam.*)

represent the crystallisation out of the previous strategy of a new policy: a locally-focused defence, involving a 'guerrilla' strategy of harassing, ambushing and dogging invading raiders, designed to stymie all but the largest forces and to prevent both the pillaging of the countryside and the economic dislocation which followed, as well as to make raiding expeditions riskier and less certain, in terms of easy booty, than before.

For the Byzantines, their most intractable foes during the tenth century were the Emirate of Aleppo, and the other independent emirates stretching along the borderlands from Cilicia east into the Caucasus and north-eastern Mesopotamia. Although the frontier had been stabilised by the middle of the eighth century, the dynamic leaders of some of these frontier commands proved to be a constant threat to the security of Byzantine Anatolia, and from the 930s for a period of over 20 years the most significant of these was personified in the figure of Sayf ad-Daulah, the Hamdanid emir of Aleppo. Eventually checked in his depredations by the campaigning emperors of the second half of the tenth century, Aleppo nevertheless remained a key point on the Byzantine frontier, and played an important role in the 980s and afterwards as a semi-independent buffer state between Byzantine and Fatimid lands, obligated by treaty to support the emperor in his wars with the encroaching Fatimid power.

The Steppes and the Rus' c. 680–1000

The collapse of the Avar hegemony in the period following the failed siege of Constantinople in 626 resulted within a few decades in their confinement to the plain of Hungary and the emergence of a number of new nomad groups in the west

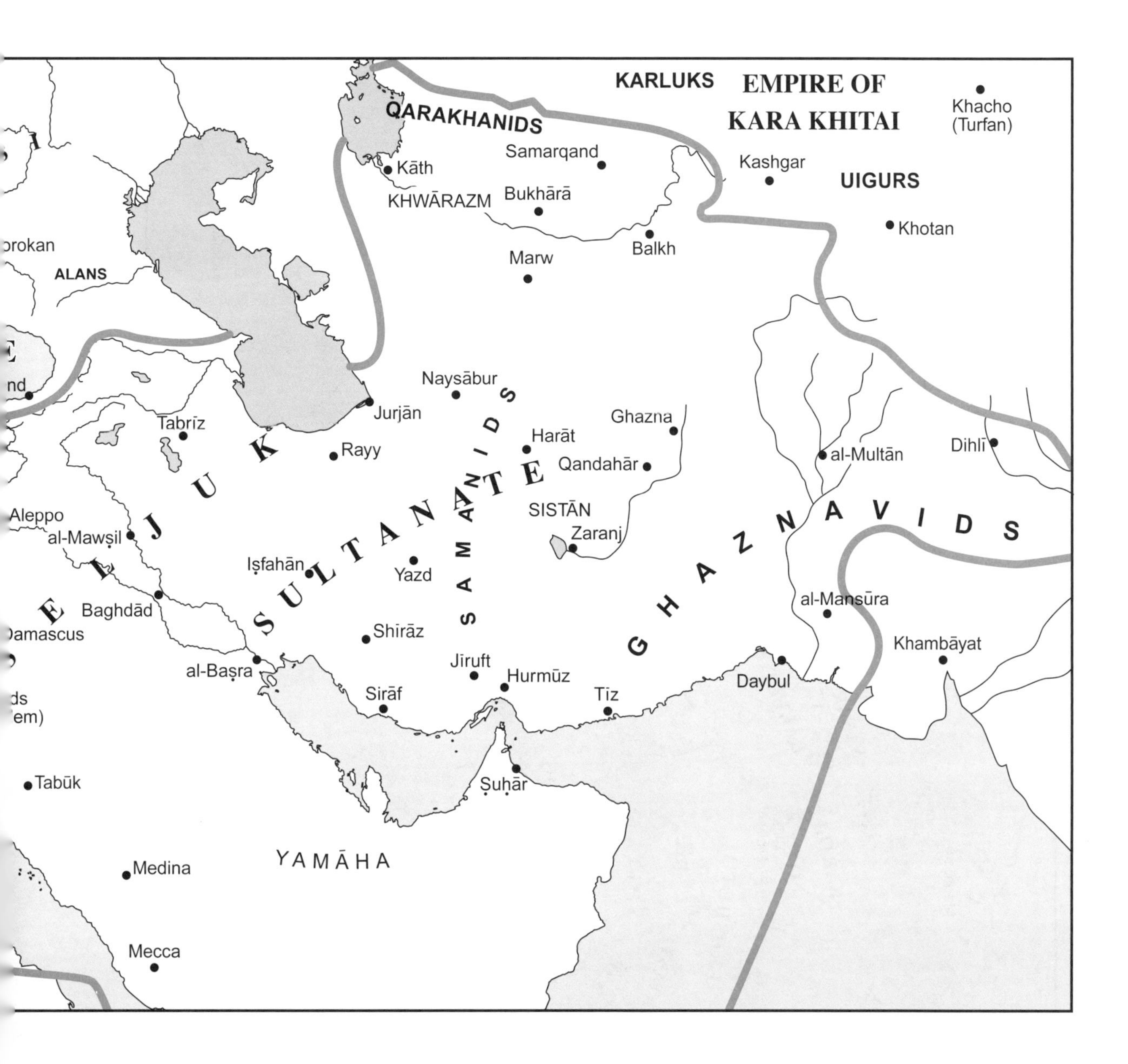

Eurasian steppe. The most important were the Bulgars and the Khazars. The former included the Onogurs who had been under Avar dominion until they split away and made a treaty with the Emperor Heraclius; and the two groups of Kutrigur and Utigur 'Huns', along with several other minor clans, who had been involved with the break-up of the Blue Turk confederacy (ca. 630s). They were in conflict with the Khazars, another faction in this fighting. Khazar victory in the 670s led to the expulsion of several Bulgar groups. Some migrated northwards (to form the later Volga Bulgar group) or westwards (as far as the Pentapolis in north-east Italy, or into Avar territory in Pannonia), and one group moved down to the Danube delta. Under their leader, Asparuch, their request to enter east Roman territory was refused. When they crossed into what was seen as Roman territory in 679/680, the Emperor Constantine IV marched to meet them, but tactical confusion led to a Roman defeat, which made possible the foundation of the Balkan Bulgar state (see pages 31–32 and 58). By the early eighth century the Bulgar khans were able to interfere in internal Byzantine politics, and during the eighth century they presented almost as grave a threat to the empire as did the Arabs in the east.

By 680 the Khazars controlled much of the western Eurasian steppe, their leading clan claiming the lineage and rights of the senior clan of the western Blue Turks. They were for Constantinople an important ally against Islam, and mounted several raids into Islamic Caucasian territories, although temporarily checked by a successful Arab offensive in 737. They were also a counterweight to the Bulgars in the Balkans. Two emperors arranged marriage alliances with the Khazars (Justinian II married the Khagan's sister; Constantine V married a Khazar princess) and other contacts flourished. Tensions over conflicting claims in the Crimea did not hinder

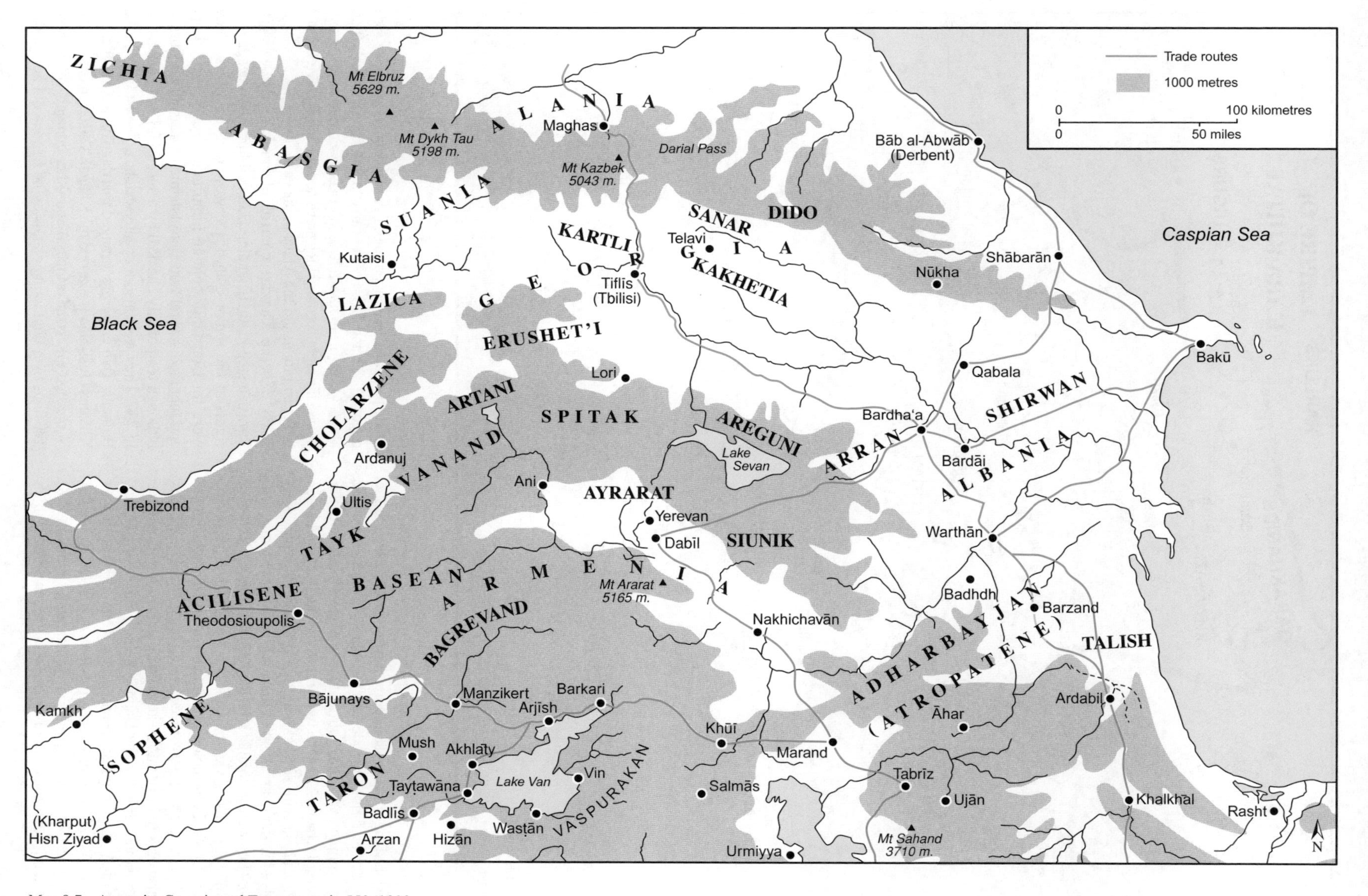

Map 8.7 Armenia, Georgia and Transcaucasia 550–1000.

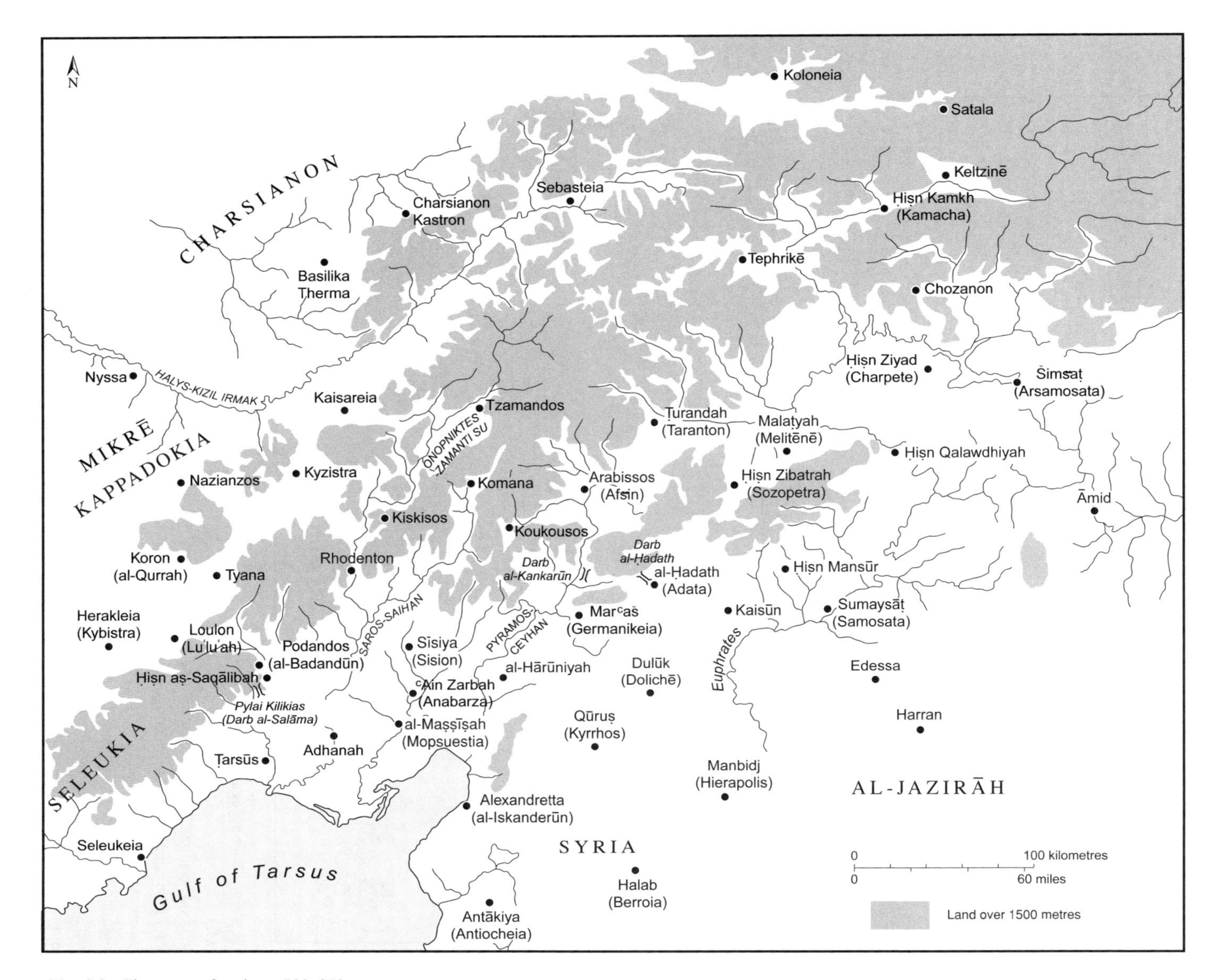

Map 8.8 The eastern frontier c. 700–950.

Map 8.9 The steppes and the Rus' c. 680–1000.

otherwise good relations, even after the leading clan converted to Judaism in the eighth century. The kingdom of Abasgia was under Khazar rather than Byzantine protection from 786; and the semi-autonomous Iranian Alans, in the northern Caucasus region, were also tributary. But the Khazar rulers slowly moved away from their steppe nomad roots to derive most of their wealth from commerce, for they controlled the major trade routes across the southern steppe zone and between the northern forests and the south. When a new threat in the form of the Pecheneg and Oğuz Turks appeared on their eastern borders, they were able to offer only limited resistance.

The Pechenegs were a Turkic group subject to the rump of the former Blue Turk Khaganate and living in Turkestan. With the dissolution of the Turkic Khaganate, another group, the Oğuz, under pressure from the expanding Uighurs, moved westwards. Some Pechenegs remained and joined with the newcomers; others moved away towards Khazaria, dislodging the Magyars, an Ugric-speaking people tributary to the Khazars, who were also pushed westwards. By the 880s effective Khazar control was confined to the steppe north of the Caucasus. The Pechenegs dominated the south Russian steppe, with the advancing Oğuz to the east and the Magyars to the west.

Byzantine encouragement to the Magyars to attack the Bulgars in 894–896, combined with Pecheneg pressure, finally pushed the Magyars into the Hungarian plain, where they founded the kingdom of Hungary during the tenth century and adopted western Christianity. Khazar power was further reduced by Russian attacks in the 960s (Itil, the capital, fell in 965), and by the emergence of an independent Alan kingdom, which by the 1030s had swallowed up the western half of Khazaria and, with the attacks of the Oğuz (or Cumans as the Byzantines also called them) driving the Pechenegs west. By the 1060s the kingdom of the Alans had absorbed also the eastern half. Independent Khazaria disappeared.

The Rus' first appear in the 830s. Scandinavian raiders and traders, they colonised the Baltic coastline in the mid-eighth century, before moving inland along the great rivers from the Baltic and down to the Black and Caspian seas. They established a series of riverine principalities at Novgorod, along the Volga, Kiev and, a little later, at Tmutorokan on the Black Sea. Asserting their control over the scattered Slav populations, raiding Byzantine territory in the 830s and 840s, one group – probably from the Volga region – launched a surprise attack on Constantinople in 860. They were brought under a single rule by Oleg of Kiev, successor to the Rurik who had first established a lordship over the Novgorod district, but were initially client tributaries of the Khazars.

Kiev increased in importance partly through the support of the Byzantine empire, which needed a reliable ally in the region to prevent a recurrence of the attack of 860. Trade agreements in the early tenth century, and the baptism of the Russian princess Olga in 960, cemented this alliance. But there were tensions. Olga's son Svyatoslav remained a pagan, and in a joint attack with the Oğuz in 965 destroyed the Khazar Khaganate and brought the Volga Bulgars and Rus' under his sway. An uneasy *modus vivendi* existed between Kiev and the Pechenegs. At Roman request, Svyatoslav attacked Bulgaria in 967, but then occupied the whole country. The Byzantine counter-attack forced the Rus' leader to terms, but he was killed on his return journey to Kiev in a Pecheneg attack. The Byzantines were able to occupy eastern Bulgaria; in Kiev there followed ten years of conflict until the prince Vladimir seized power in 980. Treaties with the empire eventually brought about the conversion of Vladimir and his circle to Byzantine Christianity (in 988: Kiev became an ecclesiastical province of Constantinople, with a metropolitan bishop at Kiev) and the establishment of the Varangian guard, a consequence of the Rus' promise to furnish troops to the Byzantine emperor when asked to do so. With the increasing threat from the Pechenegs the Rus' now became key allies of the empire in the south Russian steppe.

Part Three:

The Later Period
(c. 11th–15th Century)

9 *Apogee and Collapse: the Waning of East Rome*

The Empire in Context 1050–1204

The traditional view of the later tenth century and the reign of Basil II is that of a Golden Age, in which the empire attained the height of its power. In terms of international esteem and territorial extent this is not incorrect, but the result is that the reigns of Basil II's successors are inevitably compared unfavourably with what went before. Basil left no children and was succeeded by his brother Constantine VIII, who ruled just three years alone before he too died. Constantine's elder daughter Zoe now determined who should become emperor according to whom she married. In the period from 1028 until 1042 she had in succession three husbands who became emperor through her: Romanos III Argyros (1028–1034), Michael IV 'the Paphlagonian' (1034–1041) and Constantine IX Monomachos (1042–1055). The predominantly civilian power élite at Constantinople took the stability achieved by the end of Basil II's reign for granted, and proceeded to a radical reduction of the standing army and frontier militias – in order to limit the power and ambition of the provincial 'military' élite. In doing so, however, they did not, and perhaps could not, take into account changing circumstances outside the empire. The military élite were just as concerned with their own position within the empire and their relations with the governing clique and the emperors. Provincial rebellions in Bulgaria in the 1030s and 1040s, a result of misguided fiscal policies as well as political oppression, foreshadowed greater problems. The arrival on the Balkan frontiers of the Pechenegs, a Turkic steppe people, at about the same time, and their first incursions into imperial territory, similarly should have alerted the emperors and their advisers. The rebellion in 1042–1043 led by George Maniakes, commander of the imperial forces in Sicily, was cut short by the death of its leader, but nevertheless indicated that there was considerable unrest and discontent among leading elements of the élite. The brief reign of the Emperor Isaac I Komnenos (1057–1059), a soldier of the Anatolian military clans, indicated the direction of change. And although the empire continued to expand its frontiers in the Caucasus, efforts to re-establish imperial control in Sicily and southern Italy failed, so that by the early 1070s the empire had lost its last foothold there to the dynamic Normans.

As well as the Pechenegs in the north, a new and yet more dangerous foe appeared in the 1050s and 1060s in the east, in the form of the Turkic Seljuks, a branch of the Oğuz Turks (called Ouzoi by the Byzantines) who had already established themselves as masters in the Caliphate, and whose energies were now directed northwards from Iraq into the Caucasus and eastern Asia Minor. The Emperor Romanos IV Diogenes (1068–1071) made an attempt to stem the flow of raids and incursions, but after some initial success, a combination of treachery and tactical blunders led to his defeat in 1071 at the battle of Manzikert in eastern Anatolia. Losses following the battle were in themselves not great, and the evidence suggests that the strategic situation could have been rescued. But Romanos himself was captured, and although ransomed after a brief period as the guest of the Sultan Alp Arslan, found that he had been deposed. Civil war broke out and the empire's remaining forces exhausted and weakened themselves in the conflict that followed. Asia Minor was simply left open, and Turkic herdsmen with their flocks, families and chattels moved in to the central plateau, ideally suited to their way of life (in contrast to that of the Arabs who, in the seventh century, had signally failed to make this transition). Further civil conflict ensued, and by the time the general Alexios I (1081–1118) seized the throne in 1081, the empire had lost central Asia Minor, while the Balkans were overrun by Pecheneg raiders and the Normans had invaded the empire from their base in southern Italy.

Alexios had restored stability by 1105 through good planning, diplomacy and able generalship. The Pechenegs were defeated and settled within the empire as soldiers in the imperial army; the Normans were thrown out of Epiros; the Seljuks were checked. Alexios astutely used the armies of the First Crusade to help as they passed through Byzantine territory in 1097–1098. He carried out administrative reforms and he allied himself through marriage and the imperial system of offices with other élite clans in order to re-establish an effective central administration. Under Alexios' son John II (1118–1143) substantial tracts in western Asia Minor were recovered, while under Manuel I (1143–1180) the imperial position in the Balkans was strengthened and imperial armies began slowly to push into central Anatolia again. But tactical misjudgement led to the defeat at the battle of Myriokephalon near Ikonion in 1176, ending Byzantine efforts to recover the central plateau. The empire was never again in a position to mount such a campaign in the region, and the 'Turkification' – and Islamisation – of central Asia Minor were firmly under way.

Along the Danube the Hungarians were coped with, but the empire was becoming an increasingly European state. At the same time a number of relatively new political powers entered the historical stage.

First, the maritime power of Venice (followed by Genoa, Pisa and Amalfi) introduced a new element into the economic as well as the political relations between Byzantium and the west, and the imperial government made several concessions in respect of customs and trading privileges to these city-states. Secondly, the empire had to deal with a complex situation in its relations with the German emperors and with the nascent power of the Sicilian Norman kingdom. The Crusades transformed the political situation: while the Crusaders themselves established a series of fragile principalities around the Kingdom of Jerusalem, they also brought with them the vested interests of the western powers, whose interest in what had been an exclusively Byzantine sphere of influence grew. John II and Manuel were able to maintain good relations with the German emperors, playing them off, for example, against the Normans of Sicily.

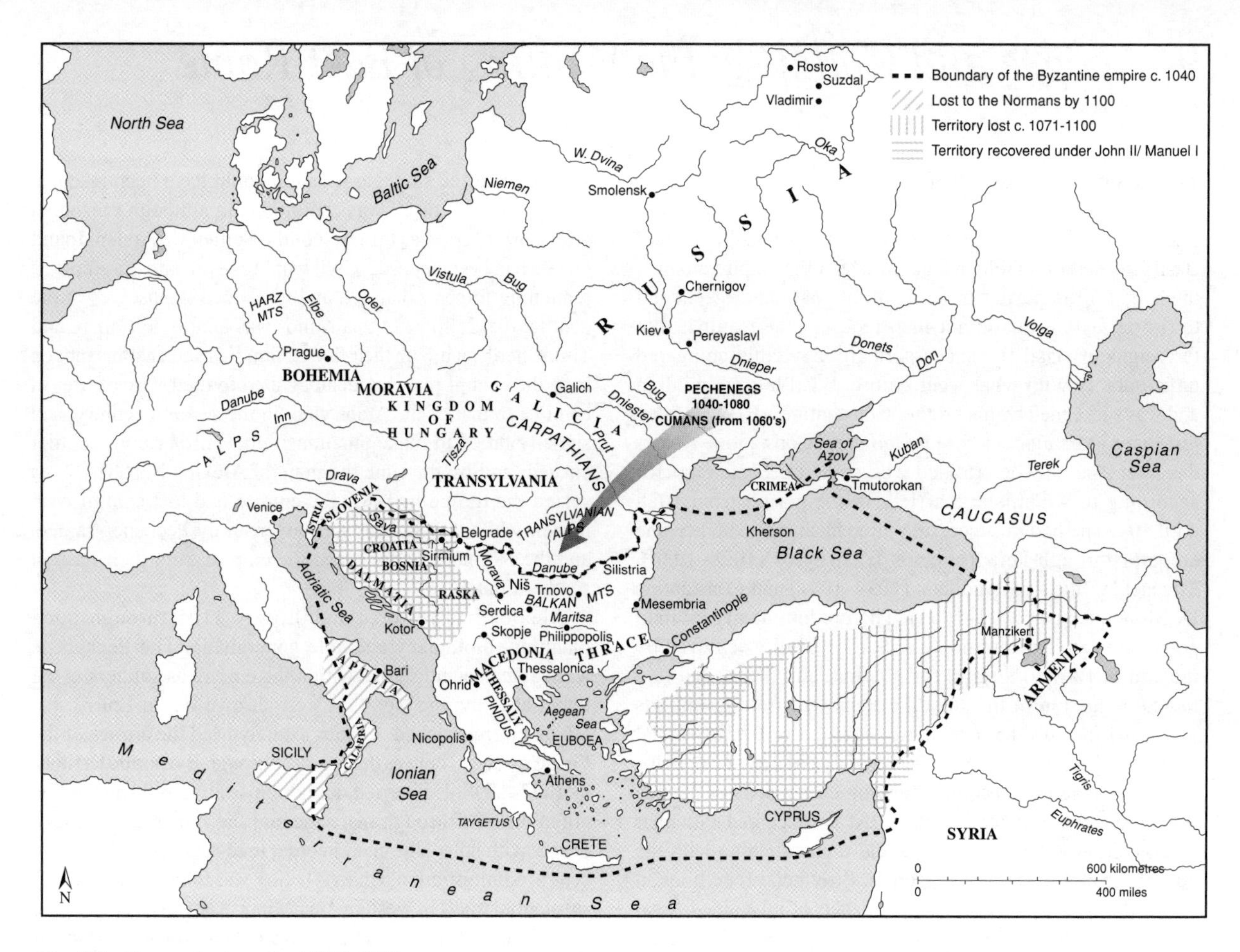

Map 9.1 The empire in context 1050–1204.

Yet on the death of the Emperor Manuel I, the system he had built up quickly fell apart, and the fragility of the imperial position soon became apparent.

Manuel left no heir competent enough to deal with this inheritance, and the empire was riven by petty dynastic squabbles. The hostility of the Italian cities and other western powers was provoked by attacks on westerners – in 1182, for example, a massacre of western merchants was orchestrated by Andronikos I. In 1185 the Normans exploited the situation by attacking and sacking Thessalonica, the second city of the empire. Andronikos was deposed and killed in 1185 and succeeded by Isaac II Angelos (1185–1195), a relation of the Komnenos clan; he was replaced by his brother Alexios III Angelos in the years 1195–1203. In this latter year, however, the armies of the Fourth Crusade appeared before Constantinople. Alexios III was removed, to be succeeded by Alexios IV.

A rebellion in Bulgaria in 1185 led to the defeat of imperial garrisons and the re-establishment of an independent Bulgarian state. Isaac II was himself heavily defeated in 1190 in an effort to check the revolt; while Serbia, which had occupied the position of a vassal state for some time, began also to distance itself from the empire, partly a result of the empire's wars with Hungary. Isaac was able to stabilise the situation through a successful military campaign and a marriage alliance. But it was apparent that imperial power and authority were in decline on every front; by 1196 Serbia had turned to Rome rather than Constantinople for political support, while the situation in Bulgaria was hopeless; and in 1189 Cyprus was also lost to the English under Richard I in the course of his crusading campaign.

The Crusades 1096–1204

Military expeditions organised or inspired by the papacy and aimed at the protection of pilgrims and Christian holy places or to help fellow Christians in their struggle against unbelievers had begun already in the 1060s when knights from Burgundy and Languedoc marched to fight with the armies of Castile in Spain. The reform movement in the western church, which affected the papacy and its policies, coincided with a renewed threat, in western eyes, to the Holy Land from the Seljuks, whose capture of Jerusalem and Syria from the Fatimids directly affected western pilgrims. Unarmed pilgrimages were a well-established tradition, and Seljuk interference caused considerable anger. At the same time, the notion of holy war against the heathen was growing in popularity and in 1095 at the Council of Clermont the Pope, Urban II, responding to a request from the Emperor Alexios I for military aid, preached that knights should join a crusade to recover Jerusalem. In 1096 an unorganised crusade of adventurers led by a certain Peter the Hermit arrived at Constantinople and was transferred to the Anatolian side, where it rapidly disintegrated under Seljuk attacks. Between 1096 and 1099, however, a much more effective expedition was conducted under the leadership of several leading nobles. The northern French contingent was led by Duke Robert of Normandy, those of Lorraine and the Flemish lands by Godfrey of Bouillon, Baldwin of Boulogne and Robert II of Flanders, and those of southern France by Raymond of Toulouse. The Normans of southern Italy were led by Bohemond of Tarento and his nephew Tancred. The forces of Godfrey marched across Germany and down through Hungary to enter Byzantine territory (some of the Crusaders also attacked several Jewish communities en route), while those of Raymond of Toulouse marched along the Adriatic shore lands of Croatia to enter Byzantine territory north of Dyrrhachion (mod. Durrës). The Normans crossed the Adriatic and entered the empire further south on the coast of Epiros. The Emperor Alexios was able to provide supplies for them and, on the whole, to keep the peace between his own soldiers and the subjects of the empire and these substantial alien armies, eventually – after extracting oaths of fealty from the leaders – helping them to cross the Bosphorus and engage the Seljuks. Although initially outwitted by Seljuk tactics, the Crusader leaders quickly adapted to the new conditions. Defeating the Seljuk Sultan at Dorylaion, Antioch was taken after a seven-month siege, and eventually Jerusalem itself was captured in July 1099.

The result was the foundation of a number of Crusader states. The most important was the Kingdom of Jerusalem (whose first king, with the title 'protector of the Holy Sepulchre', was Godfrey of Bouillon), followed by the Principality of Antioch (claimed, however, by the Byzantines), taken by Bohemund, and the Counties of Edessa and Tripolis. Weakened by internal fighting and conflict as well as by conflict with Byzantium, the city of Edessa, which had become an important Crusader stronghold, was taken by the Zengid emir of Mosul in 1144. This led directly to the preaching of the Second Crusade (1147–1149), led by the German Emperor Conrad III and the French King Louis VII. But the effort was weakened by the conflicting foreign policy interests of the two (the German emperor allied with Byzantium against the interests of the Norman King of Sicily, and the latter allied with the French). The result was defeat at the hands of the Seljuks and unsuccessful expeditions against the Muslim strongholds of Ascalon and Damascus. Continued Crusader rivalries and factionalism as well as an essentially untenable strategic position led to the capture of Jerusalem in 1187 by the general Saladin, founder of the Ayyubid dynasty, who took over the empire built up by the Zengid emir Nur ad-Din.

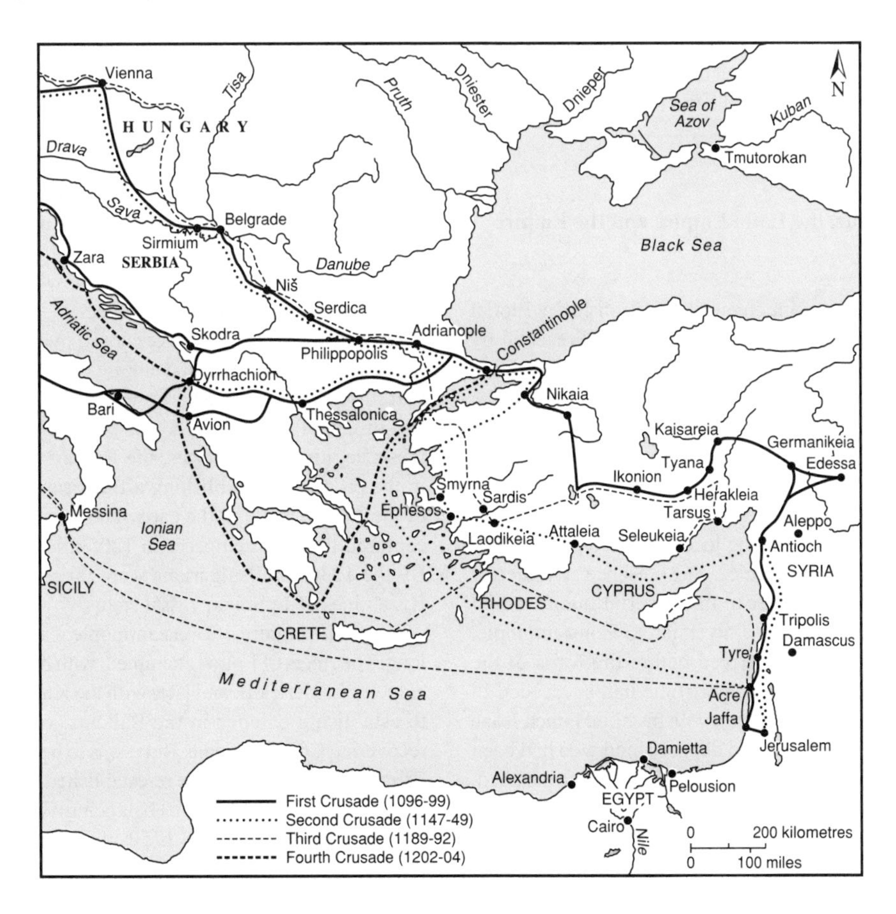

Map 9.2 The Crusades 1096–1204.

The Third Crusade which followed (1189–1192) was led by the German Emperor Frederick I Barbarossa, and consisted of three separate columns, his own, across the Balkans and via Constantinople; that of Richard I of England, which sailed via southern France, the toe of Italy and Cyprus (which was seized from the Byzantines en route); and that of Philip II Augustus of France which sailed from Genoa via southern Italy to the Holy Land. After a victory over the Seljuks at Ikonion Frederick drowned while crossing the river Kalykadnos (Salef) and his son, Duke Frederick of Swabia, took over. His army reached Acre, but Frederick died in 1191; Richard and Philip succeeded in taking Acre, and a treaty was eventually concluded with Saladin, whereby Tyre and Jaffa were ceded to the Crusaders, and pilgrimages to Jerusalem were to be permitted without hindrance. Cyprus was given as a fief to Guy of Lusignan.

The Crusades represented what was to be, in the last analysis, a strategically impossible task. Apart from the financial and resource problems posed by the need to defend such a long and exposed frontier, the Crusader states were perpetually short of manpower, had problems with maintaining their stock of warhorses, and were split by internal rivalries. But they were given a certain strength and resilience by the efforts of the military religious orders of knights, the Knights of St John and the Knights Templar. The latter, formed in 1120 with their mission the protection of pilgrims and securing the conquest of the Holy Land, formed a core of highly trained and effective soldiers. The former, originating in a fraternity attached to the hospital of St John in Jerusalem, became a military order in 1120 also and, like the Templars, acted as a key support of the Crusader states until forced to leave the Holy Land in the late thirteenth century.

The Fourth Crusade, the Latin Empire and the Empire of Nicaea

The fatal blow to the Byzantine empire in its established territorial form came in the form of the Fourth Crusade. Preached by Pope Innocent III and with Egypt the objective, the Crusading leaders hired ships and obtained some of the finance for the expedition from the Venetians, to whom they rapidly became heavily indebted. While not opposed to a Crusade, Venice was also interested in consolidating its commercial position in the Adriatic and the eastern Mediterranean, and in return for financial assistance the Crusading leaders agreed first to seize the city of Zara, claimed by Venice. The presence at Venice of Alexios IV Angelos, pretender to the imperial throne, could likewise be used to legitimate a diversion to Constantinople. In 1203, the Crusader army arrived before the walls of the Byzantine capital and within a short time had succeeded in installing Alexios IV as co-emperor, with his blind father, Isaac II, whom his uncle Alexios III had deposed, and who had been brought out of prison after the latter fled the city. Once installed, Alexios IV found it impossible to pay the promised rewards, and found himself increasingly isolated. Early in 1204 he was deposed and murdered by Alexios V Doukas. Although the new emperor strengthened the defences and was able to resist an initial Crusader attack, the city fell on 12 April. The booty taken was immense. The city had amassed a store of precious objects, statues, liturgical and ceremonial vestments and objects since its refoundation by Constantine I, and had never before fallen to violent assault. Now it was mercilessly sacked for three days, during which countless objects were destroyed, while precious metal objects were melted down or stolen. Some of the most spectacular late Roman objects can still be seen in Venice today.

Alexios V fled but was captured shortly afterwards and executed, and Baldwin of Flanders was elected emperor. The empire's lands were divided among the victors, according to a document known as the *Partitio Romaniae*, drawn up during 1204, and probably based on imperial tax registers. According to this, the Latin emperor at Constantinople would receive a quarter of the empire, the others three-eighths each. Venice received the provinces and maritime districts it had coveted, while Greece was divided among several rulers: the Principality of Achaia (the Morea) and the Duchy of the Archipelago were subject to the Latin emperor at Constantinople. A kingdom of Thessalonica was established, to whose ruler the lords of Athens and Thebes owed fealty; while the county of Cephalonia (which – along with the islands of Ithaca and Zante – had been under Italian rule since 1194) was nominally subject to Venice, although it was in practice autonomous, and after 1214 recognised the prince of Achaia as overlord. The lord of Euboea (Negroponte) was subject to the authority of both Thessalonica and Venice.

The Byzantine empire continued to exist. Despite the parcelling out of imperial territory, a number of counter-claimants to the imperial throne asserted their position. One branch of the Komnenos-Doukas clan established an independent principality in the western Balkans, with its focus in Epiros in north-western Greece, which lasted almost to the fifteenth century. From the 1240s its ruler was referred to by the title *despotes* ('lord'). The dynasty of the Komnenoi governed a more-or-less autonomous region in central and eastern Pontos, where the 'empire' of Trebizond now appeared. Members of the Laskaris family continued to exercise effective control over much of Byzantine western Asia Minor, and the empire of Nicaea evolved around that city. Theodore Laskaris was crowned ruler, and as the son-in-law of Alexios III, he had some legitimacy. Apart from these 'legitimist' territories, the Bulgarian Tsar Kalojan was in the process of establishing a Bulgarian power to rival that of the Tsar Symeon in the early tenth century, and he actually captured the Latin emperor in 1205 after crushing his army. By the 1230s the Bulgarians were threatening to reduce the Byzantines of Epiros to vassal status.

The Latin empire of Constantinople was not destined to last long. The rulers of Epiros attempted, with help from the German emperor Frederick II, and later with the King of Sicily Manfred, to establish a balance in the Balkans, with the intention of recovering Constantinople. But it was to be from Nicaea that the Byzantine empire was to be re-established. The emperors allied themselves with Genoa, thus balancing Venetian naval power. Throughout the 1240s and 1250s they extended their lands in Europe. With an imperial court and household rooted in that of the emperors before 1204, and with an effective administrative framework inherited from that evolved under the first three

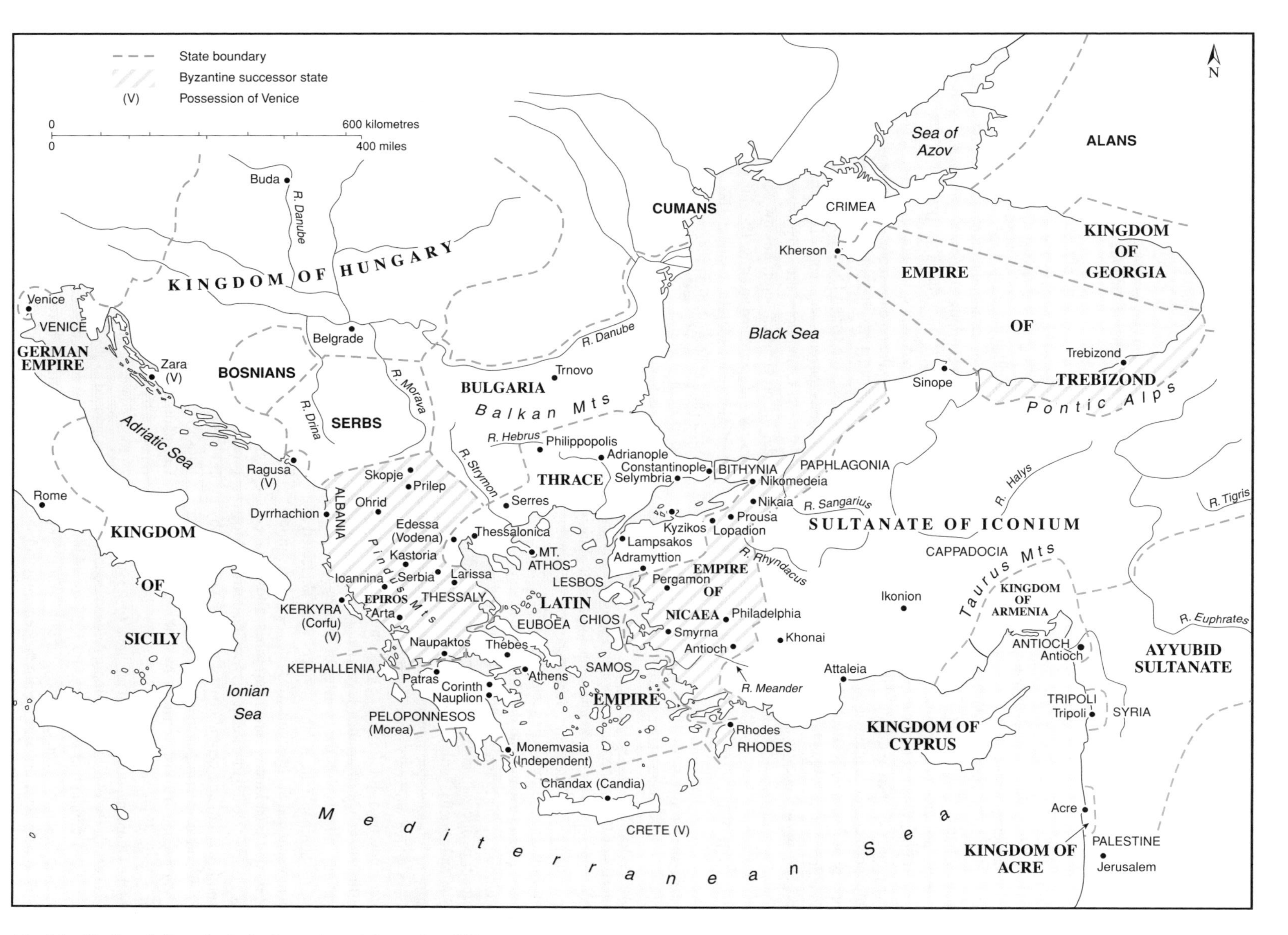

Map 9.3 The Fourth Crusade, the Latin empire and the empire of Nicaea.

Komnenoi, they had a sound logistical and strategic basis from which to operate. By the end of the thirteenth century, parts of central Greece were once again in Byzantine hands: the Byzantine 'despotate' of Morea controlled much of central and south-eastern Peloponnese, although the Latin principality of Achaia remained an important power to its north. In Asia Minor, the policy of rapprochement with the Seljuks pursued by the emperors at Nicaea permitted a temporary stabilisation of the frontier. In 1261, taking advantage of the absence of most of the Latin garrison of Constantinople on an expedition, a small Nicaean force was able to gain entry to the city, reclaim it for the empire, and drive out the remaining western troops and Latin civilian population. Constantinople was the capital of the east Roman empire once more.

The 'empire' of Nicaea was short-lived, but its strategic position, its sound economic base and its intelligent diplomacy and strategic policies enabled its rulers to re-establish a Byzantine empire. Whether this was in the medium and longer term a wise move is open to debate. The beleaguered position of the refounded empire in international political terms and the impossibility of dealing effectively with western, Balkan and Anatolian threats from such a limited territorial resource base made its very survival questionable.

Recovery, Civil War, Contraction 1261–1351

In spite of the recovery of the imperial capital by the emperors of the Nicaean state, and the consolidation and temporary recovery of some territory in Asia Minor, the last two centuries of Byzantine rule represent a story of slow but inevitable contraction. By the middle of the fourteenth century the empire had been reduced to a dependency of the growing Ottoman Sultanate. Under Michael VIII (1259–1282), the empire was able to expand into the Peloponnese and to force the submission of the Frankish principalities in the region; in international politics alliances with Genoa and the kingdom of Aragon and, briefly, with the papacy, enabled Byzantium once more to influence the international scene and to resist the powers which worked for its destruction and partition. But under Andronikos II (1282–1328), the empire was unable to meet the challenge of a much less friendly international environment. Indeed, with the transfer of imperial attention back to Constantinople the Asian provinces were neglected at the very moment that the Mongols weakened Seljuk dominion over the nomadic Türkmen (Turcoman) tribes, allowing them unrestricted access to the ill-defended Byzantine districts. Most of the south-western and central coastal regions were lost by about 1270, and the interior, including the important Maeander valley, had become Turkish territory by 1300. Independent Turkish principalities or emirates, including the fledgling power of the Ottomans, posed a constant threat to the surviving imperial districts. By 1315, the remaining Aegean regions had been lost, and Bithynia had succumbed by 1337. In addition, the mercenary Catalan Grand Company, hired by Andronikos II in 1303 to help fight the Turks and other enemies, turned against the empire when its demands for pay were not met and, after defeating the Burgundian duke of Athens in 1311, seized control of the region, which it held until 1388. Other mercenary companies behaved similarly. The empire's resources were simply insufficient to meet any but the smallest hostile attack on its territory, and it was rapidly losing even the possibility of hiring mercenaries.

By the time of Andronikos' death in 1328, the empire held only a few isolated fortress-towns. With the loss of the semi-autonomous region about Philadelphia in 1390 to the Ottomans, the history of Byzantine Anatolia comes to an end. The situation was greatly exacerbated by the civil wars which rent the empire in the period after the death of Andronikos II, wars which were for the greater part inspired by the personal and family rivalries of a small group within the Byzantine aristocracy. Andronikos III rebelled against his grandfather Andronikos II in 1321, and after four years of squabbling, marching and countermarching, succeeded in becoming co-emperor in 1325. Conflict flared up again in 1327, but Andronikos II died in 1328 leaving the empire to his grandson, who ruled until 1341. It broke out again in 1341, when the regency representing the young John V, son of Andronikos III (dominated by the dowager Empress Anna of Savoy and the Grand Duke Alexios Apokaukos) declared John Kantakouzenos an enemy of the empire.

John was Grand Domestic and had been the leading minister under the previous emperor; he now had himself proclaimed emperor and assembled an army. The situation was complicated by religious and social divisions. The rise of hesychasm, a mystical and contemplative movement which found particular support in monastic circles, had polarised opinion within the church, since much of the regular and higher clergy, including the patriarch, were fiercely opposed to its teachings. The regime at Constantinople thus found that the hesychasts, particularly on Mt Athos, the centre of Byzantine monasticism, allied themselves with John Kantakouzenos, who thus gained the support of Gregory Palamas, one of the greatest theologians of the last centuries of Byzantium, a leading proponent of hesychasm, and a powerful speaker. At the same time, the regency whipped up discontent in the provincial towns against the fundamentally aristocratic party led by Kantakouzenos, so that popular movements sprang up and expelled many of the latter's supporters, with the result that Kantakouzenos fled to the Serbian king Štefan (Uroš) IV Dušan (1331–55). This alliance suited Serbian expansionist interests, but did not last long; for John then allied himself with the governor of Byzantine Thessaly, John Angelos, a practically autonomous region. In response, Dušan negotiated an alliance with the regime at Constantinople. Using Turkish allies, Kantakouzenos prosecuted the war until 1345; and following the collapse of the regency and the murder of Alexios Apokaukos, he had himself crowned John VI at Adrianople in 1346 and entered Constantinople the following year, where his coronation was repeated by the patriarch.

John's victory meant also the victory of hesychasm, which was opposed to any compromise with the western church. From 1351, when hesychast doctrine became the official doctrine of the Byzantine Church, its conservative anti-western values came to the fore and proved an important influence on the last century of Byzantine culture and politics, especially in respect of attitudes to western culture and Christianity. Politically and

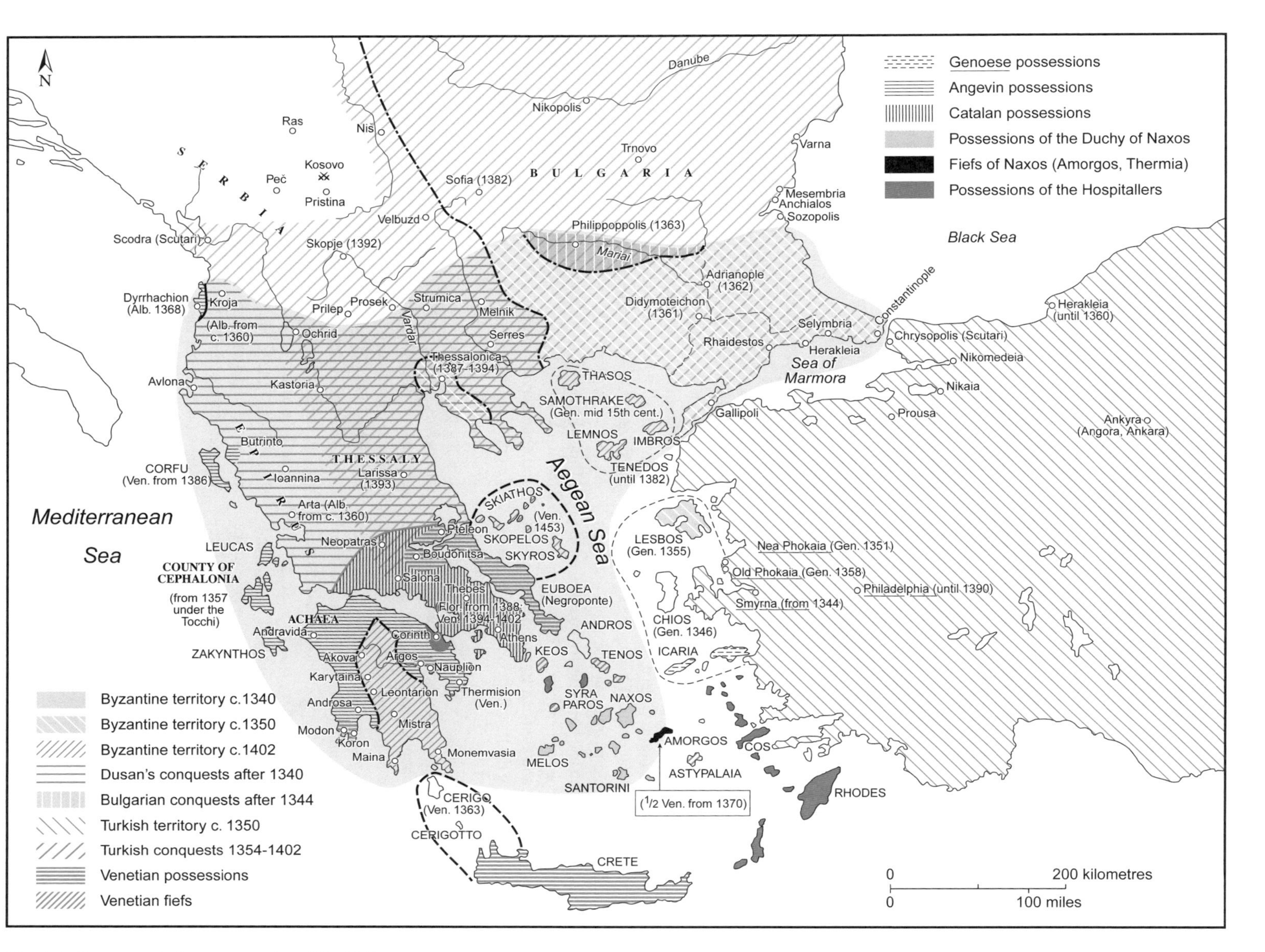

Map 9.4 Recovery, civil war, contraction 1261–1351.

economically the empire was now in a desperate situation. The Serbian ruler Štefan IV Dušan had exploited Byzantine weakness to swallow up Albania, eastern Macedonia and Thessaly. The empire was left with Thrace around Constantinople, a small district around Thessalonica, surrounded by Serbian territory; together with its lands in the Peloponnese and the northern Aegean isles. Each of these regions was, in practice, a more-or-less autonomous province, constituting together an empire only in name and by tradition. But the civil wars had wrecked the economy of these districts, which could barely afford the minimal taxes the emperors demanded, a point well illustrated by the fact that the Genoese commercial centre at Galata, on the other side of the Golden Horn from Constantinople, had an annual revenue seven times as great as that of the imperial city itself.

Decline and Fall 1350–1453

From the 1350s a new European enemy appeared, for the Ottomans had expanded into Europe during the civil wars, following a request in 1344 from Kantakouzenos for Ottoman help against John V. As a result, they began permanently to establish themselves in Europe, taking the chief towns of Thrace in the 1360s, and Thessalonica in 1387. By about 1400, and with the exception of imperial territory in the southern Peloponnese (the despotate of the Morea) and certain Aegean islands, the empire had no lands in Greece.

The complex territorial and political history of the empire after 1204 reflects the vested interests and factional divisions of the western powers which had been directly involved on the Fourth Crusade. Much of the southern part of the Aegean came under Venetian authority; and although Byzantine power was re-established briefly during the later thirteenth century, Naxos remained the centre of the Latin 'duchy of the Archipelago', established in 1207 in the Cyclades by Marco Sanudo, a relative of the Venetian Doge. Initially under the overlordship of the Latin emperor at Constantinople, the duchy later transferred its allegiance to Achaia (in 1261), and to Naples (in 1267), although Venice also laid claim to it. The Sanudo family was replaced in 1383 by the Lombard Crispi family, which retained its autonomy until after 1550, when the duchy was absorbed into the Ottoman state. The remaining islands were held at different times by Genoa, Venice, the Knights of St John and, eventually, the Turks. Rhodes played a particular role in the history of the Hospitallers' opposition to the Ottomans from 1309 until its fall in 1523. The Knights were permitted under treaty to move their headquarters to Malta. In the northern Aegean, Lemnos remained Byzantine until 1453. Thereafter it came under the rule of the Gattilusi of Lesbos (independent until the Ottoman conquest in 1462). In 1460, it was awarded to Demetrios Palaiologos, formerly Lord of the Morea, along with the island of Thasos (which had fallen to the Ottomans in 1455). Lemnos was eventually occupied by Ottoman forces in 1479. Most of the Aegean islands had equally checkered histories – Naxos and Chios fell only in 1566, while Tenedos remained under the Venetians until 1715.

By 1371 the Ottoman forces had defeated the Serbs on the Maritza. In 1388 Bulgaria became a tributary state, and in 1389, following their victory (although the battle seems in fact to have been a draw, the Ottoman superiority in troops and resources gave them the advantage) at the battle of Kosovo, Ottoman forces were able to force the Serbs to accept tributary status. The Ottoman advance caused considerable anxiety in the west. A crusade was organised under the leadership of the Hungarian king, Sigismund, but in 1396 at the battle of Nikopolis his army was decisively defeated. The Byzantines, caught between the Ottomans and the western powers, attempted to play the different elements off against one another. One possible solution, the union of the two churches – with the inevitable subordination of Constantinople to Rome which this entailed – was espoused by some churchmen and part of the aristocracy. Monastic circles in particular, and much of the rural population, were bitterly hostile to such a compromise, to the extent even of arguing that subjection to the Turk was preferable. Hostility to the 'Latins' had become firmly entrenched in the minds of the majority of the orthodox population, both inside and outside the empire's territories. Hesychasm, which represented an alternative to the worldly politics of the pro-westernists, thrived on this ground and further exploited the alienation between the two worlds. Neither party was able to assert itself effectively within the empire, with the result that the western powers remained on the whole apathetic to the plight of 'the Greeks'.

In 1401 the Ottoman Sultan Bayezid began preparations for the siege of Constantinople. But the end was not yet reached. As the siege was under way, Mongol forces under Timur (Timur Lenk – Tamburlane) invaded Asia Minor, where, at the battle of Ankyra in 1402, the Ottomans were defeated and themselves forced to accept tributary status. In the Peloponnese, the Byzantines were able to use the opportunity to bring the remaining Latin princes under their authority. But the respite was of short duration. Timur's death shortly afterwards brought with it the fragmentation of his empire and the revival, indeed strengthening, of Ottoman power. The Sultans consolidated their control in Anatolia, and set about expanding their control of the Balkans. The Emperor John VIII travelled widely in Europe attempting to muster support against the Islamic threat, even accepting ecclesiastical union with the western church at the Council of Florence in 1439, but this did little to hinder the inevitable. A last effort on the part of the emperors led to the western Crusade which ended in disaster at the battle of Varna in Bulgaria in 1444, and in 1453 Mehmet II set about the siege of Constantinople. After several weeks of the siege the Ottoman forces, equipped with heavy artillery, including cannon, were able to effect some serious breaches in the Theodosian walls. In spite of a valiant effort on the part of the imperial troops and their western allies, who were massively outnumbered, the walls were finally breached by the élite Janissary units on 29 May 1453. The last emperor, Constantine XI, died in the attack and his body was never found. Constantinople became the new Ottoman capital; the surviving Aegean isles were quickly absorbed by the Ottoman state; the despotate of Morea was conquered in 1460, and

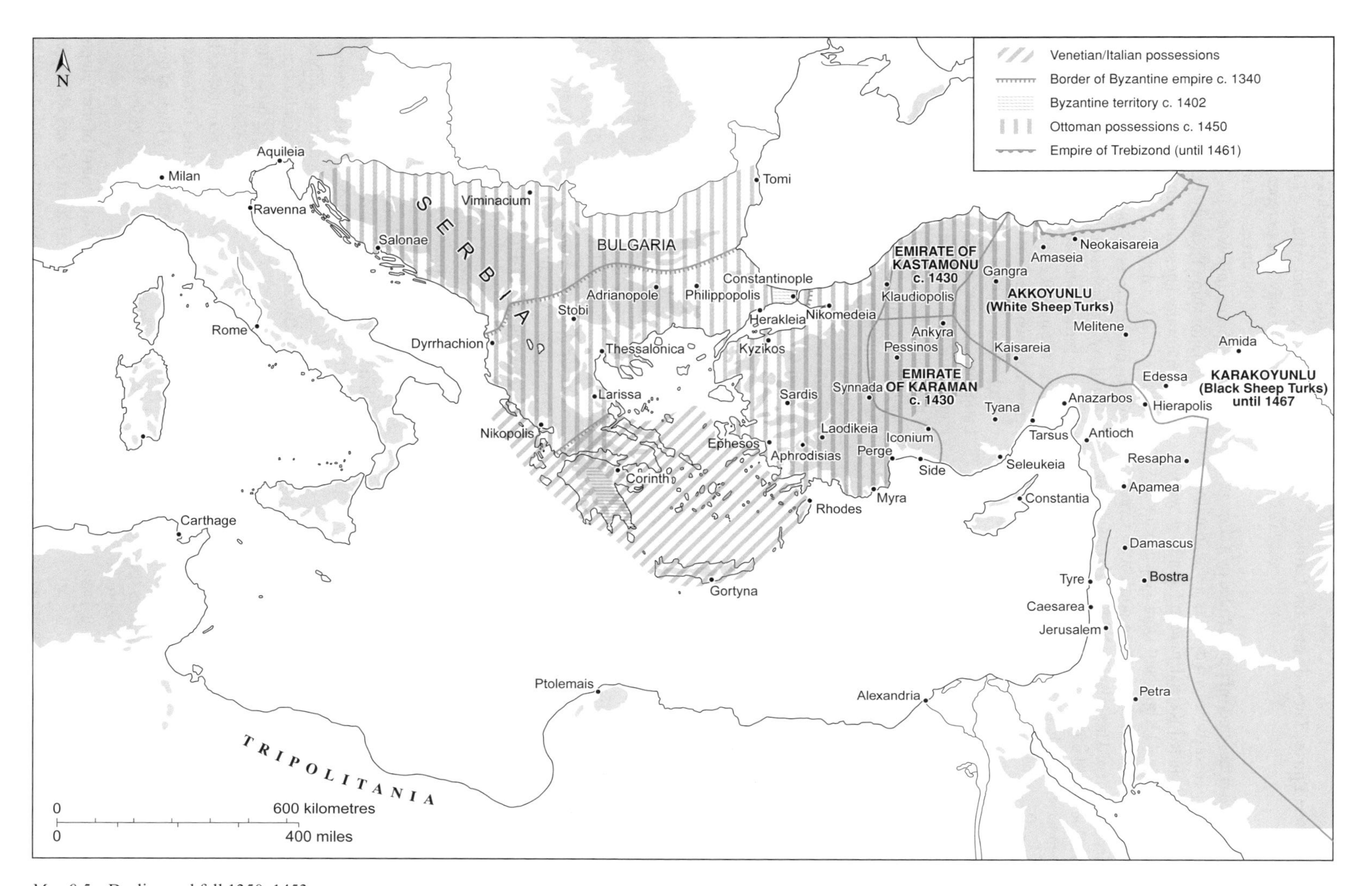

Map 9.5 Decline and fall 1350–1453.

Trebizond, capital of the empire of the Grand Komnenoi, fell to a Turkish army the year after.

Competing States: Epiros, Thessaly and the Latin Territories

The small state based in the region of Epiros (ruled by a *despotes*, 'lord', and hence generally referred to as a 'despotate'), although it usually included Kephalenia/Cephalonia as well, was established by Michael I Komnenos Doukas, who imposed an effective control after 1204 throughout north-west Greece and a considerable part of Thessaly. His brother and successor Theodore was able to retake Thessalonica in 1224, where he was crowned as emperor, thus challenging the emperors of Nicaea who also claimed legitimacy as true heirs to the imperial throne. But in 1242 John III Vatatzes of Nicaea compelled Theodore's son and successor John to abandon the title of emperor; and by 1246 Thessalonica was under Nicaean rule. The armies of Nicaea extended their control over much of Epiros after their victory at the battle of Pelagonia in 1259, fought because the alliance between Epiros, the Frankish principality of Achaia under William II Villehardouin, and Manfred of Sicily was intended to thwart the rise of the Nicaean ruler Michael VIII Palaiologos. This control was temporary, however, and after 1264 Epiros was ruled by independent despots (*despotai*) until 1318. Its geographical situation, cut off between the spine of the Pindos range and the Adriatic, facilitated a degree of political separatism and independence from Constantinople until the Ottoman conquest. Because the emperors at Constantinople always insisted on their rights to confer the title of *despotes*, the rulers of Epiros were viewed as rebels for much of the fourteenth and fifteenth centuries.

From 1318 until 1337 Epiros was ruled by the Italian Orsini family; and after a short Greek recovery, it was taken by the Serbs in 1348. Ioannina and Arta were the main political centres. From 1366 to 1384 Ioannina was ruled by Thomas Komnenos Palaiologos, also known as Preljubović, the son of the *caesar* Gregory Preljub who had been Serbian governor of Thessaly under Štefan Uroš IV Dušan. Thomas was able to assert Serbian control over northern Epiros and fought with the Albanian lords of Arta in the south, eventually defeating them with Ottoman help. In 1382 his title of *despotes* was confirmed by the Byzantine emperor at Constantinople. He was assassinated late in 1384, probably by members of the local nobility who objected to his rule. His wife, the Byzantine Maria Angelina Doukaina Palaiologina, remarried the Italian nobleman Esau Buondelmonti, who ruled as *despotes* until about 1411. As a result of the connections between Esau and the Florentine Acciajuoli, the despotate came under the house of Tocco thereafter, whose rulers were also able to recover Arta from the Albanians. But in 1430 the Ottomans took Ioannina, and Arta fell in 1449. Henceforth Epiros was to be part of the Ottoman empire. Kephalenia was taken in 1479, but Venice seized it in 1500.

After the partition of the Byzantine empire in 1204, eastern Thessaly was ruled by the Franks, while the western regions were disputed by the rulers of Epiros and Nicaea. In about 1267 John Doukas (known as John the Bastard, an illegitimate son of Michael II of Epiros) established himself at Neopatras as an independent ruler, with the Byzantine title *sebastokratôr*. As he attempted to expand his control eastwards, however, he came into conflict with the Emperor Michael VIII, whose attacks he was only able to repel with difficulty together with the assistance of the dukes of Athens and Charles I of Anjou. Venetian support following the conclusion of a favourable trading relationship (Thessaly exported agricultural produce) helped maintain Thessalian independence until the arrival in 1309 of the Catalan Grand Company, which occupied the southern districts from 1318.

Hired by Andronikos II in 1303 against the Turks, the company turned against the empire when its demands for pay were not met. It established itself initially in the Gallipoli peninsula, then plundered Thrace and Macedonia, and seized control of the duchies of Athens and Thebes, expelling their Latin lords. Under Aragonese protection they dominated the region until the Navarrese Company, temporarily in the service of the Hospitallers, took Thebes. This opened the way for the Florentine Acciajuoli, lords of Corinth, to take Athens in 1388. The latter then ruled all three regions until their defeat at the hands of the Ottomans in the 1450s. The northern regions of Thessaly remained independent until 1332 under the ruler Stephen Gabrielopoulos. At this point they were taken by John II Orsini of Epiros. In 1335 Thessaly was retaken by the Constantinopolitan ruler, and from 1348 acknowledged the overlordship of the Serbian ruler Štefan IV. After his death (1355), the self-styled emperor Symeon Uroš, *despotes* of Epiros and Akarnania, was able to seize control of both Epiros and Thessaly following the death of Nikephoros II of Epiros in 1358/9, and rule independently. He was succeeded by his son John, who adopted the monastic life in 1373, upon which the *caesar* Alexios Angelos Philanthropenos took control, governing as a vassal of the Byzantine emperor John V. In 1393 the conquest of Thessaly by Ottoman forces put an end to its independence.

In the Peloponnese the main rival to continued Byzantine authority was the continuous struggle with the Latin principality of Achaia. The principality was at its most successful under its prince William II Villehardouin (1246–1278); but after the battle of Pelagonia (see above) in 1259, he had to cede a number of fortresses, including Mistra, Monemvasia and Maina, to the Byzantines. Internecine squabbles weakened resistance to Byzantine pressure, especially from the 1370s, when one claimant to the principality hired the Navarrese Company to fight for him, which from 1381 exercised effective political control over the Frankish territories. In 1401 the last Navarrese Prince joined with the Ottomans against the Byzantines, but by 1430 the remaining lands of the principality had passed to the Byzantine *despotes* of the Morea through a marriage alliance.

While the Byzantines lost control of the Peloponnese after 1204, after 1259 imperial territory was slowly expanded at the expense of the princes of Achaia, and in 1349 the Emperor John VI Kantakouzenos appointed the first *despotes*, his son Manuel, whose capital was the hilltop fortress of Mistra near

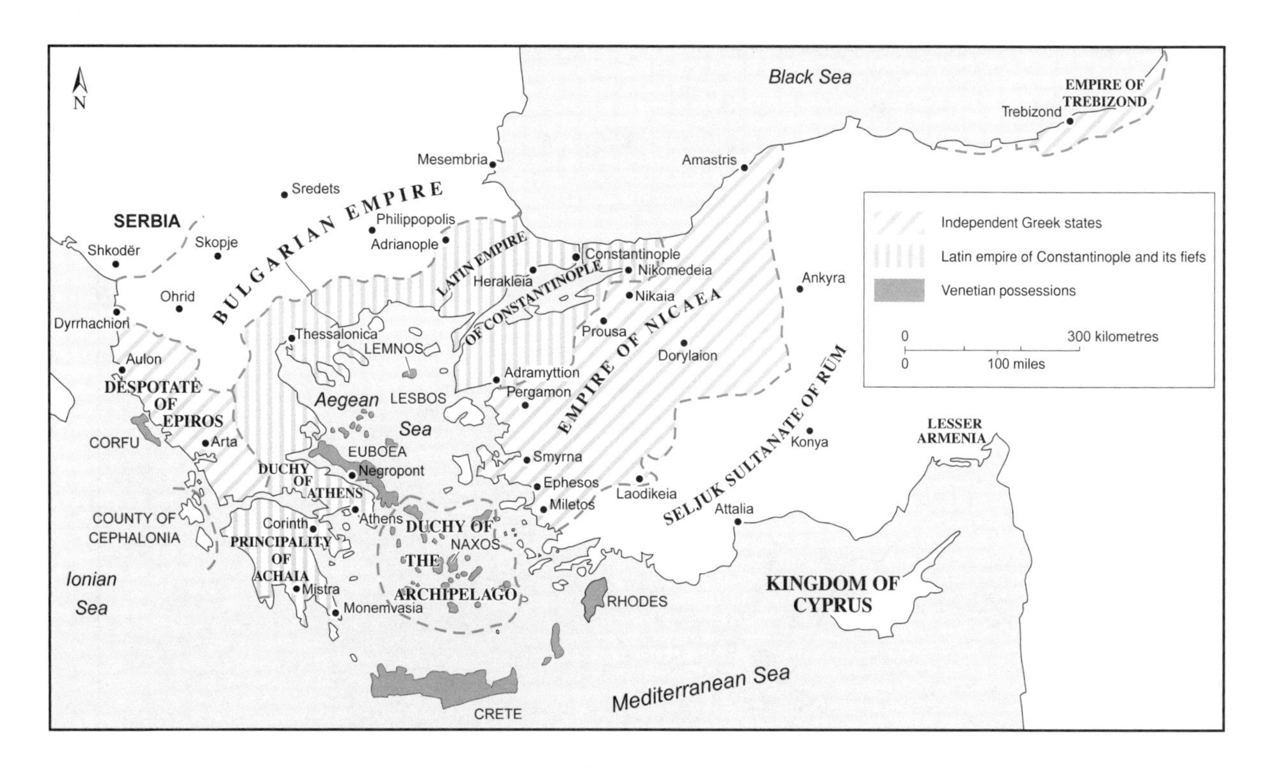

Map 9.6 Competing states: Epiros, Thessaly and the Latin territories.

Sparta. Manuel ruled effectively until 1380, restoring order and encouraging a degree of prosperity to the region. By 1430, as a result of intelligent diplomacy, marriage alliances and effective warfare, the Byzantine despot of the Morea controlled virtually the whole peninsula. The Ottoman presence and the fall of Constantinople to the Sultan Mehmet II in 1453 effectively ended this final period of Byzantine rule. The Morea resisted until 1460.

The Empire of Trebizond

The so-called empire of Trebizond is the longest-lived of the Byzantine successor states which arose in the years following the Fourth Crusade. But in fact its origins lie in the civil conflict which engulfed the empire after the overthrow of the Emperor Andronikos I Komnenos in 1185. Shortly before the Fourth Crusade took Constantinople, the grandsons of Andronikos, Alexios and David Komnenos, had established themselves as independent rulers in Trebizond with the aid of their relative, Queen Thamar of Georgia. True to the claims of their family, their successors refused to recognise the emperors at Constantinople as having a superior claim and refused to renounce their own titles. Although restricted largely to the coastal zone along the Pontic coast and reaching inland as far as the high pastures separating coast from inland plateau, the 'empire' flourished for over 250 years through a combination of strategic good fortune, represented by the Pontic Alps, a natural defensive wall which was ably exploited by the Komnenian rulers, intelligent and careful diplomacy with all its neighbours, whether Muslim or Christian, the strong defences of the capital city, and the lack of a strong and united foe – until, that is, the Ottomans finally decided to extinguish Trapezuntine independence in 1461.

While retaining a tenuous independence, the Grand Komnenoi, as they styled themselves, depended very heavily on their status as vassals or allies of stronger neighbouring powers. Given the changing circumstances of the geopolitics of Asia Minor throughout this period, however, their flexibility in judging which side to support and when to offer diplomatic or military aid, limited though the latter may have been, served them well. Thus between 1214 and 1243 the Grand Komnenoi were tributary to the Seljuk Sultans of Konya, then to the Mongols, who invaded briefly in 1243, the Timurid Mongols after 1402 (when they defeated the Ottoman Sultan Bayezid at the battle of Ankara), and then the Ottomans after 1456. Marriage alliances with the Georgian kings on the one hand, and with the different, and competing, Türkmen rulers of the plateau (a number of imperial daughters and sisters were thus despatched as brides to Muslim lords) were a major element in the strategy of survival. But the empire also had a certain economic advantage, maintaining significant trade and commercial contacts with the Genoese and others, and serving also as a major entrepôt in the trade between east and west. This brought in resources and gave the Grand Komnenoi a degree of flexibility which they would not otherwise have had. The Komnenoi may have had an additional advantage, insofar as they were effectively local rulers who had the support and loyalty of the local aristocracy as well as the church and the mass of the ordinary population in an area which had traditionally been somewhat separate. It is notable that traditional pre-1204 institutional arrangements in respect of provincial government and administration survived in Trebizond in a more conservative form than

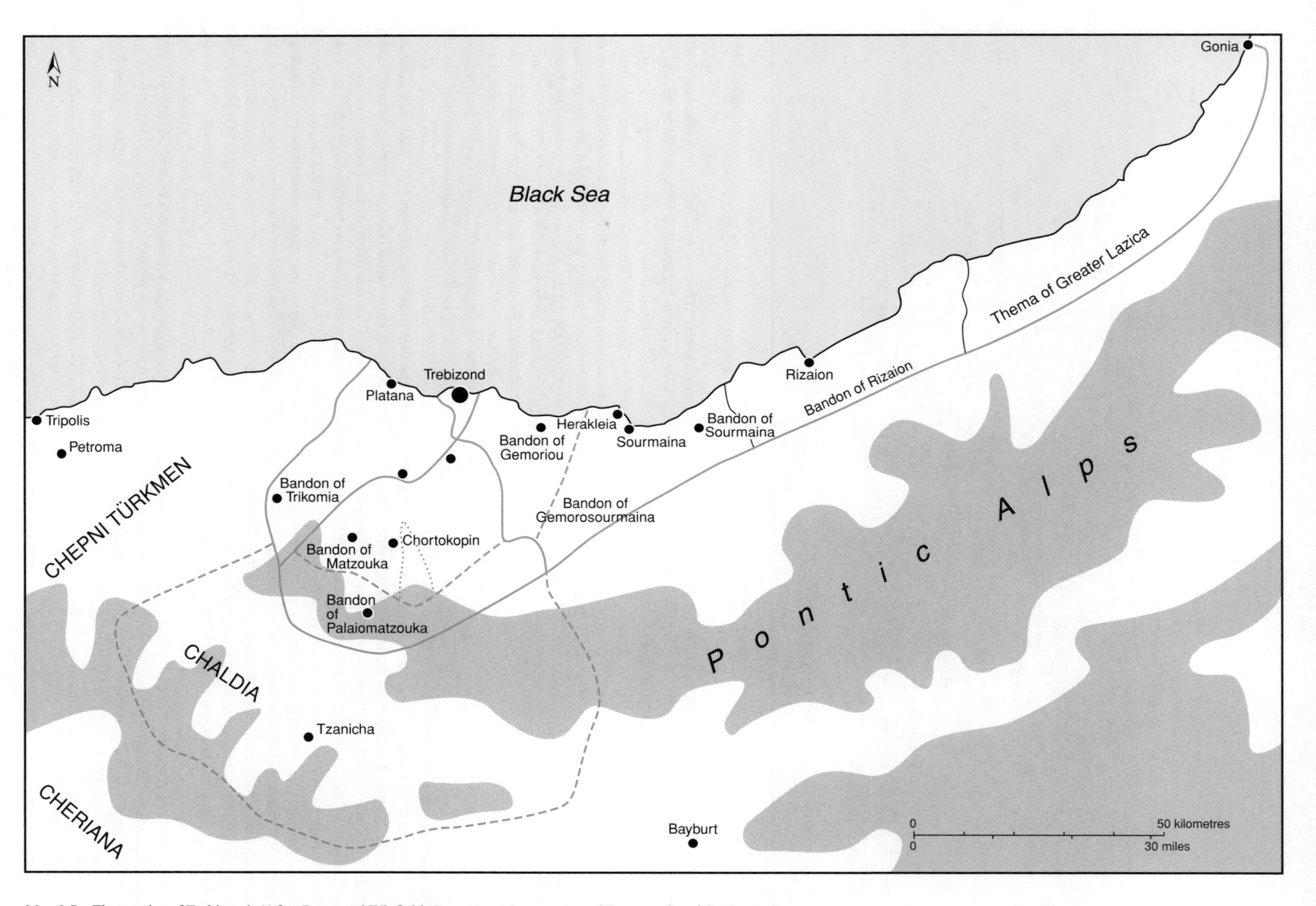

Map 9.7 The empire of Trebizond. (After Bryer and Winfield, *Byzantine Monuments and Topography of the Pontos*.)

elsewhere. At the end of its existence the empire still had *banda* organised for defence with a locally-recruited militia playing a key role, and with defence concentrated on a series of key strongholds controlling access to and egress from the fertile valleys which formed the agricultural heartlands of the empire in the hill regions between coastal plain and plateau.

The political bounds of the empire varied. In the period between 1204 and 1223 it controlled part of the Crimea, lost when pressure from the Tatars to the north became too great to resist effectively (although the Pontic family of the Gabrades remained important there in the following centuries). To the west it stretched at first as far as Sinope (until 1214 when it was taken by the Turks, and again from about 1254 to 1265); to the east as far as the coastal city of Bathys (Batumi). Fluctuation in territory was the norm: progressive incursions by Türkmen lords constantly ate away at the western borders, but were often made temporarily good by marriage alliances. This led both to a dual presence in many areas, and to a progressive assimilation of the Byzantine and some of the local Turk dynasties, so that while the western districts of the empire of Trebizond were already under Turkish overlordship by the 1380s, members of the ruling families of these regions could at a later date remain Muslims and be members of the Trapezuntine imperial court.

Administratively and militarily Trebizond retained in an evolved form much of the Komnenian system of the twelfth century, as did the other successor states. For much of the fourteenth and fifteenth centuries it was divided into seven *banda*, territorial units which reflected local geography, and concentrated resources in the hands of a local élite which could organise defence as well as a civil administration. A small imperial fleet existed until the later fourteenth century. The limited extent of the empire's territory meant that central control, reinforced by frequent visitations by the court, was readily maintained. Only on the inland fringes of the empire were local lords more independent, co-existing in an uneasy relationship with their dangerous neighbours on the plateau, some of them surviving well beyond the political transformations around them as the power of first one and then another local ruler waxed or waned. But the Grand Komnenoi retained power until the end, when the power of the Ottoman Sultan Mehmet II, invigorated by the successful capture of Constantinople in 1453, demanded the surrender of the city, which passed peacefully into Ottoman hands in September 1461.

10 Economy and Administration

Defence and Provincial Administration: the Komnenian System

The system of defence as re-established ultimately under Alexios I Komnenos (1081–1118) was a continuation of the methods he had found to be most successful in his wars to repel the Pechenegs, Normans and Seljuk Turks in the years 1081–1094. Strategy in the broader sense in the opening years of his rule did not exist: the emperor had to respond to a series of emergencies in different parts of the empire on an entirely reactive basis, although it is apparent that the Balkan theatre preoccupied him in the opening years of his rule. But imperial political control in the Balkans was achieved by 1094. The Normans were hemmed in to a small enclave on the Illyrian coast; a little before this, the Pechenegs were crushed at the battle of Lebounion and placed under treaty or incorporated into the imperial armies. The stabilisation of the situation in this theatre brought a return to the administrative arrangements of the middle of the eleventh century, and it was now the Balkan provinces which provided the resources with which the emperor could begin to reassert imperial authority in the east. Manuel I placed a great deal of emphasis on defending imperial interests in the Balkans, on protecting the hinterland behind the frontier zone, and on maintaining a firm control of the Danube frontier with its constituent fortresses, and this demonstrates the recognition by the imperial government that the resources of the area were essential to the empire's financial and political survival. The areas to the south of the Danube were kept more or less depopulated, in order to discourage raids from either the Hungarians or the Galician Russians to the north.

Imperial control in western Asia Minor was virtually non-existent when Alexios I seized the throne in 1081. By skilfully exploiting the armies of the First Crusade, however, Alexios began a slow recovery of imperial territory. A new frontier was established both to mark out the key points to be defended and to establish a safe area from which resources could be extracted and within which economic life could safely be carried on. Under Alexios numerous new commands were established in both eastern and western theatres to consolidate this progress: in the west, Abydos (1086), Anchialos (1087), Crete (1088–1089), Philippopolis (1094–1096), Belgrade (1096) and Karpathos (c. 1090–1100); in the east, Trebizond (1091), Nikaia, Ephesos, Smyrna (all in 1097), Cyprus (1099), Korykos and Seleukeia (1103), Korypho (1104/5) and Samosata (1100).

Since the tenth century the high command had been divided into an eastern and a western section. These sectors were under the supreme command of a *megas domestikos*, or Grand Duke, of east and west respectively. Defence was placed in the hands of local lords and their retinues, or specific groups of landholders with military obligations of one sort or another. Foreigners continued to be settled on imperial lands under an obligation for military service. Pechenegs were given lands in Macedonia by Alexios I; Serbs and Pechenegs were given lands in Anatolia under John II, and Cuman soldiers were given military estates in Macedonia during the reign of Manuel. This tradition lasted until the end of the empire. Resources for particular purposes were organised territorially. The *chartoularios* of the stable (also known as the *megas chartoularios*) was responsible for providing pack-animals and horses for the armies, and under Alexios and his successors managed five major estates in the Balkan regions of the empire. These estates were known as *chartoularata*, and were in effect the equivalent of the older *aplekta* and *metata* of Asia Minor, under the logothete of the herds, which had had similar functions before the 1070s.

The imperial navy remained important, in spite of increasing reliance on Venetian or other Italian warships through treaty arrangements. Commanded by the *megas doux*, or Grand Duke, the fleet was supported by revenues drawn from specific estates set aside in the provinces of Hellas-Peloponnese, the Aegean and Cyprus, and the collection of these revenues came under the Grand Duke's authority.

Substantial lost areas were recovered in the period from the death of Alexios I in 1118 to the 1160s, and already by the 1140s the empire could push onto the central Anatolian plateau itself. New *themata* were established. These were military and civil districts which replaced the older thematic regions of the pre-Seljuk years. Under John II a *thema* of Thrakesion was re-established, geographically smaller than its predecessor, as well as the new *thema* of Mylasa and Melanoudion (made up from the northern parts of the old Kibyrrhaiot *thema* and the southern sections of the old Thrakesion). Under Manuel I, the *thema* of Neokastra was established to the north, based around Atramyttion, Pergamon and Khliara; while many small forts to cover major routes from the Anatolian plateau were built and garrisoned by locally-raised militias. The term *thema*, which meant simply a province, no longer had any direct military implications. By the 1180s there were thematic provinces from Chaldia and Trebizond in the east, on the Pontic coast, westwards through the districts of Paphlagonia/Boukellarion, Optimaton, Nikomedeia, Opsikion, Neokastra, Thrakesion, Mylasa/Melanoudion, Kibyrrhaiotai and Cilicia. The forces stationed in each of these regions were commanded by 'dukes' – Byzantine *doukes* – who were also the governors of their districts.

This policy of gradual expansion came to an abrupt end in 1176, in a strategically premature and tactically misjudged attempt to eliminate organised Turkish opposition in central Asia Minor. The imperial field army, with the emperor present, was ambushed and defeated at the battle of Myriokephalon on its way to lay siege to the Seljuk capital of Ikonion (Konya). This was an extremely expensive enterprise, and the army was accompanied by a huge siege-train which was utterly destroyed. The effort was thus wasted, and as a result of changes in the international situation and rebellion in the Balkans, the empire was never again in a position to go onto the offensive in Asia

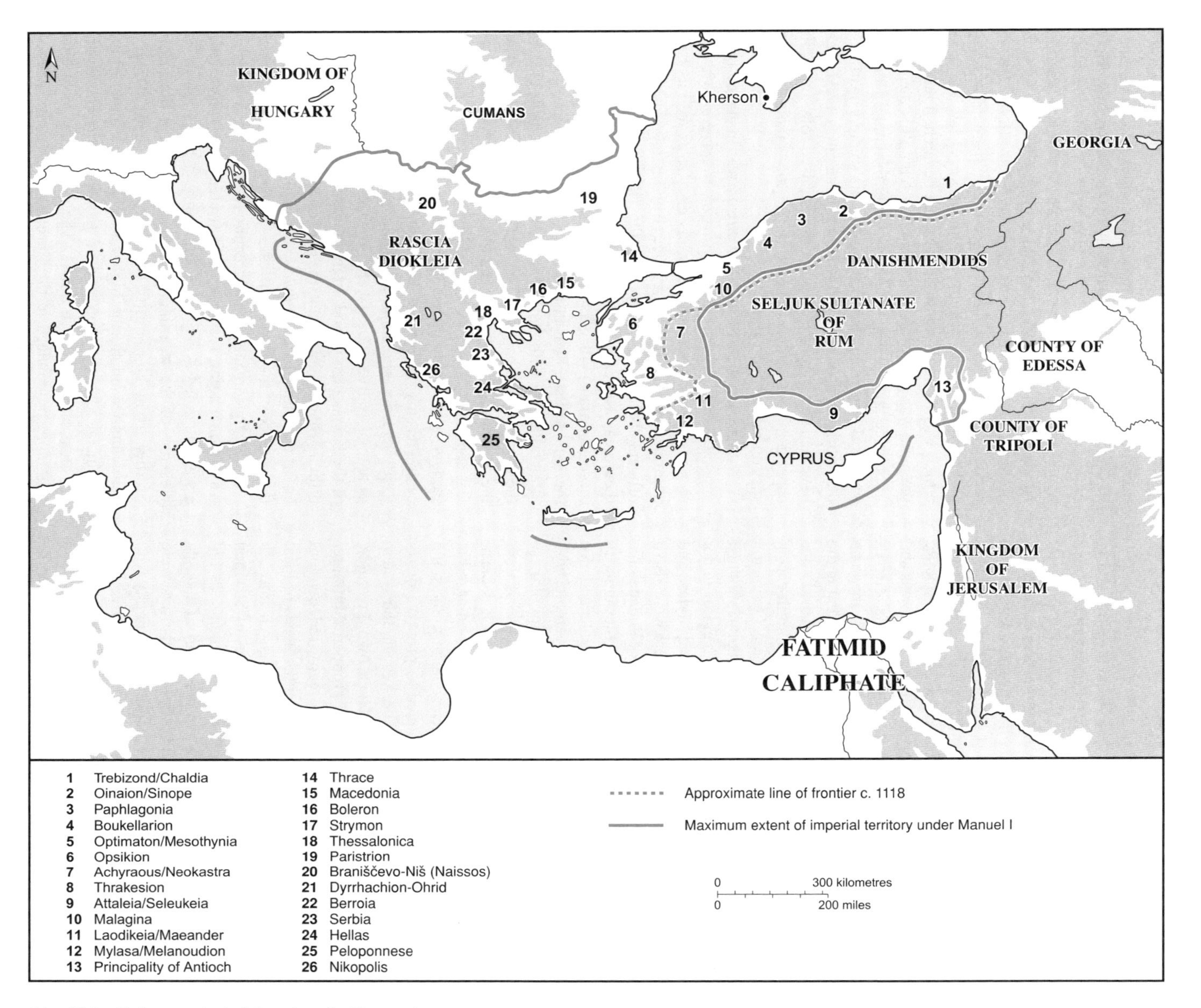

Map 10.1 Defence and administration: the Komnenian system.

Minor on this scale again. 'Turkification' and the Islamisation of central Asia Minor were already well advanced. Gradually excluded from Asia Minor over the following 150 years, the empire became an increasingly European state.

Provincial administration 1204–1453

The administrative apparatus of the Komnenoi survived in an evolved form under the rulers of Nicaea, and formed the basis of the last two centuries of Byzantine imperial administration, although there were important differences between the European and the Anatolian areas. In Asia Minor, and as with the arrangements of the twelfth century, the theme was the chief unit of provincial administration, commanded by its *doux*. The themes were themselves groups of smaller geographical units based around key fortresses and towns and called *katepanikia*, in turn divided into smaller units referred to by the terms *chora* or *enorion*, 'district', generally made up of several villages and their lands. At its greatest extent in the years just before the recovery of Constantinople from the Latins in 1261, there were some seven *themata*. From the north-east in an anticlockwise direction these were: Paphlagonia, Optimaton opposite Constantinople, Bithynia and then Opsikion (although it was probably referred to as Troados or Skamandros under the Nicene emperors), Neokastra, Thrakesion and Mylasa and Melanoudion. In addition to these provinces there were also a number of frontier districts such as the *themata* of Philadelphia and that of the Maeander, both technically under the authority of the *doux* of Thrakesion, but having a certain independence in military affairs and placed under their own governors, with the title of *stratopedarches*. No doubt the boundaries of these themes were different in many respects from those of their like-named predecessors in the twelfth century – it is clear, for example, that the theme of Thrakesion was enlarged at the expense of that of Mylasa and Melanoudion to the south – but the emperors at Nicaea attempted to preserve or re-establish the earlier arrangements. In the Aegean, the system seems less clear, since most islands had their own governors, although for fiscal purposes they were grouped into larger circumscriptions. The presence of so many independent military and civil commanders may reflect the needs of local defence against both pirates and more dangerous enemies. In the European regions of the empire, the themes of Strymon and Thessalonica reflected the earlier arrangements, as did that of Boleron, although all three had been part of a single unit before the Fourth Crusade. The frontier districts were generally very small, consisting of a single fortress or city with its immediate hinterland, and the governors were effectively military commanders, usually called 'heads' (*kephalai*) after 1261. Before this date such officers seem to have a had a good deal of autonomy and were effective senior officials. After the recovery of Constantinople they shared power with officials appointed directly by the emperor, such as the *prokathemenoi* in the larger cities and the local *kastrophylakes* in charge of fortresses.

The governors or 'dukes' – *doukes* – of the provinces were all members of the imperial household under the Nicene emperors, although after the recovery of Constantinople this was not always the case. The majority were members of the upper or middling aristocracy also. The most important officer below each duke was the stratopedarch or military commander (a post sometimes held together with the position of duke), probably responsible both for military and tactical administration as well as for liaising with the financial departments in respect of military supplies and recruitment. In addition to the stratopedarch there were also commanders of fortresses and towns. The centre of the administrative apparatus was the duke's own household and headquarters. He had secretarial staff headed by a *grammatikos*, a financial manager or *logariastes*, in charge of fiscal affairs. He was himself responsible for supervising the administration of justice and the maintenance within his jurisdiction of law and order, which was achieved through the stratopedarch. Technically the duke was also commander-in-chief within his theme, but in practice the imperial field army under the Grand Domestic (*megas domestikos*) absorbed most of his soldiers apart from the militia garrisons of the forts and frontier cities, and he may have had little more than an administrative role, including supervising the holders of *pronoiai* (revenue grants in land to support military service).

Within each theme, the *katepanikion* was managed by an appointee of the *doux*, called a *praktor* or *energon*, who was responsible for fiscal affairs in his district as well as basic administration and the local level of justice. The district officers also liaised closely with the village communities in their areas, each represented by a group of senior (wealthier, more important) villagers (including the village headman and the local priest), both in respect of fiscal as well as judicial affairs. Parallel with the *katepanikia* were the major cities of the theme, which had their own separate administration. Here the governor, generally referred to by the title *prokathemenos*, was appointed directly by the emperor, a reflection in part of the strategic and economic as well as political importance of such centres as Smyrna, Ephesos, Philadelphia and, of course, Nikaia itself. Such governorships, which incorporated judicial as well as administrative duties, were probably also a development of the Komnenian period. In many cases, especially where a military role was important, the governor would be assisted by an officer entitled *kastrophylax*, in charge of the garrison and the defence of the city or fortress.

While this organisational structure represented an effective way of managing resources and strategy, the larger cities always retained a degree of autonomy, a reflection of the presence of the wealthier local landowners in the affairs of the urban community, of the presence of some of their number among the higher levels of the imperial and provincial administration, and of the role of the church and the local bishop in such affairs. The role of the local élites was more pronounced in the European provinces, where a stronger tradition of political independence had developed in the context of the wars of the later twelfth century and the conflict over Thrace and Macedonia between Epiros, the Latins, Nicaea and other interests in the region – Latins and Bulgars, for example; although the élite of Trebizond provides an important exception.

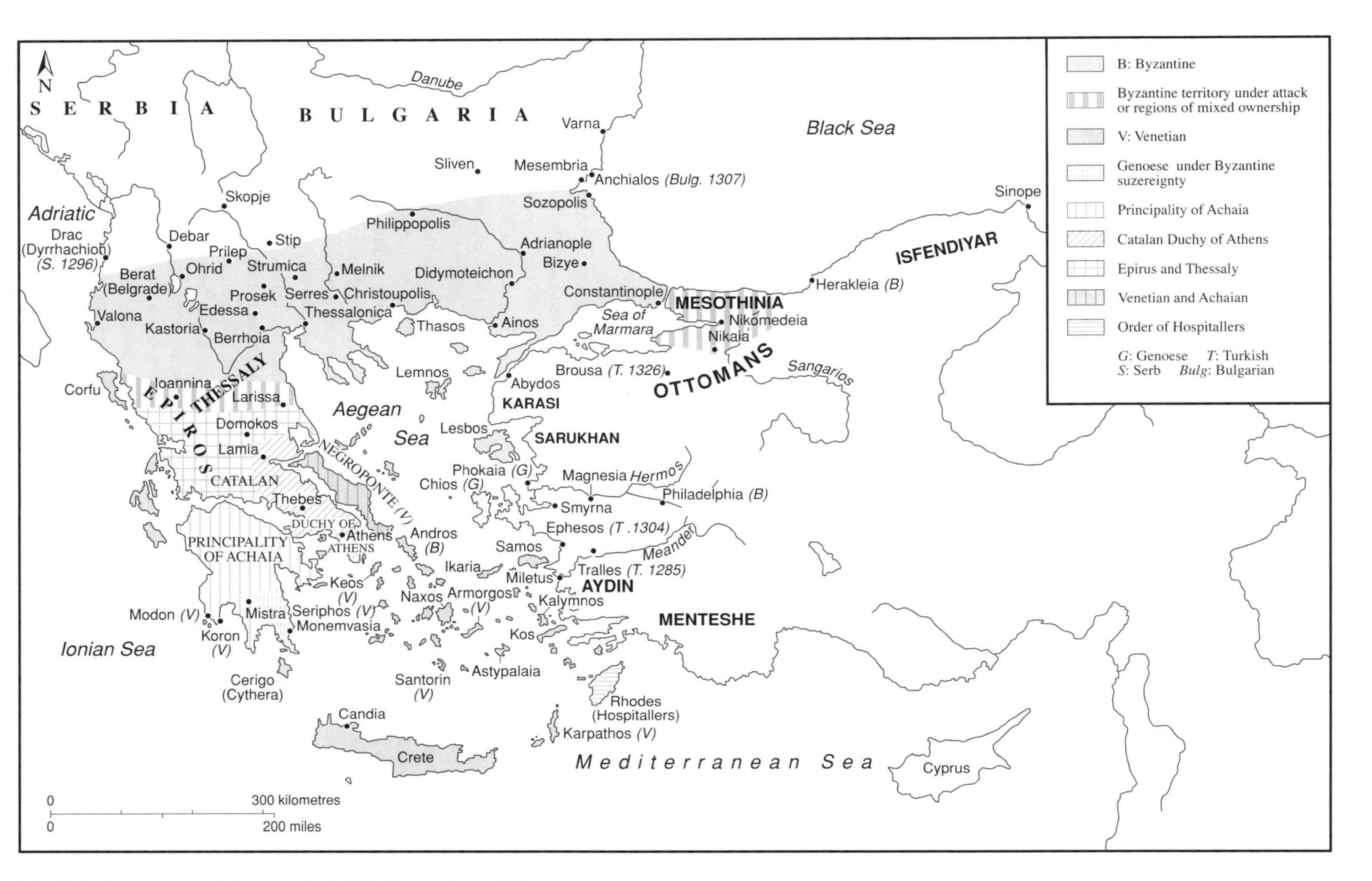

Map 10.2 Provincial administration 1204–1453 (the empire c. 1330).

Central Government and Court 1081–1204

The administrative structures of the empire underwent a series of fundamental changes during the reign of Alexios I Komnenos, changes which set the pattern for the government of the empire until its demise in 1453, much as the changes of the later sixth and seventh centuries determined the pattern of imperial administration until the eleventh century. Alexios' changes were very much in the direction of greater centralisation. The whole fiscal and financial administration of the court and empire was placed under a single senior official, the *megas logariastes*, or Grand Logariast. The *vestiarion* or public wardrobe, and the *oikeiaka*, the department responsible for state fiscal estates, with several sub-sections, now became the central fiscal administrative departments, although the old *genikon*, or general, treasury survived with provincial fiscal responsibilities. The great estates managed by various charitable institutions, such as the imperial orphanage (*orphanotropheion*) under its director, the *orphanotrophos*, became increasingly important resources to the government, and their enhanced role is clear in the status of their senior officials during the twelfth century. The *dromos* or public postal and transport system, essential to the logistics of the imperial armies as well as to the business of the state, continued to exist, managed now through the imperial chancery of which the logothete of the *dromos* was a member. Associated with this department, however, were also the imperial stables under their *chartoularios*, to which were attributed now large estates known as *chartoularata* and as *oria* along the coats, under the authority of the *megas doux*, the Grand Duke in command of the imperial navy. These were fiscal lands whose revenues or produce were intended for the support of the armies and navy. The *chartoularata* appear to be the successors of the older *aplekta* or base camps and estates of the previous period, under the logothete of the herds.

The remaining administrative departments, or *sekreta*, were placed under a single senior supervisor or manager, the *logothetes* of the *sekreta*. Imperial control was exercised through a series of palatine officials and bureaux, the most important of which was the imperial chancery, headed by an official entitled *protasekretis*. Next in importance to him (although importance and influence also depended upon personality) came the *mystikos*, the imperial private secretary, the official in charge of petitions (*epi ton deeseon*) and the 'master of the inkwell' (*epi tou kanikleiou*), all involved in issuing imperial documents of various types, but all in close regular contact with the emperor and thus of very great influence, since they formed in effect a small cabinet of close associates of the emperor.

A particularly important official was the official usually referred to as *mesazon*, literally 'intermediary', a personal assistant to the emperor who acted in effect as a representative of the emperor in dealing with day-to-day business and as private secretary to the ruler. The *mesazon* was generally drawn from among the senior officials in the chancery or related departments. As such he exercised great influence also and has been referred to as the 'chief minister'.

Justice was administered by a series of central courts also headed, by the middle of the twelfth century, by the *protasekretis*. He was accompanied by a new official created by Alexios I, the *diakaiodotes*, who had his own court, and by the Grand *Droungarios*, who continued to preside over the court of the Velum, or Covered Hippodrome. Originally commander of one of the imperial palace units, the *vigla*, or Watch, the Grand *Droungarios* had become by the middle of the eleventh century one of the key judges at Constantinople and his court, originally a lower court dealing with matters of palace security and military jurisdiction had become one of the most important higher benches at Constantinople. Another important judge was the *parathalassites*, again originally a fairly humble court dealing with maritime and port affairs at the capital, but with a much wider competence by the middle of the twelfth century.

The emperor's security was in the charge of the palace guard units, which also formed the core units in any imperial military expedition. The most important older units were the *Hetaireia*, under the command of a *megas hetaireiarches* or Grand Hetaireiarch, the Varangians (since the 1080s largely composed of English as well as Russian or Norse soldiers), under their *akolouthos*, and two small units of guards for the imperial treasuries, the *vestiaritai* (from *vestiarion*, wardrobe). Under Manuel I new units appeared, the *Vardariotai*, originally recruited from Macedonia and Thrace, and commanded by a *primmikerios*. In addition, the emperors also attached to their retinue smaller and more temporary groups of soldiers, often foreigners – chiefly Turks and 'Latins'. The Grand Domestics (*megaloi domestikoi*) of east and west, attached to the court but frequently in the field, were the commanders of the eastern and western units, or *tagmata*, of the field armies.

Central Government and Court 1204–1453

After 1204 the successor states attempted to salvage the remains of the structures of the twelfth century with which they were familiar. Most successful in this respect appears to have been the Empire of Trebizond which maintained an effective separate existence until 1461. But the emperors at Nicaea likewise reconstructed an effective imperial administration based on the Komnenian arrangements, although in a somewhat simplified and reduced form, consistent with its reduced territorial extent and administrative complexity. The most significant change was the increasingly personal, household nature of imperial administration, a result of several factors. First, the sack of 1204 appears to have destroyed the bulk of the central records in the palace archives and government departments, so that while provincial copies in all probability survived, the emperors were heavily dependent upon the know-how and knowledge of the system of their closest advisers. Second, the emphasis under the Komnenos dynasty had already been tending towards government through senior officials connected directly, through marriage or other relationships, with the imperial family. This was then given new emphasis by the central role of the small group of senior officials and the *mesazon* under the new circumstances, which meant that expertise was available, but in a greatly concentrated form, through which new methods of administration and central records had to be created. Government thus became even more than before a matter within the imperial household, more akin to the governments

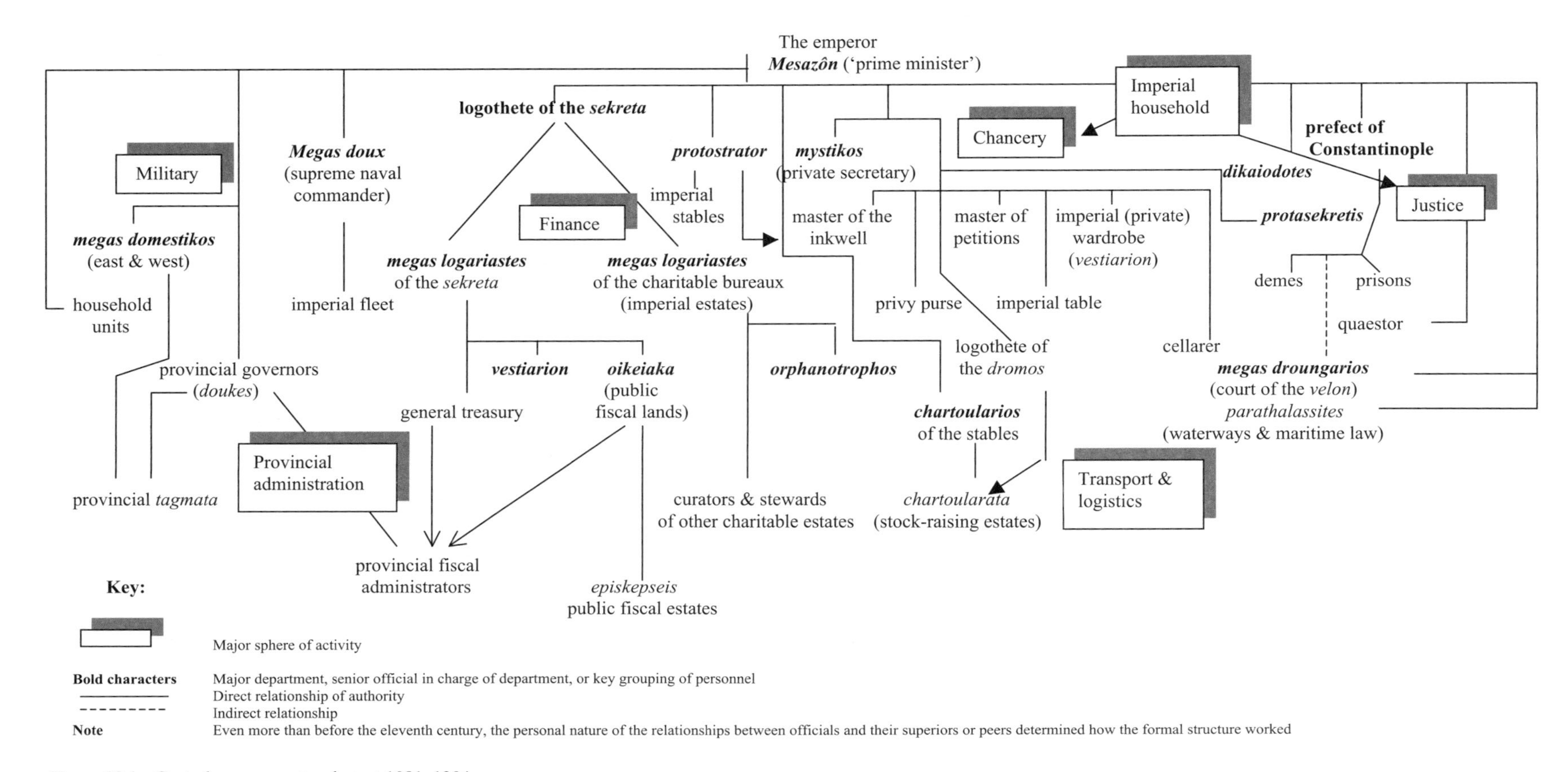

Figure 10.1 Central government and court 1081–1204

of some of the western powers such as Angevin England than the formerly impersonal and bureaucratic eastern Roman tradition.

When the Nicaean emperor Theodore I Laskaris thus came to organise his regime, he was heavily dependent upon this small group of senior household officials, developing the remaining elements of an imperial administrative apparatus piece by piece thereafter. The key figure was the *mesazon* who acted as co-ordinator of government operations and had a much more formal position in the system than under the Komnenoi. Financial affairs were centred on the imperial wardrobe, the *vestiarion*, and the rest of the imperial administration was managed through the different departments of the household and chancery. One result of this slimming down was that the older bureaucracy, with the different departments or *sekreta*, did not reappear, and government was in effect reduced to the imperial household and its secretarial staff.

After the restoration of the empire with the recovery of Constantinople in 1261 this became the pattern of imperial administration until 1453. But it was constantly evolving. There was no formally-constituted superior court under the Laskarids, for example, justice being administered on an *ad hoc* basis through the imperial court. Michael VIII Palaiologos established a special judicial tribunal known simply as the imperial *sekreton* to fulfil this function. By the same token the imperial chancery officials such as the *epi tou kanikleiou* or the *mystikos* appear to have served in a purely personal capacity until well into the reign of Theodore I, when a more formal organisation of an imperial chancery appears to develop, largely modelled on the Komnenian structure but with duties and functions more suited to the new conditions. But it is not known to what extent Michael reintroduced the arrangements which had been current before 1204 and to what extent they survived under the Laskarids of Nicaea. The imperial household dominated government and military administration. There were a number of senior officials endowed with particular responsibilities at court but to whom the emperor regularly entrusted provincial military commands, command of the field army, or some other special duty, including that of provincial governor. All the leading household officials – the *protovestiarios* (associated in fact with court ceremonial rather than with the wardrobe, which was a treasury), the *parakoimomenos* (chamberlain), the palace butler (*pinkernes*), the *protostrator* (chief military official) or *protasekretis* might be thus seconded away from the court for particular tasks.

There was always a considerable overlap in actual duties as the emperor entrusted particular individuals with tasks for which he felt them especially suited. The military organisation of the Nicaean empire was based on that of the preceding arrangements, but with substantial changes which were in turn carried over into that of the post-1261 period. The imperial retinue, consisting of the Varangian and Vardariot regiments, was commanded by a Grand Archon. The Grand Domestic (*megas domestikos*, commander-in-chief of all the armed forces of the empire below the emperor) and his deputy, the *protostrator*, was generally given overall command of campaigns, although other officers suited to a particular expedition might be appointed. Below the senior officers came the commanders of particular divisions or units such as the *allagatores* (commanders of *allagia*, regiments) or *tzaousioi*.

Towns and Local Élites 1100–1453

From the twelfth century onwards, and perhaps beginning much earlier, there developed a difference in the focus of economic activity of towns between the Asia Minor territories remaining under imperial control, and the Balkans, especially Greece and Thrace. In Asia Minor the economic basis of urban centres seems to have remained predominantly agricultural, with towns serving chiefly as market centres for livestock. In the Balkan lands, while all towns depended on agriculture for their basic needs, small-scale industrial and artisanal activity and cash-crop agriculture for market demands appears to have been a more prominent feature of life after the eleventh century. One of the reasons for this difference may be the fact that the empire had still to invest substantial resources from the Anatolian districts in defence and then in the maintenance of armies as well as the building of fortifications. The nature of the Anatolian market for agrarian and other produce was somewhat differently accented, and in eleventh- and twelfth-century Asia Minor, just as in the seventh and eighth centuries, much of the surplus that might otherwise have ended up in the market place was required by the state in the form of military supplies and related expenses. This was to some extent the case in the Balkans, but less so, and this different emphasis in demand and patterns of consumption explains some of this difference. In addition, there is also some evidence that the government offered some fiscal inducements to some towns and communities which had formerly lain outside imperial control, in order to maintain their loyalty to the empire, and thus encouraged investment in a wider range of economic activities. The difference might also lie in the pattern of communications and the fact that state fiscal structures were much more deeply embedded in the network of production, distribution and consumption of resources in Asia Minor than in the southern Balkan region. In this respect, the role of Italian traders and the demand for Byzantine agricultural produce in the west seems also to have played a particularly important role in the Balkan region. Here it was that Italian traders were most active before the thirteenth century, partly a reflection of the fact that southern Greece and the Peloponnese lay between Italy and Constantinople. At the same time, it should be recalled that Constantinople continued to dominate and thus to distort the pattern of supply and demand in the regions around it, in both the north-western parts of Asia Minor, as well as in the Balkans.

Just as significantly, the changing nature of the Byzantine provincial social élite played a significant role. Economic and political stability meant that small fortresses grew into flourishing market centres, attracting small-scale industries and trade, and promoting the development of a provincial gentry, the *archontes* – 'lords' – who were able both to invest in these new developments and draw profits from them, and who were more concerned with local affairs than with those of the imperial capital. While they were dependent in part on more powerful patrons among the higher aristocracy, their regional location

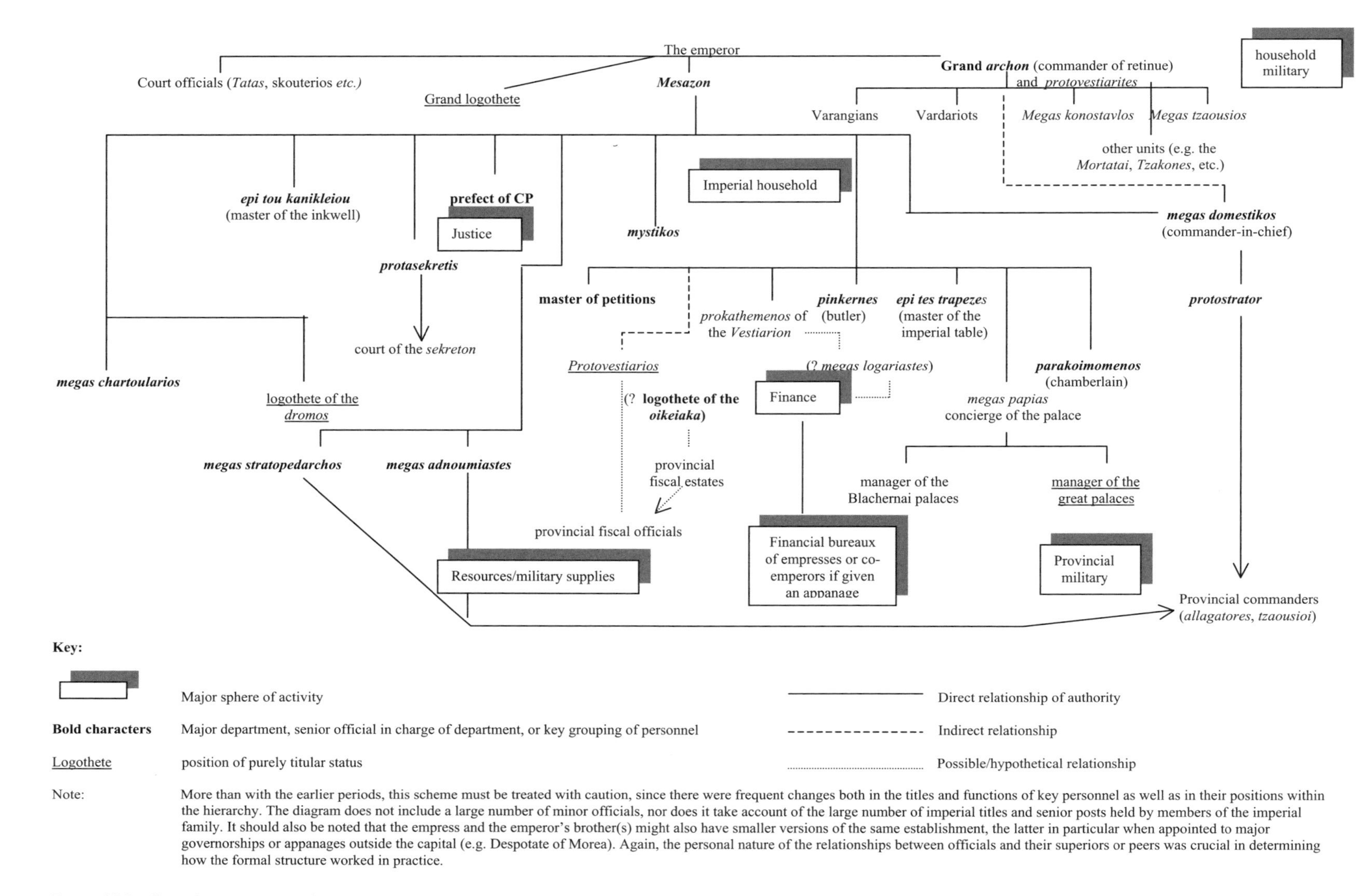

Note: More than with the earlier periods, this scheme must be treated with caution, since there were frequent changes both in the titles and functions of key personnel as well as in their positions within the hierarchy. The diagram does not include a large number of minor officials, nor does it take account of the large number of imperial titles and senior posts held by members of the imperial family. It should also be noted that the empress and the emperor's brother(s) might also have smaller versions of the same establishment, the latter in particular when appointed to major governorships or appanages outside the capital (e.g. Despotate of Morea). Again, the personal nature of the relationships between officials and their superiors or peers was crucial in determining how the formal structure worked in practice.

Figure 10.2 Central government and court 1204–1453

meant a greater investment than hitherto in local economies, especially in commerce and small-scale manufacturing, such as silk, pottery, glass, olive oil and wine production. Byzantium rapidly became an attractive and lucrative market for outsiders, in particular the Venetians and Genoese. One consequence of this was that the provinces, in particular the south Balkan regions and those districts which had access to the Aegean Sea, became much more involved in trade, whether local or international. Cities such as Nikaia, Trebizond, Ephesos and Smyrna in Asia Minor, or Thessalonica, Athens, Corinth, Mistra and Arta in Greece and the Balkans all shared in these new developments, so that paradoxically the weakened Byzantine empire of the thirteenth and fourteenth centuries included far more regional centres of significance than did the empire at the height of its power in the early eleventh century. Yet the warfare and disruptive international situation of the late twelfth century onwards, exacerbated by a demographic decline in the later thirteenth and fourteenth centuries also meant that towns could not develop beyond their traditional limits, and thus were largely unable to compete with their more successful eastern and western rivals.

Social and political structures contributed to this. Local *archontes* developed a sense of local identity and invested in their towns, and could exercise influence on their behalf. But these *archontes*, and the magnates with whom they were associated, exploited the towns for their own economic and political ends, which inhibited the development of communal urban institutions such as evolved in Italy and western medieval Europe. Local *archontes* did reside in their provinces and towns, however, whereas the great magnates generally dwelt, for political as well as cultural reasons, in Constantinople. This gave the provincial élites an advantage in respect of local commerce and industry, insofar as they were located where production took place. The distinction between local 'middling' élites and the great 'imperial' magnates of the various aristocratic clans who dominated the political and the wider economic scene survived until the end of the empire.

The loss of central and eastern Anatolia brought with it important changes in the structure of the aristocracy, as the balance between the Anatolian and Balkan magnate clans shifted. The great majority of powerful families up to the eleventh century came from Asia Minor. Since imperial control in the Balkans had been very limited until the ninth century, powerful families from these regions were fewer and newer. With the loss of much of Asia Minor the importance of the Balkan magnates increased. Further shifts in this balance

Map 10.3 Towns and local élites 1100–1453.

occurred after 1204 and the Latin occupation of the central lands of the empire. And after the Fourth Crusade the rulers of the successor states appointed members of their own families to most key positions. In its last 150 years, the empire was held together by family relationships between more or less autonomous aristocratic groups, reinforced by the still-powerful belief in a unified imperial state, as much as by any coherent administrative structure.

Commerce, Trade and Production

The fiscal interests of the Byzantine state and the non-state sector of private merchants, bankers, shipping and so on always co-existed in a state of tension. The priorities for the government were embedded in the fiscal machinery, ways and means of regulating the extraction, distribution and consumption of resources, based upon a strongly autarkic relationship between consumption and agricultural production. Thus until the later eleventh century the export of finished goods, the flow of internal commerce between provincial centres, as well as between the provinces and Constantinople, and the movement of raw materials and livestock, were determined to a large extent by three related factors. First, the needs of the army and treasury for materials and provisions; second, the need for cash revenues to support mercenary forces and the imperial court; and third, the needs of the city of Constantinople, which dominated regional trade in its hinterland. The pattern of supply and demand had been heavily slanted towards Constantinople since the fifth century, a pattern which was even more accentuated after the loss of central Asia Minor in the 1070s. Until the development of the western economies from the later tenth and eleventh centuries, trade in the Byzantine world had been largely inward-looking, from the provinces and from the empire's neighbours to Constantinople and between the provinces, or with the Islamic world – textiles and finished items of clothing, metalwork, and luxury items such as spices, for example. After the later ninth century this commerce was a flourishing aspect of the internal economy of Byzantine society, and large numbers of traders and entrepreneurs were associated with it. And although the fiscal priorities of the state continued to dominate, the number of trading ports around the Black Sea (from which Italians were excluded before the Fourth Crusade) suggests that long-distance trade by Byzantine merchants before 1204 must have been substantial.

In addition, social constraints played a role. Most well-off Byzantines derived their wealth largely from agricultural production. The possession of land bestowed social status, along with membership of the imperial system. Wealth from trade and commerce was, in comparison with that derived through rents and state positions, of less importance, so that while merchants were an active element in urban economies and playing an important role in the distribution of locally-produced commodities, they occupied a subordinate position in the process of wealth creation as a whole, and in particular in the perception of society in general in respect of the maintenance of the social order as it was understood. For the social élite, they were simply suppliers of luxury items or disposers of the surpluses from their estates, whether in local towns or fairs, or the capital. The government also inhibited enterprise to an extent through the means it employed to control and tax the movement of goods.

In this context, the longer-term results for the Byzantine economy and state of the rise of the Italian maritime cities – especially Venice and Genoa – were unfortunate. The naval weakness of the imperial government throughout the twelfth century, particularly in respect of the threat from the Normans in Sicily, directly promoted reliance upon Venetian assistance, purchased through commercial concessions. The role played by Venice, Pisa, Genoa and other cities after the First Crusade paved the way for Italian commercial infiltration of the Byzantine economic and exchange sphere during the twelfth century, culminating in the concessions made by emperors after Manuel I. It was because Italian commerce was on a small scale, and regarded as unimportant to the economic priorities of both state and aristocracy, that it was enabled to prosper. Demographic expansion in Italy stimulated the demand for Byzantine grain and other agrarian produce, which meant that Venetian and other traders slowly built up an established network of routes, ports and market bases, originally based on carrying Byzantine bulk as well as luxury goods and Italian or western imports to Constantinople, later expanding to a longer-distance commerce to meet the needs of an expanding Italian market.

The much more complex Mediterranean-wide market that evolved during the twelfth century was a market upon which cities such as Venice and Genoa depended very heavily for their political existence and the power and wealth of their ruling élites. The expulsion of Venetians from Constantinople under Manuel I in 1171 had serious effects on Venice, for example, but encouraged a much more direct interventionist approach in the Byzantine sphere. Internal strife in Genoa at the same period reveals similar concerns, as competing factions struggled for pre-eminence in the making of policy in respect of trade with east and west.

After 1261, Byzantine merchants and the Byzantine state were unable to compete with Italian and other commercial capital and shipping. In the mid-fourteenth century the Emperor John VI attempted to exploit the political situation in the Black Sea at the expense of the Genoese and to bolster the position of Byzantine merchants. Genoese military and naval power soon re-established their pre-eminence. While the emperor's plan reveals the importance of commercial revenues to the much-reduced empire, it was now too late to change the pattern. Although some Byzantine aristocrats took an active interest in commerce, Byzantines or 'Greeks' played a generally subordinate role to Italians, sometimes as business partners, often as small-time entrepreneurs, as middlemen, and as wholesalers; frequently as small-scale moneylenders/bankers; rarely as large-scale bankers (although there were some), or major investors, still more rarely in major commercial contracts. The market demands of Italian-borne commerce began also to influence the patterns of production, consumption and taste within the empire, while in its final century or so the state itself had lost any effective role in managing or directing the production of wealth.

Map 10.4 Commerce, trade and production c. 1200–1400.

Coinage, Mints and Money

The imperial gold coinage had been stable since Justinian I, varying according to circumstances from between 99% to 90% purity. But the system was inflexible in respect of commercial demand and market exchange, with the result that the Emperor Nikephoros II Phokas had introduced a lower-value gold coin, the *tetarteron*, as a means of responding to this demand. The increase in international as well as regional trade and industry during the later tenth and eleventh centuries placed further pressures on this system. During the reign of the Emperor Constantine IX Monomachos a series of devaluations took place, reducing the exchange value of the gold *nomisma* and thus answering to the needs of the market. Further devaluations followed, so that by the end of the reign of Michael X Doukas in 1068 the 'gold' *nomisma* contained about 25% silver. This devaluation seems to have been a response to two sets of pressures, first on the government to pay its armies and maintain its apparatus in a period when state expenditures were high, and second from the demands of the market, since the volume of money that could be minted was insufficient to meet the demands of commercial exchange in a context in which prices and the velocity of circulation remained relatively stable.

The crisis which enveloped the state after the events of 1071, however, meant that devaluation also became a means of saving increasingly limited resources, and the result was a more-or-less complete collapse of the system. By the early 1090s the *nomisma* contained only some 10% gold, and it was only with the reforms carried out by that emperor in the early years of the twelfth century that stability was restored, by a combination of revaluing old issues and restoring stability to a high-quality gold coin. Alexios' system involved several smaller denominations that could be used in ordinary day-to-day transactions, illustrating an awareness of the commercial role of the coinage through the minting of a smaller mixed gold and silver denomination, the *aspron trachy*, at 7 carats (as against the high-value gold *hyperperon*, at 19–20 carats), as well as a billon coin with a silver content of only 6% (confusingly also referred to as *aspron trachy*), and a copper coin known as the *tetarteron* (because of its similarity to

Table 10.1 The coinage system after the reform of Alexios I c. 1092–1204

Hyperperon (gold)	*Aspron trachy* (silver/gold)	*Aspron trachy* (billon)	*Tetarteron* (copper)	½ *Tetarteron* (copper)
1	3	48	864	1,728
*	1	16	288	576
*	*	1	18	36
*	*	*	1	2
*	*	*	*	1

Table 10.2 The coinage system c. 1261–1350

Hyperperon	*Basilikon* (silver, from c. 1304+)	*Politikon* (billon)	*Trachion* (copper)	*Tetarteron* (copper)
1	12	96	384	864
*	1	8	32	72
*	*	1	4	9
*	*	*	1	?
*	*	*	*	?

Note: The *basilikon* was based on the Venetian silver *grosso*, the *politikon* appears to have been the Byzantine version of the Latin *denier tournois* current among the Latin states of Greece. The value and equivalences for, and even the name of, the *tetarteron*, which may also have been called the *assarion*, remain uncertain, while it should be borne in mind that there were constant fluctuations in value of the different coins, so that this table lends an artificial stability to an extremely volatile situation.

Table 10.3 The coinage system c. 1350–1453

Hyperperon	Large *stavraton* (silver)	Medium silver	Small silver ('*doukatopoulon*')	Large copper ('*tornese*')	Small copper ('*follis*')
1	2	4	16	192	576
*	1	2	8	96	288
*	*	1	4	48	144
*	*	*	1	12	36
*	*	*	*	1	3
*	*	*	*	*	1

Note: The coinage system of the last years of the empire and after the abandonment of a gold coinage was extremely confused. A major change took place in the first reign of John V (after 1354). Imperial coins were generally calculated in terms of their value against foreign, especially Venetian, issues such as the ducat. The *hyperperon* was entirely notional and no longer struck after the 1350s.

the older gold *tetarteron*). Although fluctuations occurred through the twelfth century, with a marked slide in value of the *hyperperon*, the high-value gold coin, under the emperors after Manuel I from 1180 – a reflection of the political crises which shook the government – it was not until after 1204 that the Komnenian system was seriously jeopardised. Under the rulers of the empire of Nicaea, the *aspron trachy* was transformed into a pure silver coinage, while the mixed silver and copper coin became a purely copper issue. The devaluation of the *hyperperon* continued, reduced from about 17 carats in the 1230s to 11 carats by 1260. The system was maintained after 1261, but the appearance of much stronger currency competitors in the form of the coinages of the Italian cities dealt a serious blow to the value of the imperial coinage. In the mid-fourteenth century production of the *hyperperon* ceased, because the empire could no longer afford to mint it, and already in the early years of the fourteenth century silver coins minted after Italian models were beginning to replace the traditional Byzantine gold. From 1367 a new heavy silver coin, the *stavraton*, became the standard, and remained so until the fall of the city in 1453, although it appears to have circulated only in Constantinople and its immediate hinterland.

The political fragmentation of the empire which followed the Fourth Crusade meant the end of a single Byzantine system. Local sub-systems grew up in Trebizond, for example, as well as in the newly-independent provinces of Serbia and Bulgaria. Although usually copying the imperial coinage, their existence acted to limit the diffusion of the latter, which in its turn became increasingly a local coinage with little international value. Although Byzantine coins continued to influence the coinages of their neighbours, including the occupying Latin and Venetian powers, the heyday of the imperial coinage in the period from the sixth to the eleventh centuries – described as the 'dollar of the middle ages' – was past, and by the later thirteenth century Venetian *ducats* and *grossi* were widely used within Byzantine territory, to the extent that Byzantine texts convert sums given in *hyperpera* or other Byzantine coins to Venetian ducats for clarity. Byzantine coin hoards of the period are regularly made up of a mix of Byzantine and non-Byzantine coins, illustrative of the openness of the international market at this period and of the penetration of Byzantine exchange relations by western commerce in particular. Although the Florentine *florin* and the Genoese *genovino* played a key role, by far the most successful of the western coinages to penetrate the east Mediterranean and Black Sea region, and the real successor to the Byzantine *nomisma*, was the Venetian ducat, also called the *zecchino* (or *sequin*), after the Zecca, the mint in Venice where it was produced. This was also the period at which international loans and banking became effective means of moving wealth around, as notes of credit and debit, guaranteed by stable governments and banking houses such as those of Venice, supplemented and in some cases replaced transactions in coin.

11 Frontiers and Neighbours

Byzantine Italy and the Balkans c. 960–1180

Confined since the definitive loss of Sicily to the Saracens in the last years of the ninth century to the southern provinces of Calabria, imperial forces began the reconquest of Sicily and southern Italy from Saracen and Lombard masters during the last years of the reign of the Emperor Basil II. By the 1020s southern Italy was firmly administered under an imperial military governor, a *katepano*, and the recovery of western Sicily was under way. Basil's death in 1025 slowed the process, however, which eventually ground to a halt in the 1030s. In southern Italy pressure from the German emperors was fended off through an alliance with the papacy, but new enemies soon appeared on the scene in the shape of the Normans, first appearing in c. 1016 in Gaeta as pilgrims en route for the Holy Land, shortly thereafter employed as mercenary troops in large numbers by both Lombards and Byzantines. Drawn by the pay and by the possibility of rich pickings through warfare, their numbers rapidly swelled, and by the 1030s some had succeeded well enough to gain local lordships and titles and establish a permanent territorial foothold. The most successful was Robert Guiscard of the Hauteville family: by 1059 he had defeated and driven out Byzantine troops from Apulia and Calabria, and had defeated and captured the Pope, Leo IX, and had been awarded the title of Duke of Apulia and Calabria.

From this base Robert planned the conquest of the Byzantine Balkans, and to this end he launched in 1081 a major assault on the imperial fortress of Dyrrhachion, modern Durazzo or Durrës in Albania. The city fell and the new Emperor Alexios I Komnenos, who had rushed to relieve the city, was heavily defeated and himself almost captured. A well-planned counter-offensive was then begun, and by 1085 the Norman threat had been defeated. But the establishment of the Normans in southern Italy ended any Byzantine efforts to recover Sicily (which itself soon fell under Norman control) or southern Italy. Bari, the last imperial fortress in the region, fell in 1071. Guiscard's son Bohemund continued his father's anti-imperial policy and was a key participant in the First Crusade, obtaining the city of Antioch, contrary to an agreement made with Alexios in 1098. His capture by Turkish troops in 1100 was followed by his release after payment of a ransom in 1103, when he continued to oppose Byzantine troops until his return to Italy in 1104. He launched a new expedition against Dyrrhachion in 1107 (having first called for a new crusade, against Byzantium on account of the emperor's supposed betrayal of the Crusaders), but was surrounded and forced to surrender. He became a vassal of the emperor, who awarded him the duchy of Antioch (he died, probably in Italy, in 1109 or 1111). While this did not end Byzantine–Norman conflict, which was strenuously pursued by the Norman King of Sicily, Roger, during the middle years of the twelfth century, there was no further successful Norman invasion.

In the Balkans the peace which followed the wars between the empire and the Bulgar Tsar Symeon in the early 920s lasted throughout the reign of his successor, Peter I (927–967). The Bulgar demand for the annual Byzantine 'tribute' in 965 soured this relationship and encouraged the Emperor Nikephoros II Phokas (963–969) to call in the Rus' under their ruler Svyatoslav against the Bulgars' northern frontier. Svyatoslav was too successful, however. He destroyed the Bulgar resistance and occupied the northern part of the territory. Ejected after a fiercely-contested campaign by the Emperor John I Tzimiskes in 971, Svyatoslav was killed on the return journey to Kiev. But Byzantine troops now occupied eastern Bulgaria up to the Danube. When Tzimiskes died in 976 a rebellion of the sons of a local leader, based around Prespa and Ohrid, challenged Byzantine control. The real leader was one of the sons, named Samuel, who became Tsar and launched a series of attacks on Byzantine troops, forcing the empire to relinquish much of the territory it had nominally controlled.

The war which followed lasted over 20 years and resulted, eventually, in the utter defeat of the Bulgars and the conquest of the whole territory excluding Croatia in the north-west, which remained independent, although subject to imperial tribute and to pressure in the west from the nascent power of Venice, still nominally a Byzantine territory but entirely independent in practice. The newly-conquered territories were organised into three major provinces, or *themata*, along the standard pattern: in the west, the *thema* of Sirmium (Belgrade) included the tributary Serb lands (which had moved in and out of the imperial political orbit over the preceding centuries); in the centre and south the *thema* of Bulgaria covered modern Macedonia; and in the east and along the Danube delta the *thema* of Paristrion was established in the provinces formerly known as Moesia and Scythia, including the Dobrudja. Bulgaria remained an important imperial territory until the 1180s.

The Croatian districts, together with the territory occupied by the Slovenes to the north and west, had been largely under Frankish political influence since the destruction by Charlemagne of the Avar Khaganate in the 790s, although Byzantine control over much of the Dalmatian coastal region meant that imperial cultural influence was also significant. Frankish influence was reduced by local rebellions from the 870s, and by the 920s a local prince, Tomislav, in alliance with the Byzantine emperor, was ruling over an expanded and powerful Croat confederacy. After his death this collapsed, however, and independent Croatian princes survived by a shifting pattern of alliances with their surrounding neighbours. During Basil II's war with Samuel, the Byzantines relied on Venice to assert imperial influence, and although Croatia became a Byzantine vassal in 1019, Venetian interest in the wealthy Dalmatian trading cities had been aroused. Although rejecting Byzantine overlordship after 1025, Croatia again became an imperial ally when the Normans posed a threat. Croatia also had to confront the Hungarian kingdom to the north, which in the 1080s was able to impose itself as the dominant power in the region. Thereafter Croatia, with Slavonia, remained effectively

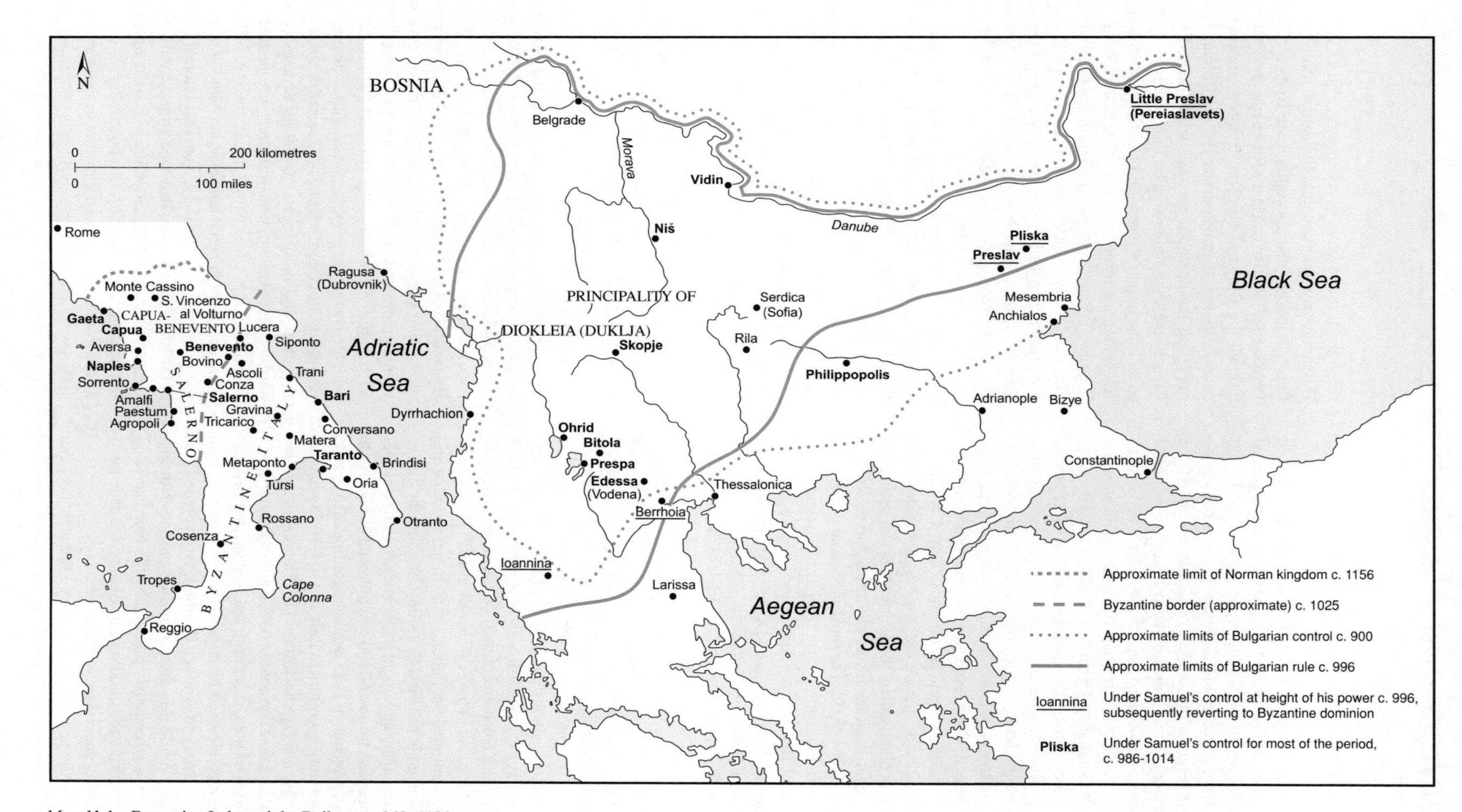

Map 11.1 Byzantine Italy and the Balkans c. 960–1180.

part of the kingdom of Hungary until the Ottoman conquest in the early sixteenth century.

The Danube Frontier and the Balkans 1050–1350

With the recovery of the eastern and central Balkans up to the Danube the empire found itself not only the dominant power in the region, but also face-to-face with neighbours with whom it had hitherto had a more distant relationship. The Magyars or Hungarians (called by the Byzantines *Tourkoi*, Turks, or *Oungroi*, Hungarians), moved into Pannonia following their service as allies of the empire in the wars with Bulgaria in the 890s, under pressure from the encroaching Pechenegs. Under the Arpad dynasty they quickly established a permanent kingdom, and although relations with Byzantium were frequently hostile, by the 950s Byzantine missionary activity in Hungary was increasing. But Latin missionary activity was also increasing, and by the later tenth century had the upper hand. The king Istvan (Stephen) (1000–1038) adopted Roman Christianity and crushed potential rivals who supported the Byzantine Church, and Hungary became increasingly western-orientated thereafter. During the eleventh century the Arpad kings became increasingly involved in the politics of the north-west Balkans, establishing hegemony over Slovenia through a marriage alliance by the 1060s. The ruler of Slovenia was a certain Zvonimir, son-in-law of the Hungarian King Bela I. Having first accepted Croatian as opposed to Hungarian overlordship, however, Zvonimir found himself crowned ruler of Croatia upon the death of the king, Kresimir, without heir, in 1075. Zvonimir also died without an heir in 1090; his widow called in her brother, the Hungarian king Laszlo I, who occupied much of Croatia; and his successor, Kalman, completed the process and made Croatia a permanent part of the Hungarian kingdom.

The annexation of Croatia and Dalmatia brought Kalman into conflict with the Byzantine empire, since the Hungarians were frequently allied against the empire with the Normans, the Serbs or the Kiev Rus'. In their turn, the Byzantines interfered in Hungarian court politics, supporting pretenders to the throne – the presence at Constantinople of Hungarian princes increased tensions – , bribing officials and spying on the king and his policy-makers. There were frequent conflicts on the Danube frontier over who controlled Serbia and Bosnia. The Emperor Manuel proposed a marriage alliance in the 1160s which would have made a Hungarian prince, Bela, heir to the Byzantine throne, but in spite of several major Byzantine military successes thereafter the plan came to nothing when Manuel had a son by his second wife, and Bela's betrothal to the emperor's daughter Maria was abandoned. When Bela succeeded to the Hungarian throne in 1172 (as Bela III) cordial relations were maintained until Andronikos Komnenos seized the throne. Bela intervened, but when Andronikos himself was deposed in 1185 friendly relations were re-established and cemented by a further marriage alliance. The Hungarian kingdom remained an ally of the Byzantine emperors – with some minor disagreements over Serbia – until the Fourth Crusade.

Bulgaria had remained firmly in Byzantine control until the 1180s. In 1185 two brothers, Peter and Ivan Asen, began a revolt against the empire. Initially unsuccessful, they solicited support from the Cumans and were able to impose a treaty on the Emperor Isaac II Angelos whereby the empire ceded control of the south Danube plain to them. In 1189 they invaded Thrace, and then heavily defeated the imperial counter-attack. An independent Bulgarian state had been re-established. Although they were themselves deposed and killed by their own élite, their younger brother and successor Kaloyan was able to maintain his independence and even extend his borders westward against Hungary and Serbia. Although firmly within the Byzantine cultural orbit – the Byzantine 'commonwealth' – Bulgaria was never again to be part of the empire. After 1204 and the establishment of the Latin empire Kaloyan was able to defeat the new Emperor Baldwin in 1205 and expand his territory to the south and west. He won papal recognition of an independent Bulgarian church. But his son and successor Ivan II Asen broke with Rome in 1232, remaining firmly in the orthodox fold. He forged alliances with the emperors of Nicaea both against the Latins and against Theodore, the Despot of Epiros, whom he definitively defeated in 1230, incorporating much of Thrace and the Latin kingdom of Thessalonica.

The expanded Bulgarian state was not to endure, however. In 1240–1241 a devastating raid by Mongol forces, which substantially weakened central power and made it a vassal of the invaders, resulted in the secession of various Vlach or Romanian populations in the north, although the neighbouring Hungarians quickly established a nominal suzerainty over these districts. Internal disarray over the succession in Hungary in the first half of the fourteenth century facilitated the rise of an independent 'land of the Vlachs' – Wallachia – by the 1340s. With Byzantine recognition of its independent status and of its autonomous orthodox church, the new principality remained independent until the Ottoman conquest in the last years of the fourteenth century.

In the 1170s in the western Balkan region Štefan Nemanja (1160s–1196) a local ruler, or *župan*, at Raška in Serbia was able to unite the neighbouring clans and territories and establish a small principality independent of both the empire and Bulgaria, which included the coastal lands around Zeta, as well as northern Albania and eastern Serbia. Initially threatened both by Hungary and by papal interference, Štefan's son, Štefan II Nemanja achieved recognition of an independent orthodox Serbian church through his brother, the monk Sava. But Serbia had to survive between the rival forces of Croats, Hungarians, Bulgaria and the Byzantines, and maintained a precarious independence until the early fourteenth century. Under the king Štefan Uroš II Milutin (1282–1321) Serbia was able to take advantage of Byzantine weakness to seize parts of western Macedonia and along the Adriatic coast as well as north-westwards towards Sirmium/Belgrade, under Hungarian control. Under his successor Štefan Uroš III, further gains were made as the empire slipped into civil war, and a victory over both Byzantines and Bulgarians in 1330 rendered Bulgaria tributary to Serbia. The pinnacle of Serbian power was attained under the next ruler, Štefan Uroš IV Dušan (1331–1355), who ousted the Hungarians from the territory south of the Danube and incorporated much of what remained of Byzantium into his domain. In 1346 he had himself proclaimed emperor.

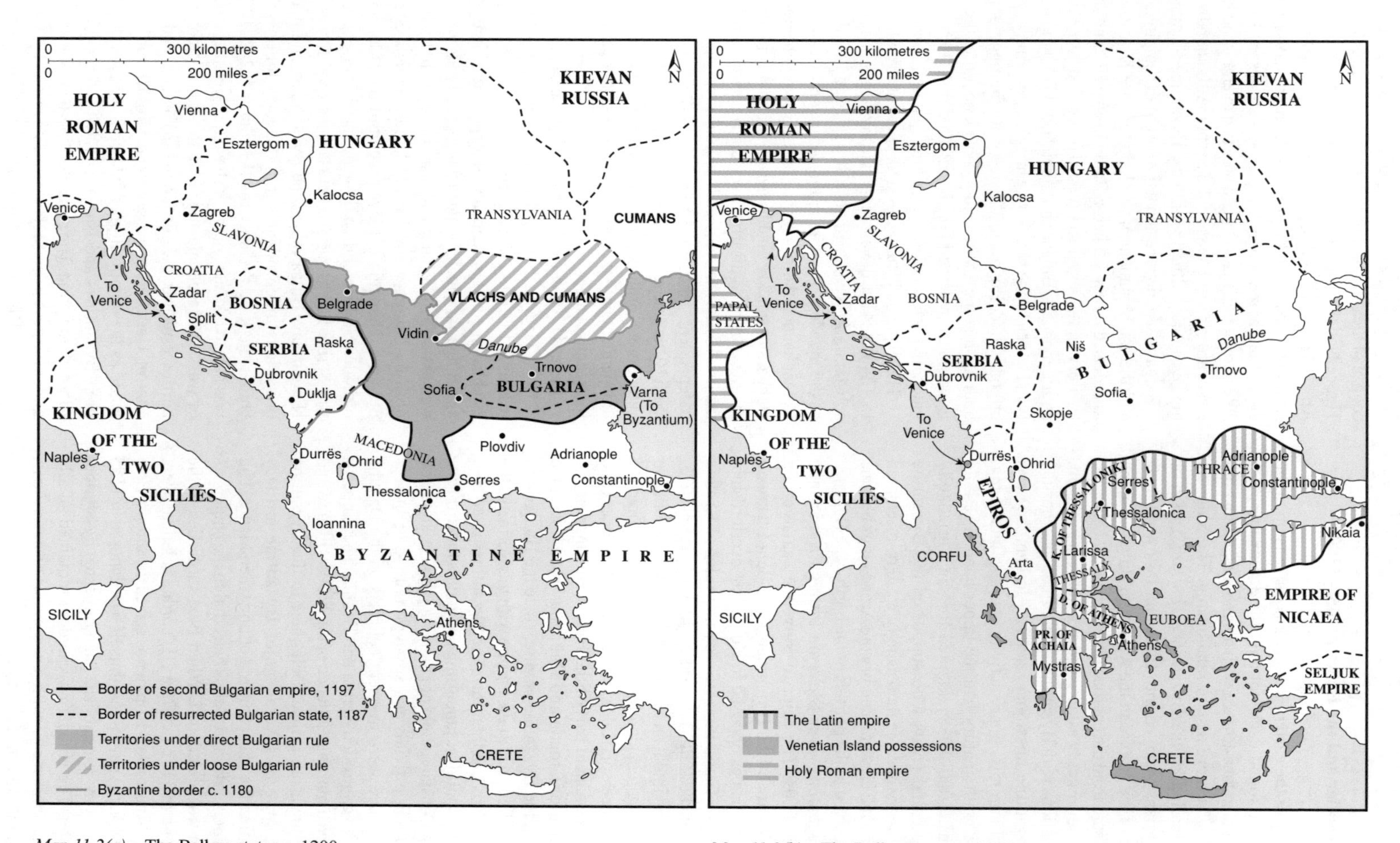

Map 11.2(a) The Balkan states c. 1200.

Map 11.2(b) The Balkan states c. 1220.

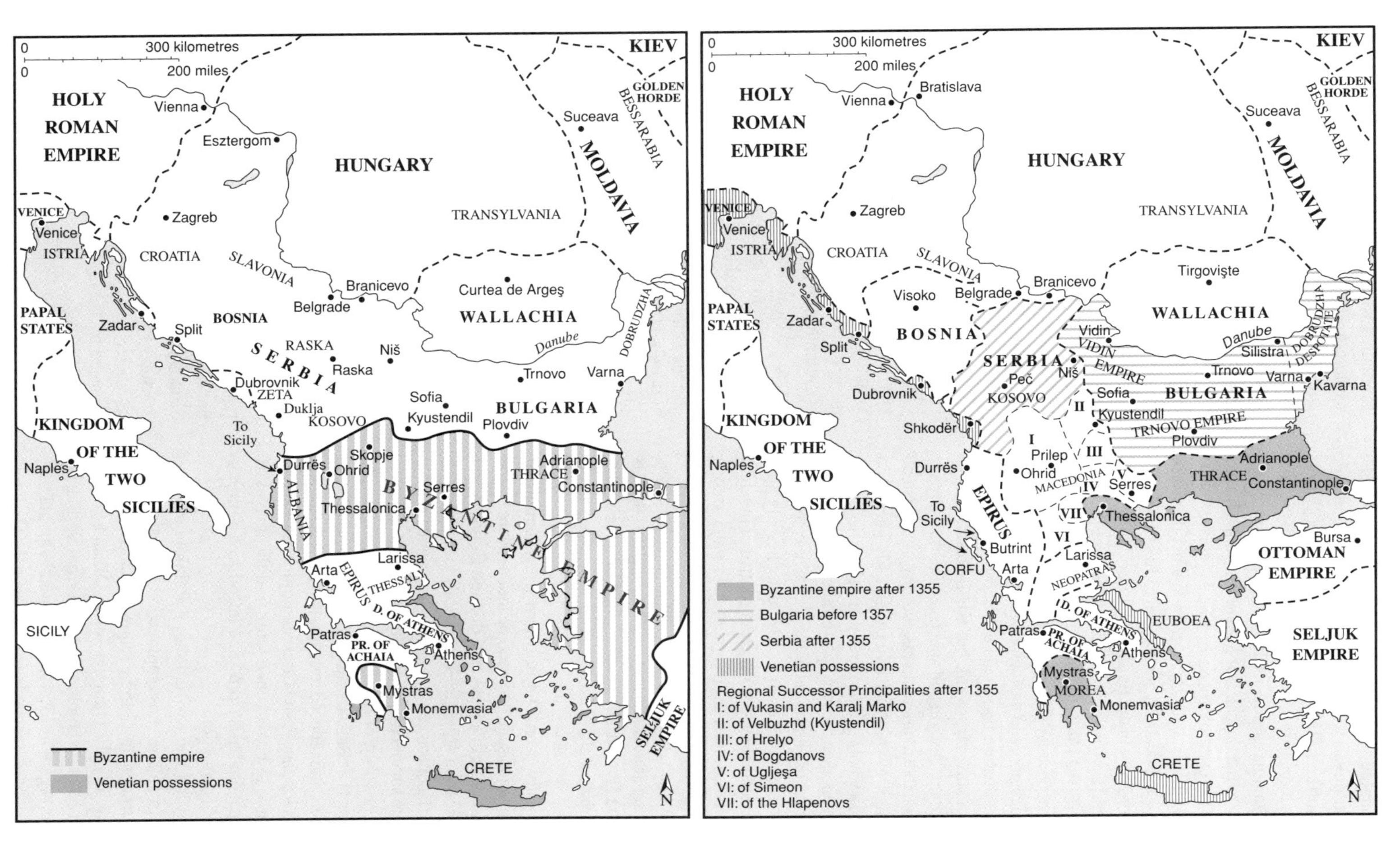

Map 11.2(c) The Balkan states c. 1320.

Map 11.2(d) The Balkan states c. 1350.

Byzantium's Balkan Neighbours 1350–1453: Serbs, Bulgars and Turks

The great raid mounted by the Mongols in 1240–1241 had brought substantial devastation and political and economic disruption to the northern Balkan region. Bulgaria collapsed into civil war and factionalism and was only reunited in the later 1320s when the Byzantine empire, itself rent by civil war, ceded substantial territory to the Bulgar Tsar Mihail Šisman, who had been elected by the *boyars* (nobles) as the best candidate to lead them. A disastrous alliance with Byzantium against the Serbs ended in defeat and the death of Šisman in 1330. Thereafter, Bulgaria was a subordinate in the Balkan scheme of things to its increasingly powerful Serb neighbour, and a victim of factional rivalries and strife. By the 1340s the Dobrudja had broken away under an independent *boyar*, Balik, whose successor, Dobrotitsa gave his name to the region, and the region around Vidin had also broken away. This situation had made the Serbian expansion under Štefan Dušan straightforward, but when Dušan died in 1355 his state broke up. His son and successor (Štefan Uroš V [1355–1371]) was unable to maintain his authority, and instead of the powerful empire which he inherited at his accession there soon appeared a whole group of petty principalities which competed with one another for local pre-eminence. While the central Serbian regions remained under the Tsar's rule, the most recently acquired Greek regions in Epiros and Thessaly split away, as did the Albanian districts. Venetian control of much of the coast served to foment further discontent and rivalry among the local lords. Autonomous Serb rulers established their own principalities in the south, in Macedonia and adjacent regions, where some seven separate statelets were established. And as all this occurred, the Hungarians again pushed into the north-western parts of Serbian-held territory, taking Belgrade and the surrounding districts.

As we have already seen, the northern trans-Danubian territories of the Bulgarian state had become independent by the 1340s in the context of both a weak and divided Bulgaria and the Hungarian succession struggle. The overall picture which thus emerges in the Balkans is one of extreme fragmentation. No major powerful state survived into the late fourteenth century, with the exception, possibly, of Hungary, which geographically does not really count as a 'Balkan state'. The Byzantine empire, wrecked by civil wars and reduced territorially to a few Aegean isles, the southern Peloponnese and Constantinople with Thrace, was no longer a force to be reckoned with, and the arrival of the Turks on a permanent basis from 1354 introduced a further complication into this situation.

In the course of his wars with John V Palaiologos, the Emperor John VI Kantakouzenos employed both western and eastern mercenary troops. In 1345 he requested, and received, military aid from his ally, the Ottoman Sultan Orhan I (1324–1360), and again in 1349, to combat the threat from Serbia, he received further aid. Again in 1354 he requested help, but this time the Ottoman troops entrenched themselves on the Gallipoli peninsula, which became a permanent base. The fragmented political situation in the Balkans after Dušan's death and the lack of any serious opposition meant that the Ottoman troops now had a free hand to raid wherever they wished; it also meant the beginnings of a permanent Ottoman presence in Europe.

The Ottoman Sultanate is named after its eponymous founder, Osman (1284–1324), a Seljuk warlord in north-western Asia Minor who prosecuted the war against Byzantium with great zeal, attracting in consequence a reputation as a *ghazi*, a fighter for the faith, along with large numbers of independent warriors, who wished to join him as much for the booty as for their religion. Early in the reign of his son, Orhan, the Ottomans took Bursa, the last Byzantine fortress in Asia Minor. Ottoman military organisation was effective, and under Orhan's successor Murad I a new phase of expansion, directed from the newly-established European bridgehead at the Serbs, Byzantines and Bulgars, was set in train. Adrianople (Edirne) was taken in 1365 and became the new Ottoman capital; new heavy infantry units, referred to as janissaries (Turk. *yeni ceri*, 'new guard') recruited from captives, bolstered the existing light cavalry of the Ottoman forces, and by the early 1370s most of Bulgaria south of the Balkan mountains had been conquered, local Serbian forces had been crushed and Macedonia incorporated, and by 1386 Ottoman troops had taken Niš and were poised to enter the heartlands of Serbia. The Serbian ruler Lazar was reduced to vassal status and, alongside many other defeated nobles and petty lords, served in the ranks of Ottoman allies in the campaigns that followed. When Lazar organised an alliance to cast off Ottoman rule both Murad and Lazar fought in their respective armies at the battle of Kosovo Polje in 1389, and both died. Serbia fell into further anarchy as a result, and by 1393 Bulgaria had been incorporated entirely into the Ottoman dominion. The unsuccessful 'crusade' led by Sigismund of Hungary, aimed at throwing back the Ottoman advance, ended in a crushing defeat at Nikopolis in 1396. The result of the Mongol (Timurid) invasion of Asia Minor and the Ottoman defeat at the battle of Ankara in 1402 was internal strife in the Ottoman court and a temporary halt to their advance in the Balkans. Serbia was able to restore some order and re-assert its territorial claims on the territories recently lost. But under Mehmet I (1413–1421) the Ottomans were able to restore the situation to their advantage, and in the reign of Murad II (1421–1451) began once more to move forward. Thessalonica and much of the Aegean were taken in the 1430s; Hungary was raided, and the last crusade, led by the Hungarian general Janos Hunyadi, was defeated near Varna in 1444. Of the Byzantine empire only Constantinople and the isles of Lemnos and Thasos remained, apart from the Peloponnese. Constantinople finally fell in 1453, and shortly afterwards the Peloponnese was also incorporated into the Ottoman lands. The final reduction of Serbia had taken place by 1458, the defeat and occupation of Bosnia was completed by 1461 (the same year in which Trebizond finally surrendered), and by 1463 Albanian resistance had been crushed. The Byzantine empire had been replaced by the Ottoman.

Seljuks, Türkmen and Mongols

Although the Seljuk Sultan Alp Arslan died while campaigning against the Karakhanid emirs east of the Caspian soon after his

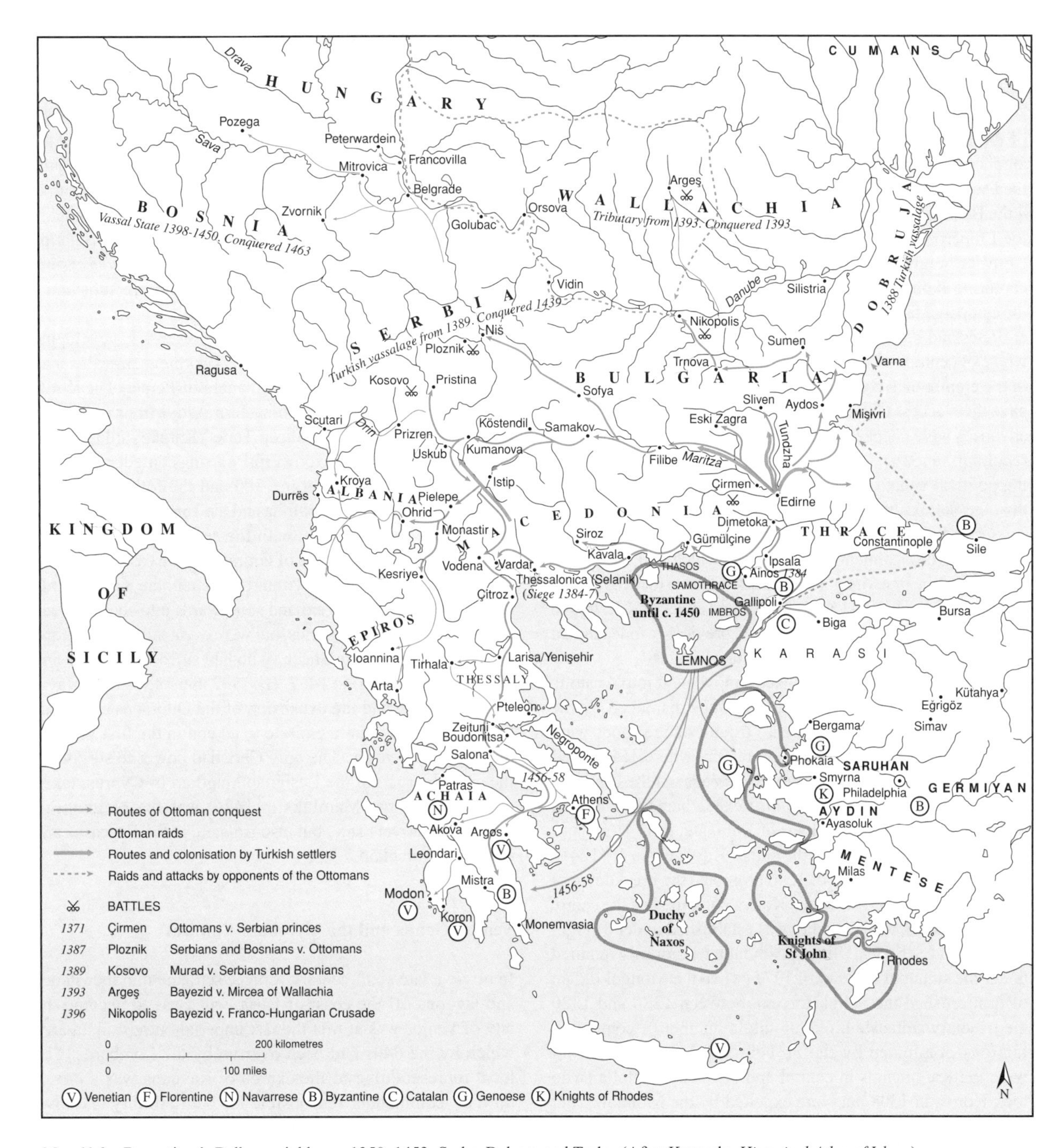

Map 11.3 Byzantium's Balkan neighbours 1350–1453: Serbs, Bulgars and Turks. (After Kennedy, *Historical Atlas of Islam.*)

victory at Manzikert, the Byzantine civil war enabled a rapid Turkish occupation of the central Anatolian plateau, much of it well suited to the pastoral nomadic lifestyle. By the 1090s, however, the Sultanate had split into three major parts, the sultanates of Merv, in the east, of Hamadan (Iran, Iraq and parts of Syria) in the centre, and of Nicaea (Iznik) in the west. The latter ruled over the greater part of the lands conquered from the Byzantine empire, although the Danishmend clan which actually controlled much of the eastern and central plateau regions barely accepted the sultan's authority. The arrival of the First Crusade effectively returned Nicaea to imperial authority, along with substantial districts around it, and made the Danishmendid emirate independent of Seljuk authority. The Seljuks, meanwhile, had withdrawn onto the plateau and established a new capital at the formerly Byzantine fortress town of Ikonion (Konya). To the east of the Danishmendid emirs the emirate of Armenia owed nominal fealty to the Seljuk sultans of Hamadan, as did their neighbours to the south, the emirs of Mosul. In practice, both were more or less entirely independent, along with a number of other petty emirs along the frontier, a factor which gave the advancing First Crusade a decided advantage in the campaign which led to their capture of Jerusalem in 1099.

By the 1150s the sultanate of Konya was the most powerful of the Seljuk states in Asia Minor, but had to contend with alliances or agreements between its neighbours to both east – the Danishmendid emirate and other minor factions – and west – the Byzantines. It faced its greatest threat in the 1170s when the Emperor Manuel I, having first isolated it diplomatically from its neighbours, set out in 1176 with the intention of capturing Ikonion and destroying the Seljuk power base. But the campaign failed. By the early thirteenth century the easterly emirates had been absorbed and the sultan at Konya ruled the whole of central and eastern Asia Minor. The westernmost parts of the empire of Trebizond fell to Seljuk forces in 1214, which gave them access to the Black Sea. Yet within a few years these advances were checked by the arrival of a substantial Mongol reconnaissance and raiding force (1221), which disrupted the political situation in the Caucasus region. In the wake of this, the Seljuks were forced to make common cause with the Ayyubids to the south to resist the expansion into this region of the revived Shahdom of Khwarizm, which had itself been defeated by the first Mongol attacks but was able to recover for a short period. Yet by 1243 a second Mongol attack had destroyed the Khwarizmian shahdom, conquered previously independent Christian Georgia, and made the Seljuks tributary.

From the 1230s there took place a movement into Anatolia of a number of Türkmen nomadic groups displaced by the Mongols further to the east. These troublesome groups were despatched to the Byzantine frontier regions where, led by their *uç beys*, marcher lords, they waged *jihad* against the Christian forces to west and north. They also exacerbated the internal political problems of the sultans of Rum, and in 1241 a major revolt had to be crushed by the sultan Kay Kusrau II (1241–1246). The Mongol attack of 1246 promoted the breakdown of central Seljuk power, however. Konya continued as the centre of one sultanate, while another was established under Mongol suzerainty at Sebasteia (Sivas). Although temporarily reunited under the sultan of Sivas until 1277 (when the Mongol Ilkhan of Iran crushed the Seljuk forces), between 1280 and 1320 the tributary sultanate broke up into a number of competing factions dominated by the *uç beys*. The most significant were the Karamanids in central and southern Anatolia (who took Konya in 1308 but were expelled by the Mongols); but the emirate of Kastamonu to the west of Trebizond and the confederation of the six emirates in the south-west competed from the early fourteenth century on equal terms until the rise of the Osmanli *beys* – the Ottomans – began to bring about substantial changes.

Although hemmed into the north-western provinces of Asia Minor, the Ottomans had the advantage of facing a Christian enemy, in the shape of the Byzantines, and of being able to exploit the situation in the Balkans when Byzantine emperors requested military aid from them. By the 1360s entrenched in Gallipoli and by the 1390s the dominant Balkan power, this provided the Ottomans with reserves of wealth and manpower which makes their eventual conquest of the independent Anatolian emirates readily understandable. Between 1390 and 1393 Bayezid I had defeated and absorbed these territories into his realm and extended the frontier of his power to the Euphrates. Unfortunately, this success was short-lived: the invasion of Timur in 1402 resulted in a crushing defeat for the Ottomans, and the re-establishment of the independence of the subject emirs. Yet in spite of a revival of Christian power in the Balkans and the emergence of revived emirates of Karaman and Kastamonu, Bayezid's successor Mehmet I was quickly able to restore Ottoman pre-eminence. By 1430 the situation before the battle of Ankara was almost restored. Only in eastern Asia Minor did the Ottoman sultan face a more substantial problem.

At the same period as the Osmanli power was developing to the west, two other powerful Türkmen emirates had evolved from the collapsing Ilkhanate of Persia. The White Sheep Turks (Akkoyunlu) in eastern Asia Minor (up to the Euphrates) and the Black Sheep Turks (Karakoyunlu) in Iran and Iraq, represented two powerful warring confederacies. The Ilkhanate itself, Islamised in 1300 and the following years, having lost control over Anatolia and the Turks, had fragmented even further by the 1330s, with the central section ruled by the Jalayrids, and a number of emirates in the eastern regions. Temporarily weakened by Timur's invasion, the Karakoyunlu were for a while able to expand southwards into southern Iraq at the expense of the Jalayrids, but were eventually defeated and absorbed by the White Sheep (who had sided with Timur and benefited therefrom) in 1467. By 1502, the rise of the Safavid Persian empire and the expansion of the Ottomans eastwards brought the Akkoyunlu emirate to an end in the first years of the sixteenth century. The only Christian power to survive in the east (apart from the Lusignan kingdom of Cyprus, taken eventually by the Mamluks in 1426) was the kingdom of Georgia, relatively safe, but also isolated, in the Caucasus and eastern Pontic plain.

Venice, Genoa and the Merchant Empires

In origin a late sixth-century refugee settlement in the islands and lagoons off the coast of Istria, what was to become the city of Venice was at first the last imperial outpost in an area which by the 640s had been overrun by the Lombards. The local representative of the exarch of Ravenna was a *dux*, a military commander responsible for both military and civil administration. The collapse of the exarchate in 751 with the capture of Ravenna left Venice entirely isolated, but under its *duces* it continued to recognise imperial authority and to assist in the defence of the surviving enclaves of imperial territory in the region. With increasing Frankish pressure from the mainland, the Venetians found it difficult to maintain their relative independence between the two great powers, on the one hand, and Slav raiders on the other. Yet in spite of family-based factional rivalry between the different centres which were growing up around the lagoons, the middle and later ninth century saw the growth of a thriving trading centre, and the early tenth saw the official adoption of the name *civitas Venetiarum* for the city and its suburbs. The position of *dux* (*doge* in Venetian dialect) tended to be monopolised by certain

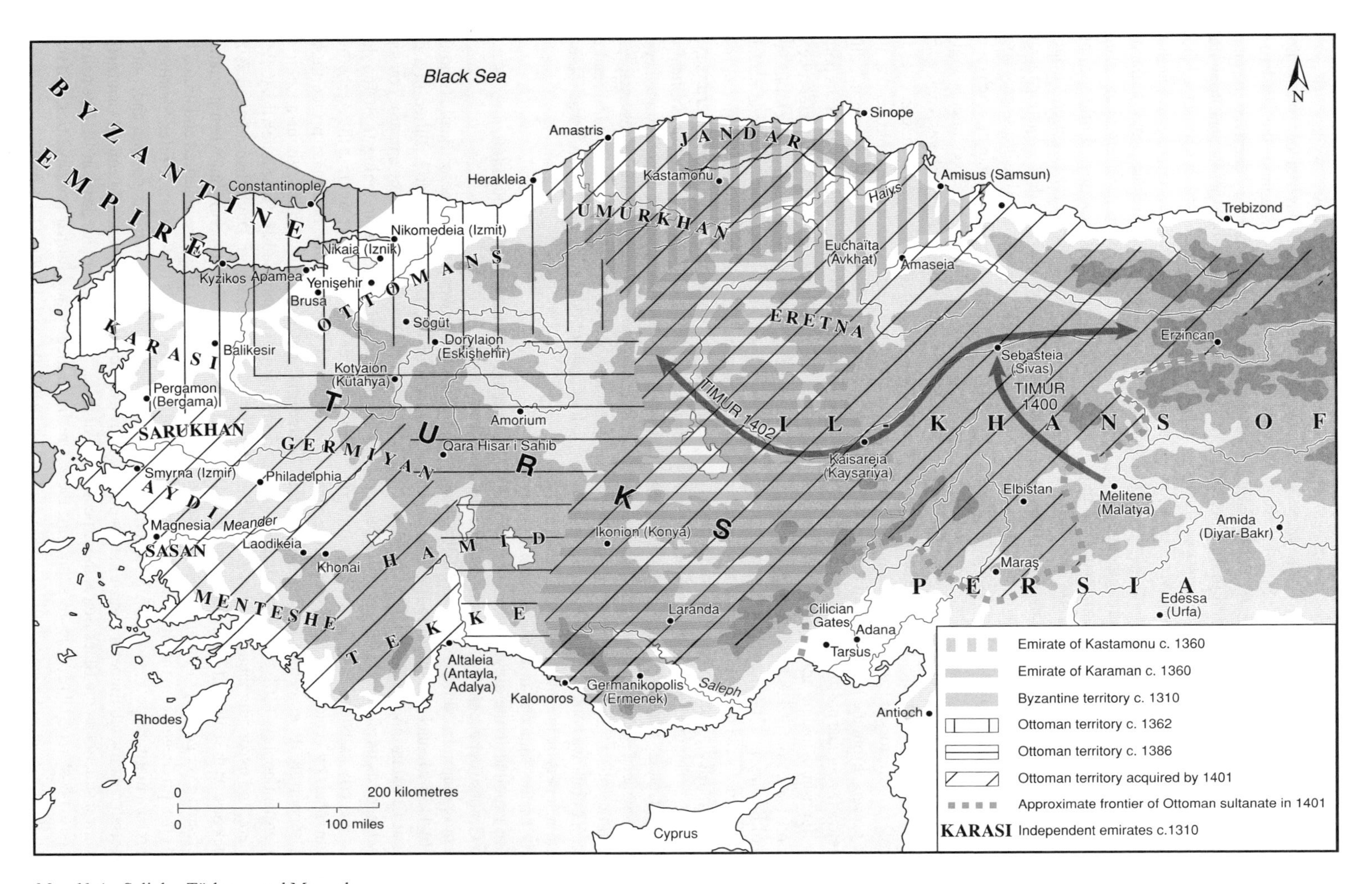

Map 11.4 Seljuks, Türkmen and Mongols

key families, but was in principle elective from the middle of the ninth century. By the late tenth century treaties with the German emperors had guaranteed the city's independence, and a treaty with Byzantium gave Venice privileged access to eastern markets and ports. By the later eleventh century Venetian military and naval power had grown sufficiently for the Venetian government to drive the Croats and other Slavs out of the Dalmatian cities (technically on behalf of the empire, but in fact a purely Venetian undertaking) and to help defeat the Normans in the 1080s. Further imperial concessions made Venice into the major mercantile power in the east Mediterranean basin by the twelfth century. Internal political reforms, the establishment of a communal government, strengthened these processes, and Byzantine resentment of Venetian maritime and commercial power became obvious in the 1170s. By this time the old alliance had been transformed into an open hostility, and Venice's exploitation of the Fourth Crusade and its results turned the small commercial city on the Adriatic into a major territorial power.

The empire attempted to thwart Venetian designs by allying itself with the rival cities of Pisa and Genoa. Pisa played a relatively brief role in the history of Byzantine relations with the Italian maritime powers. From the later eleventh century until the 1170s Pisa held a favourable position at Constantinople, for example, with a flourishing trading community there, but this ended with the anti-Latin riots of 1182 and even greater losses during the sack of the city in 1204. Rivalry with the stronger power of Genoa eventually led to an eclipse of Pisan power in the Aegean and eastern Mediterranean basin, although the city continued to be an important maritime factor until the fifteenth century. In contrast Genoa, sacked by the Lombards in the 640s, and subject to Saracen raids during the ninth century, was able to recover sufficiently by the tenth century to rival both Pisa and Venice. Sardinia was out of imperial control and in the hands of local magnets by the middle of the ninth century. Occupied briefly by the Saracens in the late tenth century, they were in turn expelled in the eleventh century, and the island became a Genoese dependency, while the Genoese war fleet could undertake major expeditions against Muslim ports such as Tunis. In 1155 Manuel I granted trading privileges to the Genoese at Constantinople. Thereafter the imperial government saw Genoa as its ally in the struggle with Venice; and the Genoese cleverly exploited this situation. In spite of the enormous gains made by Venice after 1204, Genoa was also able to profit by supporting the emperors of Nicaea and, after the recovery of Constantinople in 1261, building up their colony and fortress at Pera, and expanding their activities around the coasts of the Black Sea. At the culmination of a conflict in 1284, the Genoese defeated Pisa near Livorno and, favoured with commercial and other privileges from the Byzantine emperors, became the paramount maritime power for a while; and when, in the mid-fourteenth century, the Emperor John VI attempted to exploit the political situation in the Black Sea at the expense of the Genoese and to boost imperial revenues by supporting the position of Byzantine merchants, Genoese military and naval power soon re-established a situation more favourable to their own interests. But the Venetians also struck back at Genoa, concluding an alliance with the leaders of the Golden Horde and attacking Genoese bases, including their fortress at Galata. The war between the two mercantile empires exhausted both and an uneasy peace was arranged in 1299. As the Ottomans expanded in both Asia Minor and Europe, and then along the Black Sea coast, Genoese power waned.

Yet Venice and Genoa played a key role in the Aegean and eastern Mediterranean throughout the last two centuries of Byzantium. Between them they deprived the empire of many of its strategic resources in the Aegean; they dominated the international carrying trade; and they possessed the naval power and resources to strike at their enemies when they needed to. Their success owed much to their origins and to the emphasis their governments placed in investing in the trading activities of the shipowners and captains. While the major trading cities possessed an agricultural hinterland from which most members of their urban élites derived an income, their leading elements were at the same time businessmen whose wealth and political power was often dependent as much on commerce as on rents. The city-states themselves, increasingly dominated by merchant aristocrats and their clients, came to have a vested interest in the maintenance and promotion of as lucrative and advantageous a commerce as possible, so that the economic and political interests of the leading and middling elements were identical with the interests of the city, its political identity and its independence of outside interference. Communal government, although differently structured in both cities, reinforced such ties. State/communal and private enterprise were inseparable. The Byzantine state, in contrast, played no role at all in promoting indigenous enterprise, as far as we can see from the sources, whether for political or economic reasons, and viewed commerce as simply another minor source of state income: commercial activity was regarded as, and was in respect of how the state worked, peripheral to the social values and political system in which it was rooted.

Armenia and Georgia c. 1000–1460

The alliance between the Emperor Basil II and the Georgian Bagratid prince David of Tayk' resulted in a considerable extension of David's lands, for the emperor granted him a great tract of western Armenia stretching from Tao down to Lake Van. Successful offensives against the Arabs of Azerbaijan established David as the dominant prince in Armenia and Georgia, but upon his death without an heir in 1000 the Emperor Basil annexed the whole territory. Bagrat III, the King of Abasgia, whom David had adopted as his heir, was recompensed by the emperor, but when his father, the King of Georgia died in 1008, he inherited and united both western and eastern Georgia under a single ruler. Shortly afterwards he incorporated Kakhetia into his domain and, with the King of Armenia, defeated the Shaddadid emir of Gandza (in Caucasian Albania). By the time of his death in 1014 he had expanded his realm to incorporate further principalities, and make Georgia the paramount kingdom in Caucasia.

Internecine strife and external attacks brought about the collapse of the Armenian kingdom. Between 1018 and 1021, the country was attacked by both Daylamite raiders from

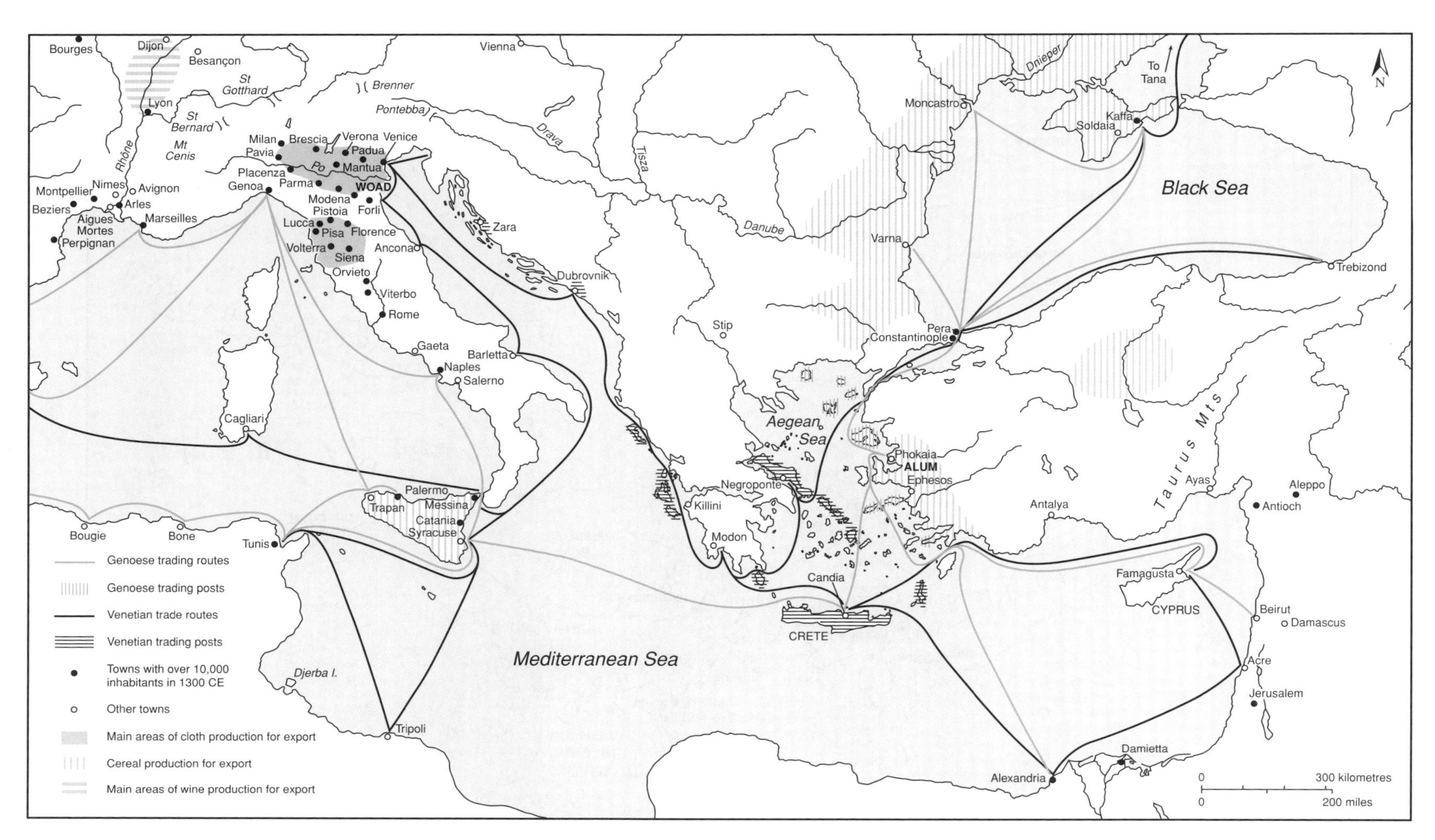

Map 11.5 Venice, Genoa and the merchant empires.

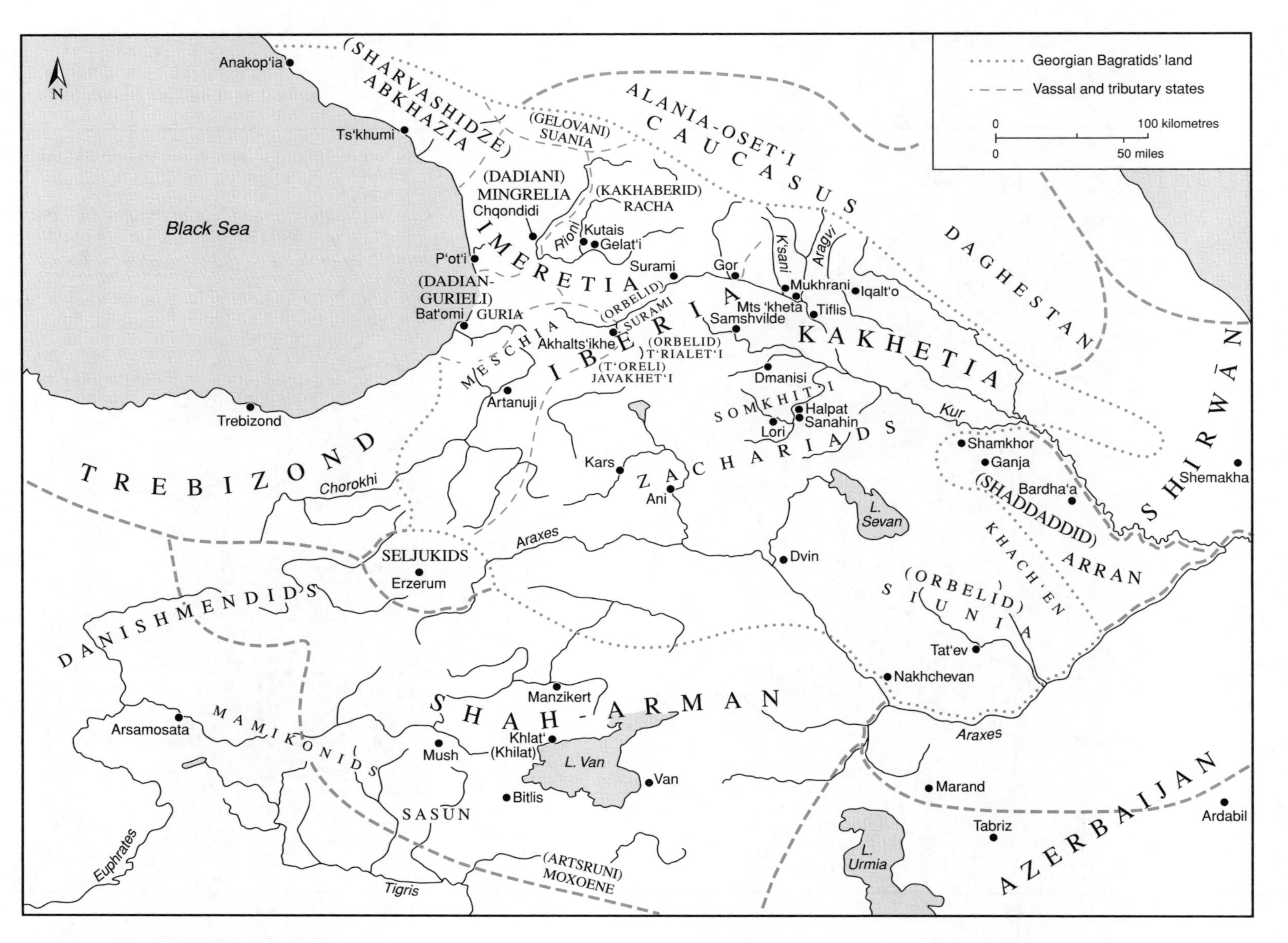

Map 11.6 Armenia and Georgia c. 1000–1460.

the south and Seljuk Turks, and in 1021 King Sennacherib of Vaspurakan ceded his kingdom to the Emperor Basil II. By 1040, again by virtue of the will of the ruler Smbat III of Armenia, Ani and Armenia were likewise to become part of the empire. But local resistance prevented annexation for a few years until 1045. Having incorporated the region, the imperial government decided to stand down the local levies, leaving the country poorly defended and easy prey to Seljuk raiders who, from 1045 onward, repeatedly attacked the country. In 1064 Ani fell to the Seljuk leader Alp Arslan, and the local kings of Siunia and Kars accepted Seljuk overlordship. The Byzantine defeat at Manzikert in 1071 sealed the fate of Armenia. A number of independent principalities – Moxoene, Arsamosata, Sasun – survived into the later twelfth century, when they were absorbed by the local Seljuk power.

As Greater Armenia suffered under Byzantine and then Seljuk pressure, the migration of Armenian nobles and their retinues, often under imperial auspices to remove the threat they posed to Byzantine influence in their homelands, called into being a Lesser Armenia, initially in Cappadocia until this region, too, fell to the Seljuks, and then in north-western Syria and Cilicia. The political origins of the kingdom lie in the refusal of its first ruler, Philaretos Brachamios, Byzantine commander of Germanikeia (Mar'aš) and Mélitene (Malatya), to accept the rule of the Emperor Michael VII Doukas after the defeat and capture of Romanos IV in 1071. By 1078 he also held Antioch, and shortly afterwards incorporated much of Cilicia, already occupied by transplanted Armenian nobles since the early part of the eleventh century. But under pressure from Seljuks on the one hand and the Franks of the First Crusade, on the other, Lesser Armenia broke up into its constituent parts. Only in Cilicia was there some element of continuity under the dynast of the Rubenids, who seem to have served the empire as generals during the earlier eleventh century. By the 1130s the Rubenids held all eastern Cilicia, and although temporarily defeated by the emperors John II Komnenos and Manuel, the Cilician principality consolidated. Under the kings Leo II (1186–1219) and Het'um I (1226–1269) Lesser Armenia became a significant local power, although they had to contend with the Ayyubids of Syria and their more formidable successors, the Mamluks of Egypt. Het'um accepted Mongol suzerainty and was instrumental in provoking the Mongol attack on Syria which culminated in the sack of Baghdad and the destruction of the Abbasid caliphate in 1258, and the battle of Ain Jalut in 1260 when the Mamluks defeated the Mongols.

This turned the tide, and although the Armenian kingdom retained its independence for a while longer, internal factionalism rendered it weak. The conversion of the Mongol Il-Khans of Persia to Islam in 1304 turned a former ally and protector into an enemy; under the Lusignan kings, who ruled the kingdom by bequest from 1329, more and more territory was lost to neighbouring Islamic powers, until by the 1370s only Sis and Anazarbos remained. The kingdom was finally extinguished by the Mamluks in 1375, when the king and his family were captured and imprisoned in Cairo. The region remained part of the Mamluk empire until taken by the Ottomans in the sixteenth century.

The Georgian kingdom fared better. Conflict with the empire marked the period 1014–1059, and from the early 1080s the Seljuks exercised a limited dominance. But by the 1120s the Shaddadids of Azerbaijan had been expelled from Tbilisi and Ani, the district of Kakhetia had been taken, and in the following decades Georgia grew to become the leading state in the region. By the time of Queen Thamar the Great (1184–1212) the kingdom spanned the whole Transcaucasus region from the Caspian Sea to the Black Sea, chiefly because royal authority was effectively asserted over the factious noble houses of the provinces. Thriving commercial centres at fortress-cities such as Dvin, Ani, Kars, Tiflis and others brought great wealth to the crown as well, the rulers built churches and decorated them, literature flourished. This was a Georgian 'golden age'. But the appearance of the Mongols heralded a change. A raid in 1220–1221 defeated a large royal army, an attack by the shah of Khwarizm devastated the land in 1225, and Mongol attacks in the 1240s reduced most of Georgia to vassal status. The kingdom remained autonomous, although it split into two, Imeretia and Georgia, as a result of further factional strife in 1258. By the 1320s, Georgian control over Armenia had been reduced, although close diplomatic ties were developed with the Komnenoi of Trebizond, and a number of marriage alliances were negotiated. The attacks of the armies of Timur between 1386 and 1403, however, permanently damaged the economy of the region, and with the Ottoman capture of Constantinople in 1453 Georgia was isolated from the western church with which efforts at a union had been attempted. Thereafter the kingdom sank into relative obscurity, the target of both Ottoman and Persian attacks, although it survived into the nineteenth century, when it was annexed by Russia.

Russia and the Steppes c. 1000–1453

On Vladimir's death in 1015, and following a civil war which ended in 1019, Russia fell into three major subdivisions, Novgorod-Kiev, Chernigov-Tmutorokan, and Polotsk. Vladimir's son Yaroslav was able to reunite them and re-establish a single principality from 1036. He defeated the Pechenegs, pushing them south-westwards into the north Danube plain whence they began raiding Byzantine territory. Although he launched a disastrous attack on the empire in 1043, friendly relations were quickly restored. Raiding from the Oğuz or Cumans (Polovtsy) weakened Kiev, and on Yaroslav's death (1054) Russia again split into autonomous and often warring principalities, Novgorod and Vladimir-Suzdal (capital at Moscow) being among the most important. In 1060 a joint attack defeated the Oğuz, but in 1067 a war between Kiev and Polotsk led to the sack of Novgorod. In spite of a major Polovtsy raid into Kiev in 1068, precipitating further internal strife (involving also the principality of Poland on one side), diplomatic and marriage alliances enabled prince Svyatoslav Yaroslavich to bring the competing principalities together again by 1076.

Although Kiev was sacked by the Polovtsy in 1093, a series of successful campaigns (1103–1116) contained them, and a degree of stability was reached, with Pecheneg and minor Oğuz groups

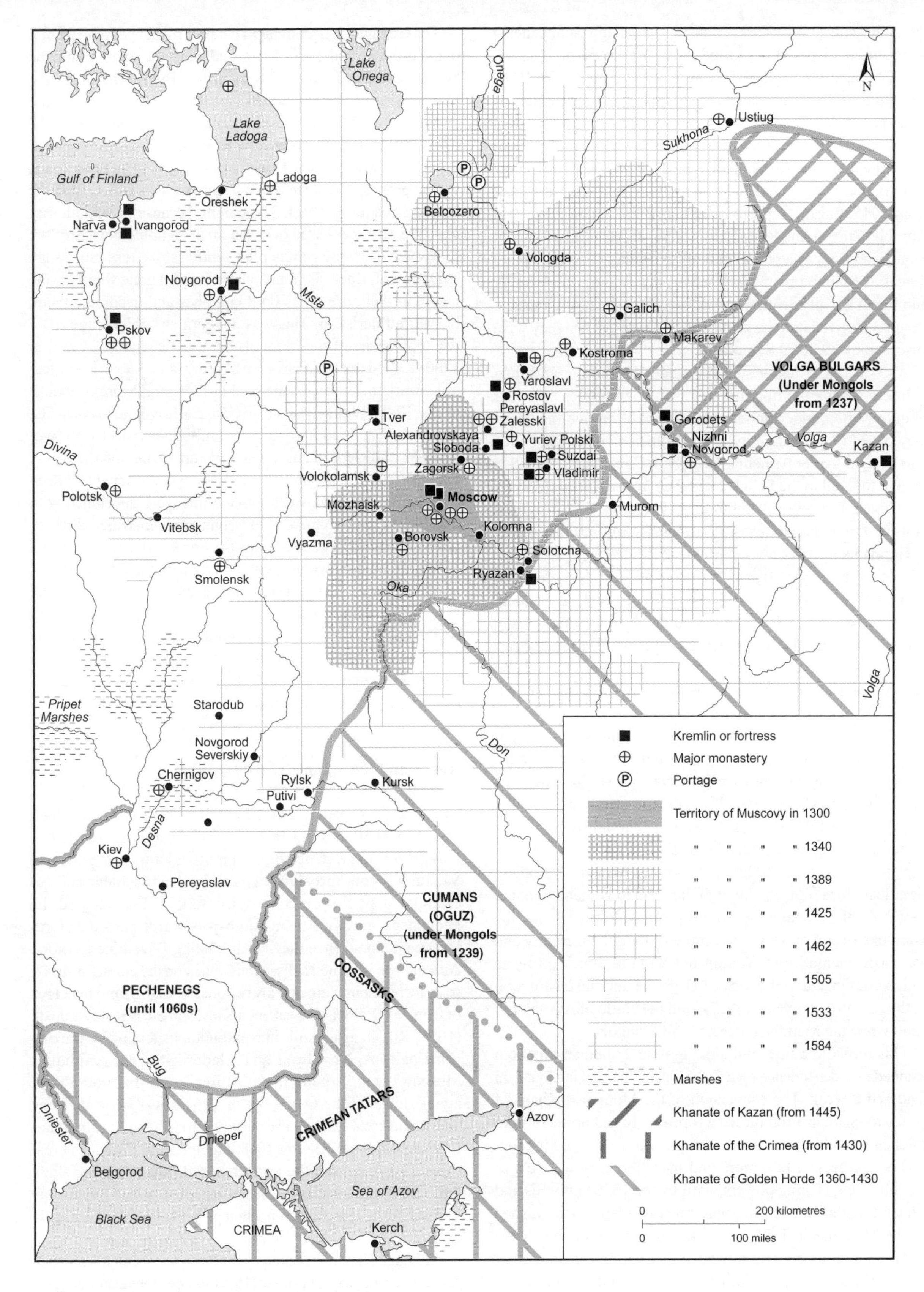

Map 11.7 Russia and the steppes c. 1000–1453.

contracted to safeguard the frontier. Fragmentation followed the death of Grand Prince Vladimir Monomakh in 1132, and after 1139, with local and long-distance trade flourishing, the Russian principalities were increasingly at odds over control of routes. When the prince of Suzdal sacked Kiev in 1169, his own territory briefly became the pre-eminent principality. A limited Mongol raid defeated a joint Russian–Polovtsy force in 1223, but withdrew and the situation appeared to return to normal. But as the prince of Suzdal prepared an attack on the Volga Bulgars (who controlled the trade route through the middle Volga basin) in 1237, the Mongols launched a second raid in strength. The Volga Bulgars were overwhelmed and the central Russian principalities were ravaged, their wealthiest centres being sacked in the process. In 1239 the southern principalities were destroyed; in 1240 Kiev was sacked; and in the north, although Prince Alexander of Novgorod (Alexander Nevsky) was able to defeat and turn back invasions by both the Swedes and the Teutonic Knights in 1240 and 1242, he was forced to pay tribute to the Mongols. On the steppe the Cumans were defeated and incorporated into the Great Khanate, and in the north Caucasus the Alans were overrun.

The 'Mongol yoke' devastated the economy of the Russian principalities, forced the peasantry into an ever greater degree of servitude and poverty, and encouraged a much more autocratic form of rule among the surviving principalities. Mongol overlordship also weakened the ability of the Russian princes to resist external aggression or interference: the territories formerly ruled from Kiev split into eastern and western regions, the latter absorbed by the princes of Lithuania and later becoming part of the kingdom of Poland, the former under tribute to the Mongols. The Russian princes attempted from time to time to throw off Mongol suzerainty – in 1382 the prince of Vladimir-Suzdal attempted to gain his independence, but was defeated and his capital, Moscow, sacked. Yet the weakening power of the Golden Horde and the interfactional strife between the different Mongol hordes enabled the Grand Princes of Vladimir to extend their power across much of the original principality between 1302 and 1420. By the 1470s they were styled Grand Princes of Muscovy, by 1478 they had incorporated Novgorod into their realm, and the way was prepared for the rapid incorporation of the other principalities and the push to the Ural river in the sixteenth century.

On the steppe to the south and east the Pechenegs were forced westwards by a combination of assaults from Kiev and the Oğuz and Cumans. Residual groups survived to ally themselves with the Russians during the twelfth century, but their place was taken by the Cumans, who seized the Crimea from Byzantine control in 1068. To the east were the Oğuz, divided from the related Seljukid clans, who had adopted Islam, by the Muslim Karakhanid Turks. In 1037 the Seljuks rebelled against their nominal overlords the Ghaznavids, whom they had ousted from eastern Iran by 1055. Penetrating into central and western Iran at the same time they subjected the Buyids, whom they replaced, establishing a Seljuk Sultanate stretching from Transoxiana to the Byzantine frontier. By 1073 they had extended their power to the north-east and conquered the Karakhanids (the first Turks to adopt Islam). The Seljuk sultanate soon broke up into a number of lesser emirates (see page 147), with rebellious Oğuz clans overthrowing the sultanate of Merv in the 1150s, and to the north the establishment of the khanate of the Karakhitai, a confederacy led by a Buddhist Mongol clan. The Cumans continued to dominate the south Russian steppe across to the Danube (the last Pecheneg attempt at re-asserting their power was defeated by the Byzantines in the 1120s), although in the period from the 1190s to 1210 the Oğuz and the Karakhitai were overthrown by the Sultan of Khwarizm, who established a new Irano-Turkish empire across eastern Persia and Transoxiana.

Between 1237 and 1241 the Mongol invasion overran the Cumans in the Russian steppe, the Alans, the Russian principalities, the Shahdom of Khwarizm, and the Seljuk sultanate. Mongol expansion westwards was only halted by the Mamluks at Ain Jalut in 1260. But in 1260 the Khanate was divided among the sons and grandsons of Chingis, creating a number of rival hordes. Although the conversion of most of these to Islam by the 1420s created an ideological unity, political and territorial rivalry remained. In 1380 the White Horde and Golden Horde fought for supremacy, with victory going to the former, although the term Golden Horde continued to describe the new formation. The conquests of Timur from the 1360s brought considerable disruption to this pattern, creating for a short period another unified Mongol empire stretching from Turkestan to Anatolia. But this in turn soon broke up into a number of smaller emirates, and by the 1460s no major Turkish or Mongol power was permanently in power in either Iran or the steppe – the Akkoyunlu in Persia had been destroyed by Ottoman and Safavid Persian power by 1502.

The Islamic Middle East c. 1100–1430

In 1092 when the Seljuk Sultan Malik Shah died, Seljuk central authority stretched from Anatolia to north-west India and from the Caucasus to the Arabian Sea. But this authority was fragile, and internecine squabbles soon destroyed it. The successes of the First Crusade may be ascribed at least in part to this lack of unity and to the rise of competing power-centres in the region, indeed the 'Franks' were frequently drawn in on one side or another in internecine Muslim conflicts. Under the emir of Aleppo and Mosul, Zengi (1128–1146), some stability was restored from the 1130s, based around the idea of unifying the disparate Islamic forces and launching a counter-attack against the Crusader states. The recovery of Jerusalem was presented as a key goal, but it remained an ideal, in view of Zengi's preoccupations in maintaining his own power in the Jazira. Nevertheless, he was able to take Edessa in 1144 and extinguish the Crusader principality of the same name, and his son and successor, Nur ad-Din (1146–1174) was able to build on this initial success. But his enemies were not just the Crusaders: he also saw the heretical Shi'a regime of the Fatimids in Egypt and North Africa as a threat to Islam, and he placed no trust in the ability or willingness of the other local emirs in Syria to support his cause. In 1154 he seized Damascus and incorporated it into his own territory, bringing all Syria under his control.

The Fatimid power was already on the brink of collapse, riven by competing factions, and the struggle for control over

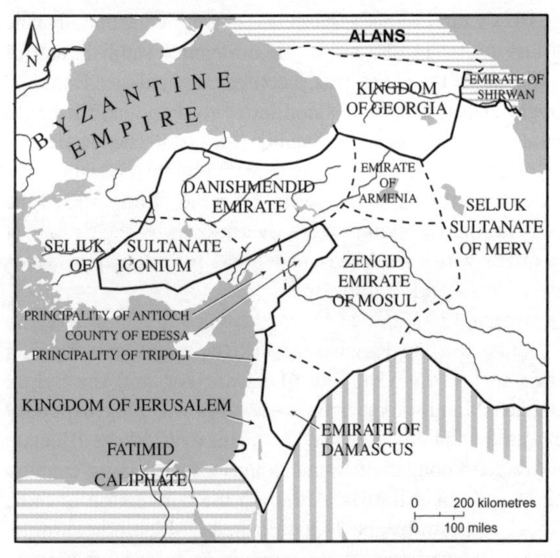

Map 11.8(a) The Islamic Middle East c. 1100–1140.

Map 11.8(b) The Islamic Middle East c. 1170–1180.

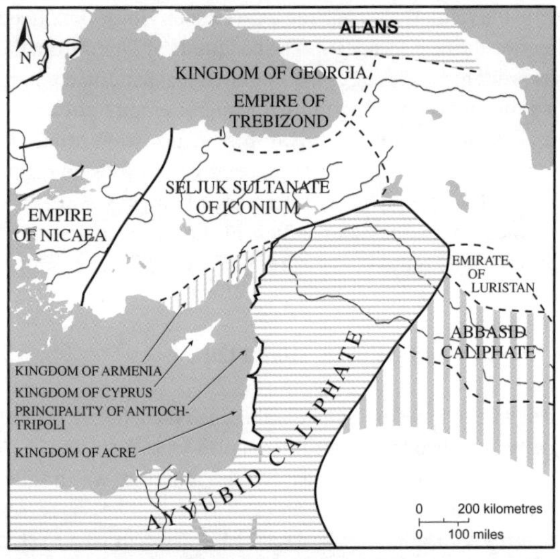

Map 11.8(c) The Islamic Middle East c. 1230.

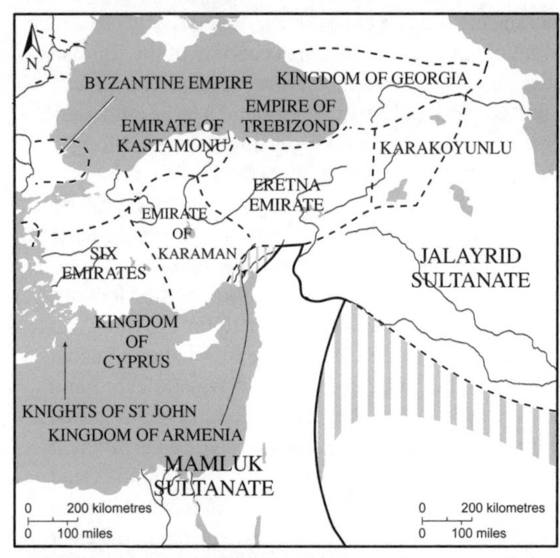

Map 11.8(d) The Islamic Middle East c. 1355.

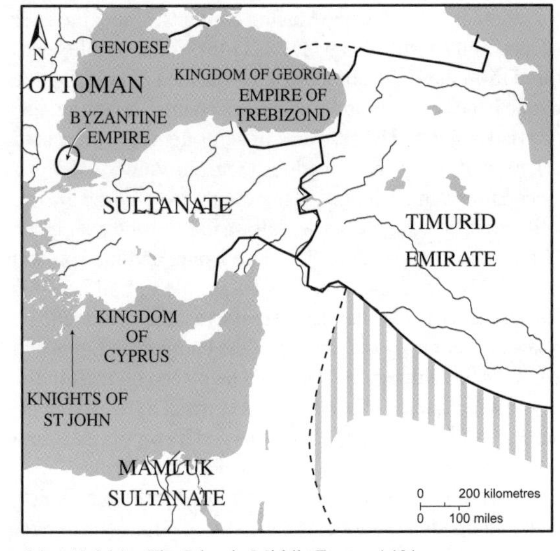

Map 11.8(e) The Islamic Middle East c. 1401.

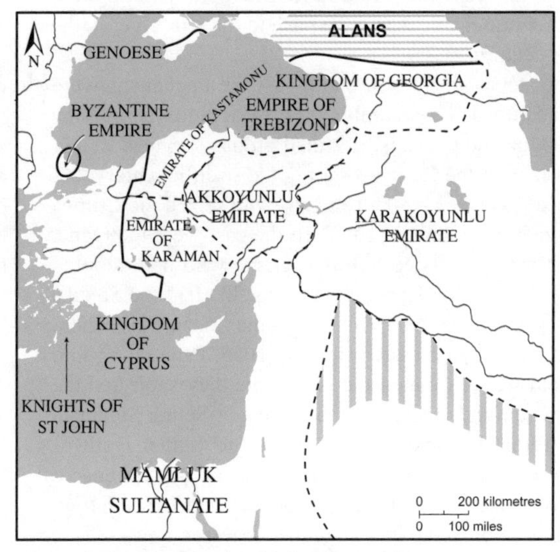

Map 11.8(f) The Islamic Middle East c. 1430.

(After Kennedy, *Historical Atlas of Islam* and McEvedy, *New Penguin Atlas of Medieval History*.)

Egypt was thus between the Crusader kingdom of Jerusalem (invited to assist one side or the other in the internal fighting) and the Zengid emir. In 1168 the Fatimids called on Nur ad-Din to help repel another Crusader attack, and the result was a Zengid administration in Egypt, formal re-establishment of the Abbasid Caliph, and the end of the Shi'a regime. The Kurdish commander Shirkuh who had achieved this on Nur ad-Din's part died in 1169, and was succeeded by his nephew, Salah ad-Din b. Ayyub, better known as Saladin. Saladin maintained only a very loose relationship with Nur ad-Din, however; and when the latter died in 1174, Saladin seized control of Syria, so uniting the two regions and establishing a new power in the Middle East, the Ayyubid Sultanate. Within 20 years he had recovered Jerusalem and reduced the Crusaders to the coastal fortresses of Palestine, but after his death in 1193 the sultanate began to fall apart, as the emirates of Hama, Damascus and Aleppo in Syria, as well as Egypt, ruled by his relatives and successors, competed amongst themselves for dominance.

The Ayyubid system was constantly under pressure. Apart from the faction-fighting between the emirs of the dynasty, there were attacks from the Khwarizmians, who were able to penetrate as far as the Syrian coast on occasion, driven west by the Mongol attack on their shahdom. In Egypt the emir Malik al-Kamil recruited large numbers of Turkish slave soldiers to support his régime; and when threatened by the Sixth Crusade in 1228–1229 (led by the Emperor Frederick II, excommunicated by the pope), he agreed to restore Jerusalem and the holy places to the Kingdom of Acre in return for their non-aggression thereafter. But this gain was short-lived, and in 1244 the city fell once more into Muslim hands, this time permanently. In 1248 the King of France, Louis IX, at the head of the Seventh Crusade, attacked Egypt from his base in Cyprus, and although successful at first – Damietta was taken – he was defeated at Mansura and captured along with most of his army. But the attack encouraged further unrest in Egypt; and in 1249 a group of military slaves (Mamluks), seized power and established a state that quickly swallowed up the remaining Ayyubid emirates. The Sultan Baybars (1260–1277) was one of their greatest rulers, inflicting a defeat on the Mongols at Ain Jalut in 1260, which put an end to Mongol attempts to conquer Syria and Palestine, and during the period up to 1291, when Acre finally fell, all the remaining Crusader strongholds were taken by the Mamluk armies. Mamluk rule over Syria as far north as the Armenian kingdom in Cilicia (the last outposts of which fell finally in 1375) was secured by the 1360s, and along with Egypt formed a polity that lasted until the defeat and conquest of their armies by the Ottomans in 1517.

12 *Church and Monastery in the Later Byzantine World*

Diocesan organisation: the *Notitiae*

The situation of the church after the Seljuk and Türkmen occupation of much of central Anatolia from the 1070s is very difficult to gauge. A series of six lists of bishops' sees (*notitiae episcopatuum*) dating from the middle of the eleventh to the middle of the thirteenth century throws some light on the extent of church authority in the provinces and in areas no longer held by the empire. The lists suggest a total of over 80 metropolitan sees, although only 50 or so had more than a handful of suffragan, or dependent, bishoprics under their authority. Many were in exposed frontier areas and the lists alone tell us nothing about the extent of episcopal control over the Christian population of the areas in question. Some 31 metropolitanates, consisting of just under 400 sees, were within the imperial frontiers. There was a great density of sees around Constantinople, in Bithynia and Thrace, and in western and south-western Asia Minor, for historical reasons – these were regions which had the most ancient traditions of church organisation, stretching back to early Christian times, and it was also a hallmark of Byzantine ecclesiastical administration that organisational conservatism was very strong – change was always to be avoided, where possible, and was certainly regretted.

Change did occur, however, and generally in response to obvious causes such as an increase in population or the expansion of towns, with a consequent demand for greater ecclesiastical supervision. New sees might thus be created, based in flourishing small towns, or at least in population centres where there was a need for the church to maintain a presence. The metropolitan bishop of Smyrna had five new suffragan sees created for it by the twelfth century, for example, and other cities or centres, such as Ephesos, were similarly endowed. On the other hand, some sees were moved in order to preserve the establishment and to protect the right of the bishop and his flock. A number of inland sees in Asia Minor, threatened by, or actually overrun by, the Turks, were transferred to the Pontic coastal region and safety. But transfers were not generally admitted, again because they disturbed the traditional order of things.

The church was collectively an extremely wealthy landlord, but this wealth was very unevenly distributed. Until the late tenth century each see was supported by revenues derived from church lands, administered by bailiffs or caretakers, by the so-called *kanonikon*, a levy on the lay communities, the priests and on all the monasteries in a see, and various miscellaneous sources – ordination fees, gifts, for example, from ordinary people as well as from landowners and, from the time of Alexios I, a part of the fines levied in court cases heard in the province. From the time of the Patriarch Sisinnios in c. 995–1000, however, the *kanonikon* on monasteries was abolished, and many bishoprics in poorer areas suffered as a result. From the end of the eleventh century the *kanonikon* or church tax exacted from the laity and the clergy was regulated so that it generated enough income to support the bishop and his establishment and the various duties he had to fulfil. The emperors paid considerable attention to the well-being of the church in the provinces – not only was it, from the point of view of the emperor's orthodox duty, a necessary part of his responsibilities, but from a much more practical standpoint the church was a key symbol and support of the empire and imperial rule, orthodox belief and identity, and thus a force for social and political cohesion. Where the church suffered, senior state and ecclesiastical officials were frequently at pains to restore its fortunes. There are several examples of impoverished sees which had fallen on hard times, either due to poor management by the bishop or his agents, external disruption – such as piracy or enemy raids – or famine, drought or pestilence, which were restored by investment in personnel, buildings and income by the patriarch or the emperor.

Bishops were often instrumental in founding monasteries in their sees, either directly or, more usually, by offering support and encouragement. But an important administrative and economic development which affected the church from the eleventh century onwards was the custom of *charistike*, whereby a layperson (known, therefore, as a *charistikarios*) was granted the administration of a monastery and its possessions in land and other forms during his lifetime, sometimes for several lifetimes. The purpose was to ensure the proper administration of the community's property, on the one hand, and its being put to appropriate use, beneficial to the community as a whole, on the other. It was also used as a means of helping monasteries that found themselves in financial difficulties. The grant was usually made by a bishop, but it could be made by patriarch or emperor, or anybody with existing formal rights over a monastic community and its property. *Charistike* grants were at first strictly regulated, and there were a series of specific conditions attached to the grant to make sure that it was not abused. In fact, many *charistikarioi* exploited the situation by concealing the extent of their monastery's property to avoid tax, for example, and thus enrich themselves.

Bishops occupied an important role in late Byzantine society. Not only were they a crucial link between local church and society and the administration at both provincial as well as central level; they were also in many cases the only protectors of the victims of an oppressive and increasingly corrupt and venal taxation system, representing to both court and local administrative chiefs the problems of the humbler taxpayers. During the final century of the empire, indeed, they also became the real protectors of all their orthodox population, acting as intermediaries between the conquering Turks and the indigenous population, for example, and arranging for the administration and taxation of the conquered communities under the changed circumstances as best as they could. Not always successful in ameliorating the condition of the conquered, they nevertheless became the effective and active replacements for the failed secular state, just as had their predecessors in the late Roman and early medieval west.

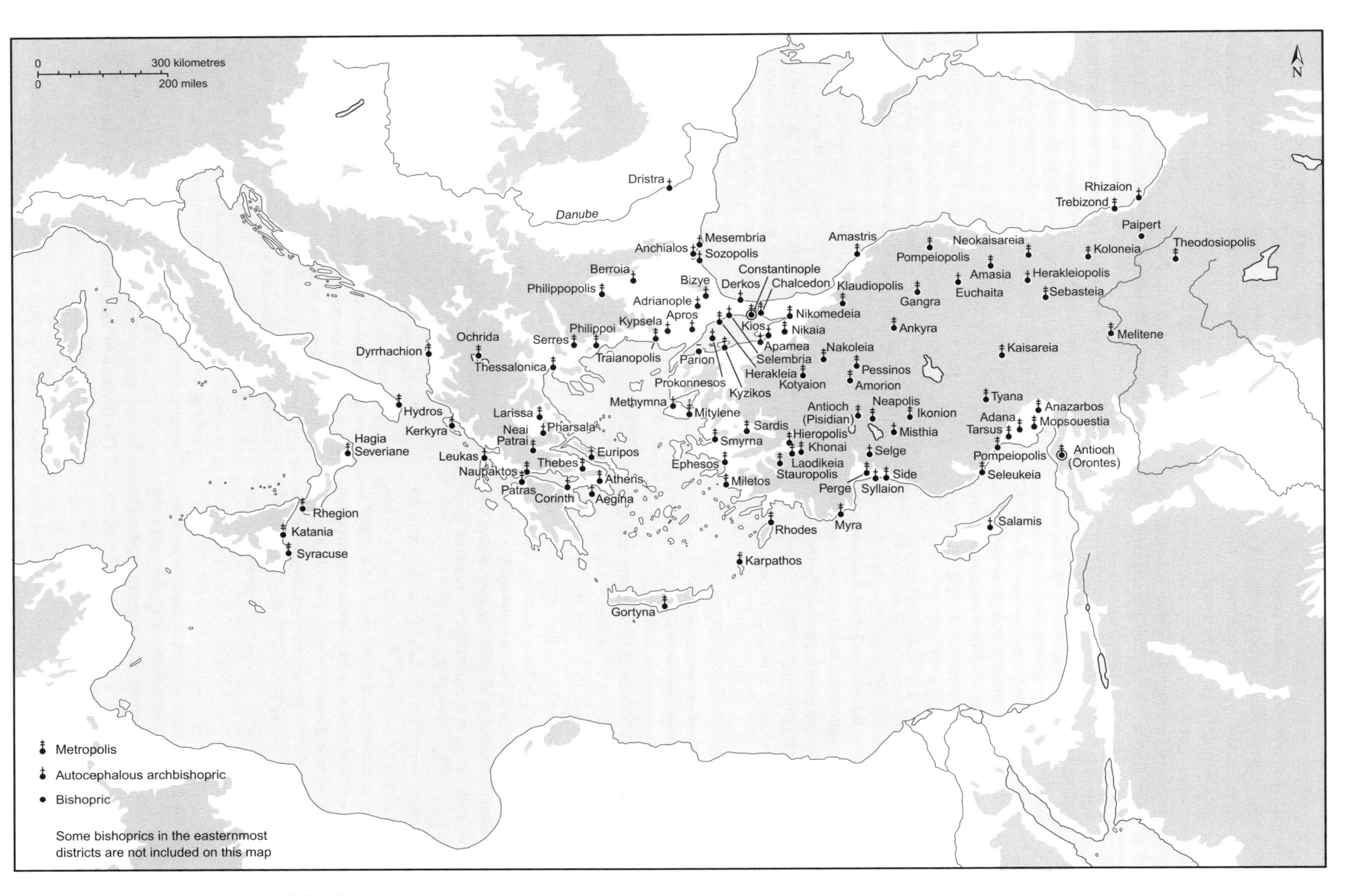

Map 12.1 Diocesan organisation c. 1070: the *Notitiae*.

Monasteries and Rules

Although monasteries remained important centres of spiritual activity and authority in the eastern Christian world throughout the history of the empire and beyond, many monastic communities were also great landlords in their own right, possessing substantial landed properties in the provinces, from which they derived considerable incomes in money and in kind. Members of all social classes, from peasant farmers to emperors, made donations to such establishments in the hope of gaining spiritual benefit in the future and especially after their death. In return, the monks would pray for the souls of the individual donors.

During the late Byzantine period, and as the empire shrank territorially, monasteries began to play an ever greater role in the secular affairs of the state. Not only did they continue to receive donations or exemptions from various state taxes and fiscal impositions; as the state's resources shrank with its territory, so monasteries became economically relatively more and more important. In some cases, they controlled so much land and resources that they became responsible for local defence and the maintenance of fortifications or the building of warships. An excellent example of this process is provided by the history of monastic property and taxation in the monastic centre of Mt Athos in the Chalkidike peninsula in northern Greece (see Map 12.2). From both Byzantine fiscal documents, imperial letters of exemption in respect of taxes and other privileges, as well as from the Ottoman detailed tax registers, it is possible to piece together the intricate picture of how the state and monasteries co-operated or competed for resources in the region, and what happened to property rights as well as the peasants and their landlords after the conquest by the Ottomans. Monasteries on Mt Athos acquired ever greater endowments from the middle of the thirteenth century until 1453, and the cumulative wealth of the richer monasteries was, by the end of the fourteenth century or not already some time before, considerably greater than that of the state. Even so, from 1371 the government was able to impose some limits, even redistributing monastic lands to support soldiers for the army, for example.

Until the Turkish occupation of much of Asia Minor in the last years of the eleventh century one of the most famous and populous centres for monastic activities was Mt Olympos in Bithynia. Others were to be found in the western coastal region – at Kyminas, Galesion and Latros, for example, but all were permanently damaged by the Byzantine–Turkish conflict of the later eleventh century and afterwards. In the north-eastern area around Trebizond a number of important monastic houses appeared in the last centuries of imperial rule, those at Soumela and Vazelon being the two best known. In Europe, already by the late tenth century Mt Athos was becoming a famous centre; there were other, much smaller, centres in Thrace; and from the thirteenth century the great monastic centre at Meteora in Thessaly began to grow. As many as a thousand different monasteries are mentioned in the documentary sources across the life of the empire, although not all of them were active at the same period. There were several important monasteries also in Constantinople, such as the famous St John of Stoudios, but several new monasteries were founded during the eleventh and twelfth centuries under imperial patronage.

The surviving *typika* or foundation charters for these establishments provide a great deal of evidence about their internal administration as well as their spiritual life and aims. Monasteries and convents varied greatly in size. Some of the larger establishments grew to have several hundred monks, but the majority remained much more modest, with tens rather than hundreds as the norm. Their functions varied. Charity and alms, the relief of the poor and the ill were key aims, and several monasteries had almshouses, hostels and hospitals attached to them, with considerable staffs of trained personnel to care for the inmates. Food might be distributed daily to the poor in an urban monastery, for example, or in particular times of dearth in the countryside. Since there were no monastic orders as such, the *typikon* of one monastery might serve as the basis for those of one or more other establishments. Thus the *typikon* of the Evergetis monastery in the suburbs of Constantinople, founded in the middle of the eleventh century, served as the model for those of several other Constantinopolitan and provincial communities.

One of the hallmarks of late Byzantine monastic development is the appearance and increasing popularity of so-called 'idiorrhythmic' monasticism, whereby the monk did not observe a common round of prayer and other duties with his brethren, but followed rather a more individual form of prayer, contemplation and work, eating alone, for example, rather than in a refectory. This was never fully condoned by the eastern church because it set up a challenge to the fundamentally cenobitic principles of the monastic life, but it nevertheless became very popular in the fourteenth century, and remained an important facet of orthodox monasticism thereafter, where it is to be found in the orthodox world today.

An equally significant development which impacted on the whole eastern orthodox world, including monasteries, in the fourteenth century was the so-called 'hesychast' movement. In the eleventh century Symeon the Theologian, building on a long tradition, argued that divine activity could be experienced both through the spirit and through the senses. On Mt Athos, the idea was developed that such experiences were open to all, provided that the right means was employed to attain them. It was argued that deep concentration and repetitive prayer, accompanied by special breathing techniques, could open the consciousness to visionary experiences, a development which contrasted strongly with traditional modes of spiritual devotion, and which divided the church as well as lay society. The debate became closely entwined with the political issues of the day. Hesychasm only came to the fore for a short period, but it left its mark on the history of orthodox spirituality, and the tradition of mysticism it entailed continued to play an important role after the Ottoman conquest.

The Division of the Churches and the Orthodox Commonwealth

The term 'Byzantine commonwealth' has been coined to describe the eastern orthodox world from the eleventh century

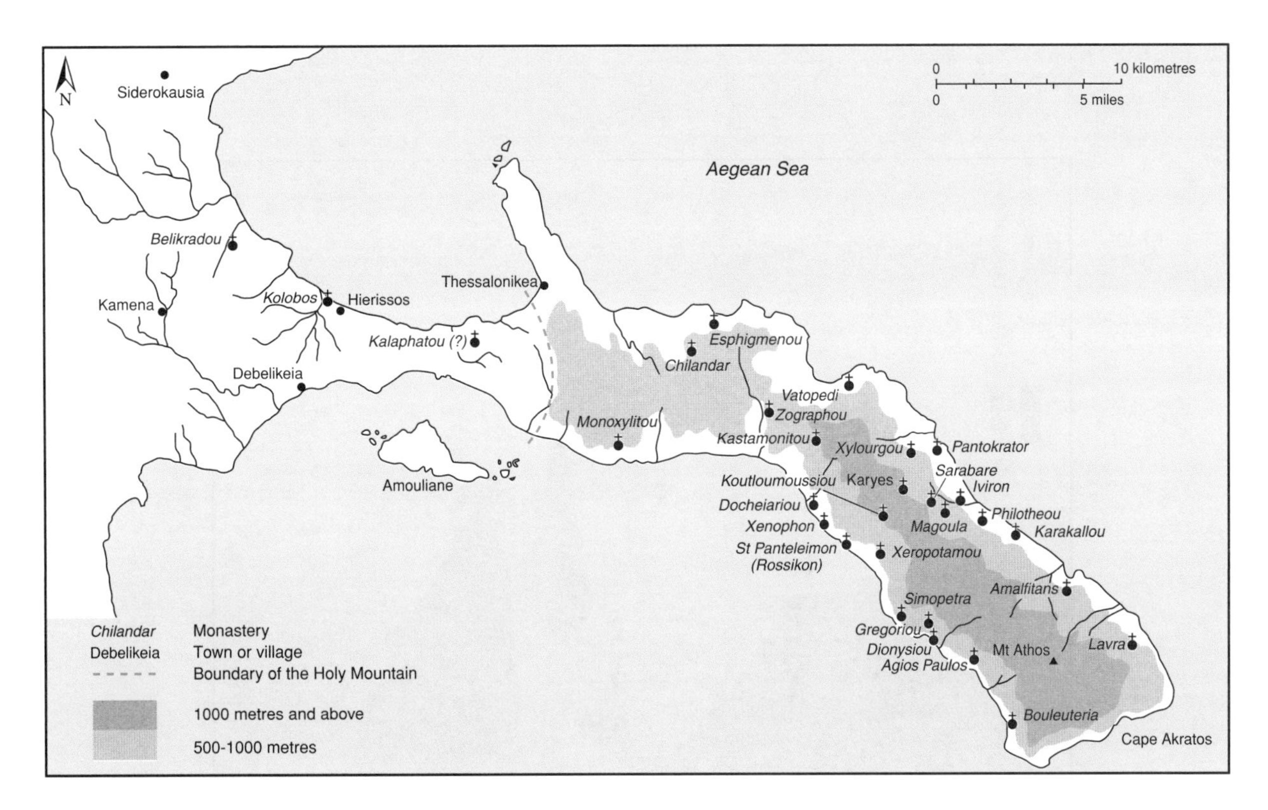

Map 12.2 Monasteries and rules: the monasteries of Athos.

onwards, and it is indeed very apt. For not only did the orthodox church outlive the secular, medieval eastern Roman empire of Constantinople, it also succeeded to many of its responsibilities thereafter. Under the Ottomans, the church was the major intermediary with the Ottoman authorities for most of the 'Greek' Christian population of the empire, and was recognised by the Ottoman authorities as the representative of the interests of the Christian communities whose voice it was. The church was not the only intermediary, but it was the most important one until after the sixteenth century. Outside the Ottoman empire the orthodox tradition dominated the lands of the Russian principalities and, as Moscow became the 'third Rome' after 1453, was employed to support the claims of the Russian rulers to be the legitimate heirs of the Byzantine emperors and thus protectors of all Christians. Quite apart from its political role, however, orthodoxy gave a common theological heritage to all those peoples whose faith it became, a heritage which, in spite of very considerable cultural differences, remains a unifying element still in the twenty-first century.

The differences between the churches of Rome and Constantinople which led to the schism of 1054 were not simply differences over doctrine or over ecclesiastical primacy, but reflected also cultural differences between the Greek eastern Mediterranean and south Balkan world, and the Latin-dominated lands of central and western Europe. Cultural alienation and misunderstandings are already apparent in the ninth and tenth centuries, first in the absurd claims made by the Patriarch Photios (but writing in the name of the Emperor Michael III) to Pope Nicholas I about Byzantine political and cultural superiority, and in the pope's learned but damning response; second in the contempt and anger for the 'Greeks' displayed by the envoy of the German Emperor Otto, Liutprand of Cremona, when he visited the court of Nikephoros II Phokas in the 960s. As western economic, political and military strength began to be a serious problem for the Byzantine empire in the later eleventh century, the situation worsened. The south Italian Normans on the one hand and the German emperors on the other posed serious threats, and a growing challenge to Byzantine maritime power from Italian merchant cities such as Venice and Genoa did not help. The Crusading movement, western prejudices about Greek perfidy, and the expansion of the Seljuk emirates in Asia Minor, transformed alienation and suspicion into open conflict. The capture of Constantinople in 1204 and the establishment of a Latin empire finalised the split, for the Latin patriarchate was not recognised by the orthodox populations of the Byzantine or formerly-Byzantine regions.

The rapidly expanding power of the Turks, especially once the Ottoman Sultanate had become established, together with the political and economic collapse and fragmentation of the Byzantine state, made reconciliation between eastern and western churches a matter of urgency. The west tended to view the Byzantines as schismatics or even heretics, and offer little or no support. Negotiations aimed at resolving this problem through a union of the churches continued throughout the fourteenth and fifteenth centuries, but at the Councils of Lyons, in 1274, and Ferrara-Florence, 1439, agreement on many issues could not be reached, even though in both cases the emperor of the time (respectively Michael VIII and John VIII) were willing to bow to western principles in order to obtain military and financial assistance. The majority of the

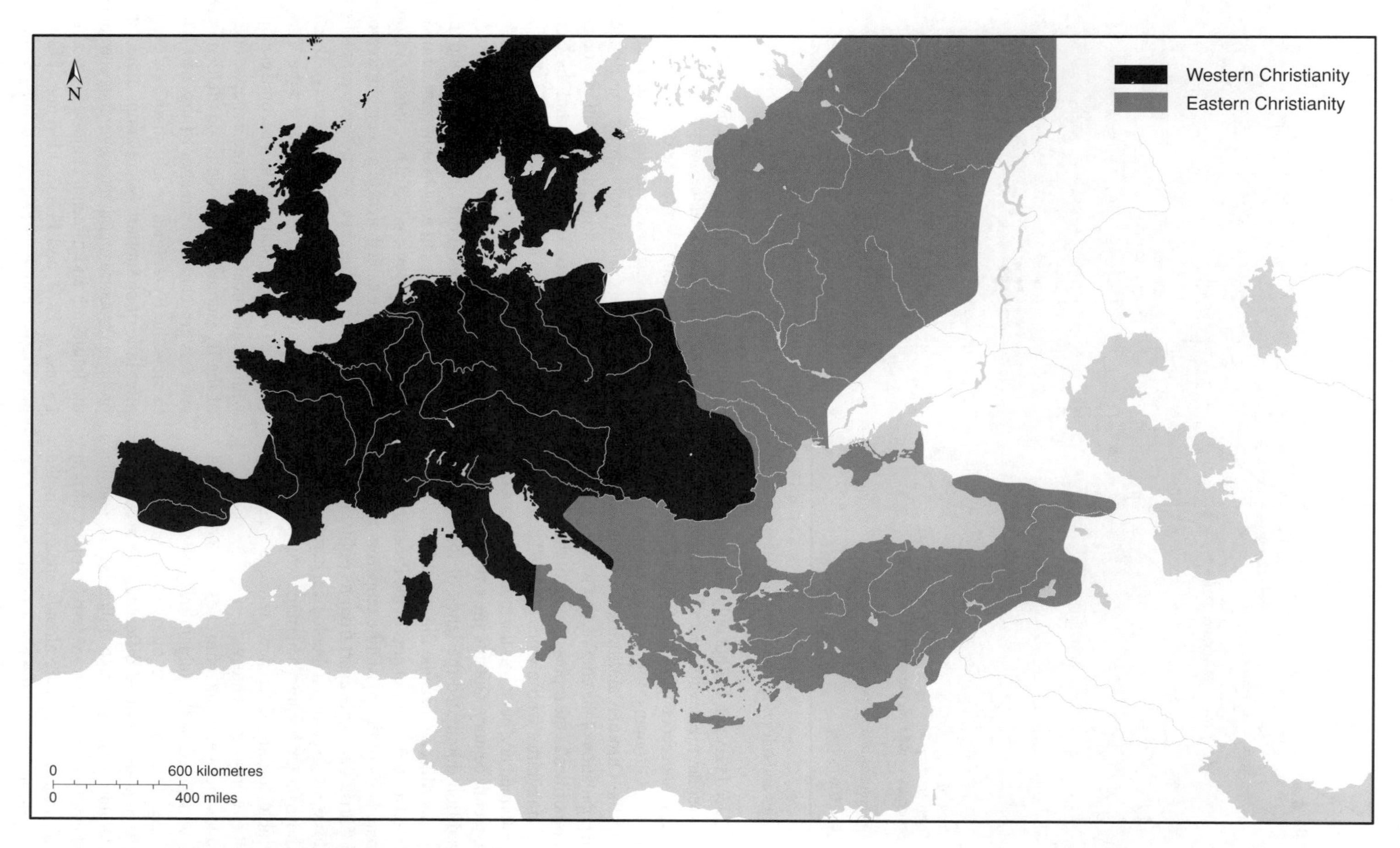

Map 12.3(a) The Roman and Constantinopolitan churches c. 1025.

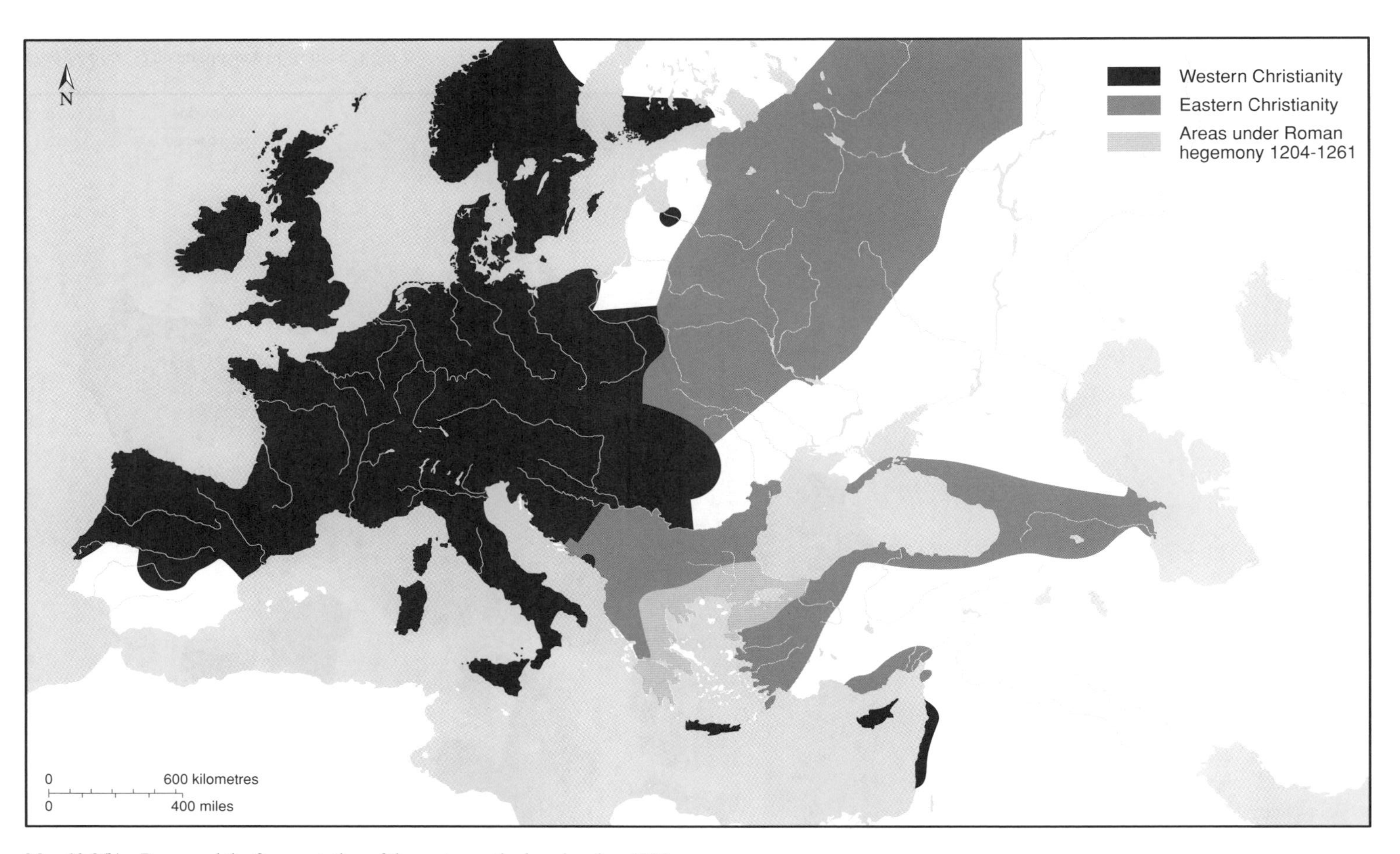

Map 12.3(b) Rome and the fragmentation of the eastern orthodox church c. 1220.

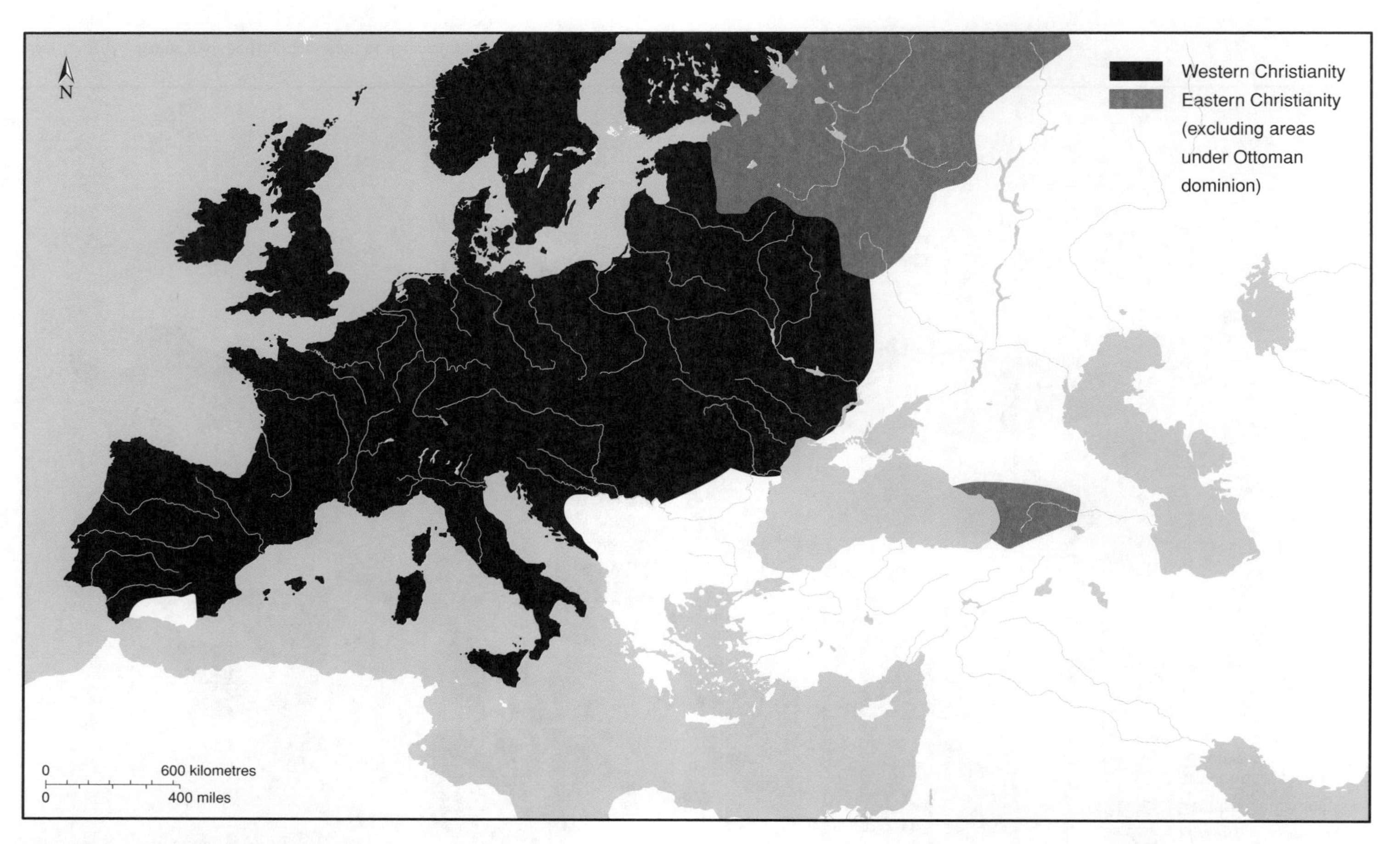

Map 12.3(c) The dominance of Rome c. 1470.

clergy and the population of the empire were utterly opposed to such concessions, however. No real progress was ever made, and no western help was ever forthcoming. The frustrations felt with regard to the attitudes and demands of the west are evident in the words of the *Megas Doux* (Grand Duke) Loukas Notaras, a leading minister of the last emperor, Constantine XI, who is reported to have said in 1451: 'better to see the turban of the Turk ruling in the City than the Latin mitre'. Notaras had been active in the negotiations with the western church, and in fact was executed two years later, with his family, on the orders of Mehmet II after the city had fallen. While not all Byzantines felt as Notaras did, there was a substantial degree of alienation, which his words make clear.

Paradoxically the decline of the secular state meant a concomitant rise in the authority and prestige of the church. By the end of the thirteenth century the patriarch held sway over a vastly larger territory than the emperor. Having finally taken the city, the Ottomans needed to find a means of ruling peacefully over the orthodox populations of the Balkan and Aegean regions. It was to the Byzantine Church and its clergy, and more particularly to the patriarchate in Constantinople, that the Ottomans turned. The differences in outlook and mentality which had become so apparent between east and west were part of an evolving cultural context, in which economic and social developments played an equally important role. The result was an increasing cultural isolationism in respect of Byzantine political attitudes, and an increasing rejection of alternatives coupled to a degree of fatalism with regard to the church and to Orthodoxy, which made planning for anything but the impossible – unconditional western support, a revival of Byzantine military might, and the turning back of the Ottoman advance – quite pointless. The rump of empire was left with little more than an ideology which no longer corresponded to reality. In day-to-day respects, Byzantine peasants, merchants and churchmen had to get on with life: when the end of the empire had become a reality, the Ottomans found that the 'Greeks', in spite of their ideological hostility, quickly settled down to a routine which was barely different from that to which they had been accustomed. The patriarch, and the orthodox church, were now the means through which Ottoman rule could be both tolerated and administered.

Chronological Overview

284–305	Diocletian and the tetrarchy
306–337	Constantine I (sole ruler from 324)
311	Edict of toleration issued by Galerius
312	Constantine's victory at the Milvian bridge
313	Edict of toleration issued by Constantine and Licinius
325	Council of Nicaea and condemnation of Arianism (first ecumenical council)
330	Consecration of Constantinople
337	Baptism and death of Constantine I
361–363	Julian the Apostate leads pagan reaction and attempts to limit the influence of Christianity
364	Jovian dies: empire divided between Valentinian I (West) and Valens (East)
378	Defeat and death of Valens at hands of Visigoths at battle of Adrianople
381	First Council of Constantinople (second ecumenical council): reaffirms rejection of Arianism; asserts right of Constantinopolitan patriarchate to take precedence after Rome
395	Death of Theodosius I and division of empire into eastern and western parts again
410	Visigoths sack Rome
413	Construction of Theodosian land walls of Constantinople
429–533	Vandal kingdom in North Africa
431	Council of Ephesus, rejection of Nestorianism (third ecumenical council)
449	Council of Ephesus ('robber council')
450/451	Council of Chalcedon, defeat of Monophysitism (fourth ecumenical council)
450	Attila and Huns defeated at Chalons
455	Sack of Rome by Vandals
476	Deposition of Romulus Augustulus by Master of Soldiers, Odoacer. End of the Western Roman empire
488	Ostrogoths under Theoderic march into Italy
493–526	Theoderic rules Ostrogothic kingdom of Italy
507–711	Kingdom of Visigoths in Spain
529	Justinian closes Academy of Athens; Codex Justinianus completed
532	'Nika' riot in Constantinople
533–534	Belisarius reconquers Africa (pacification completed in 540s); Pandects or Digest completed
534	Belisarius begins reconquest of Italy (war lasts until 553)
537	Dedication of the new church of the Holy Wisdom (Hagia Sophia) in Constantinople
540	Persian king Chosroes I takes Antioch in Syria
542+	Plague in the Byzantine world
550+	Avars establish hegemony over Slavs north of Black Sea and Danube
552	Narses defeats Totila and last Ostrogothic resistance in Italy
553	Second Council of Constantinople (fifth ecumenical council): Three Chapters condemned, concessions to Monophysites.
553+	Reconquest of South-east Spain from Visigoths
558	Treaty with Avars and agreement to pay 'subsidies'
562	'Fifty-year peace' signed with Persia
565–591	Wars with Persia
566 +	Slavs begin to infiltrate across Danube frontier; pressure on frontier fortresses from Avars
568+	Lombards driven westward from Danube, invade Italy
572	Lombards besiege Ravenna
577	Major invasion of Balkans led by Avars
584, 586	Avaro-Slav attacks on Thessalonica
591–602	Gradual success in pushing Avars back across Danube
602	Maurice overthrown, Phokas proclaimed emperor
603	War with Persia; situation in Balkans deteriorates
610	Phokas overthrown by Heraclius, son of exarch of Africa at Carthage
611–620s	Central and northern Balkans lost
614–619	Persians occupy Syria, Palestine and Egypt
622	Mohammed leaves Mecca for Medina (the 'Hijra')
622–627	Heraclius campaigns in east against Persians
626	Combined Avaro-Slav and Persian siege of Constantinople fails
626–628	Heraclius defeats Persian forces in east
629	Peace with Persia
634+	Arabs begin raids into Palestine
634–646	Arab conquest and occupation of Syria, Palestine, Mesopotamia, Egypt (636 – battle of Gabitha/Yarmuk)
638	*Ekthesis* of Heraclius: attempt to reconcile Monophysites and Chalcedonians
644+	Beginning of long-term raids and plundering expeditions against Byzantine Asia Minor
648	*Typos* of Constans II. Imperial enforcement of Monotheletism
649	Lateran synod in Rome; Maximus Confessor and Pope Martin reject imperial Monotheletism
653	Martin and Maximus arrested by exarch Theodore Calliopas and sent to Constantinople
655	Martin and Maximus found guilty of treason and exiled. Sea battle of Phoenix, Byzantines defeated
662	Constans II leads expedition through Balkans into Italy, takes up residence in Sicily
668	Constans assassinated; Mizizios proclaimed emperor in Sicily, but defeated by forces loyal to Constantine IV
674–678	Arab blockade and yearly sieges of Constantinople. First recorded use of 'liquid fire', to destroy Arab fleet
679–680	Arrival of Bulgars on Danube; defeat of Byzantine forces under Constantine IV
680–681	Third Council of Constantinople. Monotheletism rejected (sixth ecumenical council)
685–692	Truce between caliphate and Byzantium (Arab civil war)
691–692	Quinisext or Trullan council at Constantinople. Canons partly rejected by papacy
693	Byzantine defeat at Sebastoupolis
698	Carthage falls to Arabs; final loss of Africa
717–718	Siege of Constantinople; Leo, general of Anatolikon, seizes power and crowned as Leo III
726	Volcanic eruption on Thera/Santorini, leading Leo to adopt iconoclastic ideas
730	Patriarch Germanus resigns; probable beginning of public policy of iconoclasm
739/740	Leo and Constantine defeat Arab column at Akroinon

739	Earthquake hits Constantinople
741	Artabasdos, Leo's son-in-law, rebels against Constantine V and seizes Constantinople
743/4	Artabasdos defeated
746+	Plague in Constantinople
750	Abbasid revolution, removal of Umayyads from power, capital of Caliphate moved to Baghdad.
751/752	Constantine V begins publicly preaching in favour of iconoclasm
754	Iconoclast Council of Hiereia (claims to be seventh ecumenical council)
750s–770s	Constantine launches major expeditions against Bulgars and Arabs
786	Eirene attempts to hold seventh ecumenical council in Constantinople. Council abandoned due to opposition of iconoclast soldiers
787	Second Council of Nicaea (seventh ecumenical council). Iconoclasm rejected and condemned
792	Byzantines under Constantine VI defeated by Bulgars at Markellai
797	Constantine VI deposed by mother Irene; blinded and dies
800	Coronation of Charlemagne by pope in St Peter's, Rome
802	Irene deposed by chief finance minister Nikephoros (Nikephoros I)
811	Nikephoros defeated and killed by forces under Khan Krum after initially successful campaign in Bulgaria
813	Bulgar victories over Byzantine forces
815	Leo V convenes synod at Constantinople; iconoclasm reintroduced as official policy
821–823	Rebellion of Thomas 'the Slav'
824+	Beginning of Arab conquest of Sicily and of Crete
826	Theodore of Stoudion dies
838	Arab invasion of Asia Minor; siege and sack of Amorion
843	Council held in Constantinople to reaffirm acts of seventh ecumenical council. Empress regent Theodora and chief courtiers restore images; end of official iconoclasm
850s	Missionary activity in Bulgaria
860	Rus' (Viking) attack on Constantinople; mission to Khazars of St Cyril
863	Major Byzantine victory over Arabs at Poson in Anatolia
864	Conversion of Bulgar Khan and leaders
869–870	Council convoked by Basil I at Constantinople to settle Photian schism: Photios deposed, Ignatios, his predecessor, reinstated. Bulgaria placed under Constantinopolitan ecclesiastical jurisdiction (contrary to papal demands)
879–880	Acts of council of 869–870 annulled, Photios reinstated. Recognised in Rome, schism ended
900+	Final loss of Sicily; Bulgar expansionism under Tsar Symeon; war with Byzantines
917	Bulgar victory at river Achelo
920	Local council of Constantinopolitan church held in Constantinople to settle schism caused by the fourth marriage of Leo VI ('Tetragamy'), reconciling Nicholas I and his supporters, who condemned the marriage, with the Patriarch Euthymios, who had condoned it.
922	Peace with Bulgars
923–944	Byzantine conquests and eastward expansion led by general John Kourkouas
960–961	Recovery of Crete under general Nikephoros Phokas
963+	Major Byzantine offensives in east, creation of new frontier regions
965	Nikephoros II captures Tarsus and Cyprus
969	Nikephoros II captures Aleppo and Antioch
969–976	Reign of John I Tzimiskes. Continuation of eastern expansion; defeat of Bulgars with help of Rus' allies under Svyatoslav; defeat of Rus' at Silistra (971)
975	John I invades Palestine, takes several towns and fortresses, but withdraws
985+	Bulgar resistance in western Balkans leads to growth of first Bulgarian empire under Tsar Samuel
989	Conversion of Vladimir of Kiev to Christianity
990–1019	Basil II crushes Bulgar resistance; Bulgaria re-incorporated into empire, Danube new frontier in North
1022	Armenian territories annexed to empire
1034–1041	Michael IV takes first steps in debasement of gold currency
1054	Schism with papacy
1055	Seljuks take Baghdad; Norman power in southern Italy expanding
1070+	Major Pecheneg advances into Balkans; civil war within empire
1071	Romanos IV defeated and captured at Mantzikert by Seljuks; beginning of Turk occupation of central Anatolia; Normans take Bari
1081	Alexios Komnenos rebels and defeats Nikephoros III and is crowned emperor
1081–1085	Norman invasion of western Balkan provinces
1082–1084	Commercial privileges granted to Venice
1091	Seljuk–Pecheneg siege of Constantinople; defeat of Pechenegs
1092	Coinage reform carried out by Alexios I
1094	Synod held at Blachernae to decide the issue of Leo of Chalcedon, a hard line opponent of the church's decision to melt down ecclesiastical treasures to aid the imperial treasury. Deposed by the permanent synod in 1086, this council reinstated him after he was reconciled to the official church position.
1097+	First Crusade; Seljuks defeated
1098/1099	Jerusalem captured; Latin principalities and Kingdom of Jerusalem established in Palestine and Syria
1108	Alexios defeats Normans under Bohemund
1111	Commercial privileges granted to Pisa
1130s	Alliance with German empire against Normans of south Italy
1138–1142	Byzantine confrontation with Crusader principality of Antioch
1143–1180	Manuel I Komnenos: pro-western politics become major factor in Byzantine foreign policy
1146–1148	Second Crusade
1153	Treaty of Constanz between Frederick I (Barbarossa) and papacy against Byzantium
1155–1157	Successful imperial campaign in Italy; commercial and political negotiations with Genoa
1156–1157	Council of Constantinople: teachings of the Patriarch elect, Panteugenos, condemned
1158–1159	Imperial forces march against Antioch
1160+	Successful imperial political involvement in Italy against German imperial interests; Manuel defeats Hungarians and Serbs in Balkans and reaffirms imperial pre-eminence
1166–1167	Local Constantinopolitan council meets to discuss Christological issues arising from discussions with western theologians

1169–1170	Commercial treaties with Pisa and Genoa
1171+	Byzantine–Venetian hostilities increase
1175–1176	Manuel plans crusade in east
1176	Defeat of imperial forces under Manuel by Seljuk Sultan Kilidj Aslan at Myriokephalon
1180	Manuel dies; strong anti-western sentiments in Constantinople
1182	Massacre of westerners, especially Italian merchants and their dependants, in Constantinople
1185	Normans sack Thessalonica; Andronikos Komnenos deposed
1186+	Rebellion in Bulgaria, defeat of local Byzantine troops, establishment of Second Bulgarian Empire
1187	Defeat of Crusader forces at battle of Horns of Hattin; Jerusalem retaken by Saladin
1192	Treaties with Genoa and Pisa
1203–1204	Fourth Crusade, with Venetian financial and naval support, marches against Constantinople. After the capture and sack of the city in 1204, the Latin empire is established, along with several principalities and other territories under Latin or Venetian rule
1204–1205	Successor states in Nicaea, Epirus and Trebizond established
1205	Latin emperor Baldwin I defeated by Bulgars
1259	Michael VIII succeeds to throne in empire of Nicaea; Nicaean army defeats combined Latin and Epirot army at battle of Pelagonia. Fortress-town of Mistra handed over to Byzantines (Nicaea)
1261	During absence of main Latin army Nicaean forces enter and seize Constantinople
1265	Pope invites Charles of Anjou, brother of Louis IX of France, to support him militarily against Manfred of Sicily and the Hohenstaufen power in Italy
1266	Manfred of Sicily defeated at battle of Benevento by Charles of Anjou; Angevin plans, supported by papacy, evolve to invade and conquer the Byzantine empire
1274	Gregory X summons second Council of Lyons; representatives of Byzantine Church present; union of the churches agreed, under threat of papally-approved invasion led by Charles of Anjou. Union not accepted in the Byzantine empire
1280–1337	Ottomans take nearly all remaining Byzantine possessions in Asia Minor (Ephesus 1328, Brusa 1326)
1282	'Sicilian vespers'; Death of Charles of Anjou and end of his plans to invade Byzantium
1285	Council of Constantinople ('second synod of Blachernae'): discussed and rejected pro-western interpretation of the Trinity as enunciated by the Patriarch John XI Bekkos. Also rejected decisions of Council of Lyons (1274)
1303	Andronikos II hires Catalan company as mercenary troop
1321–1328	Civil war between Andronikos II and Andronikos III
1329	Turks take Nicaea
1331–1355	Štefan Dušan Kral (King) of Serbia
1337	Turks take Nicomedia
1340+	Height of Serbian empire under Štefan Dušan
1341	Synod in Constantinople to discuss the issues raised by the traditionalist orthodox views (defended by Barlaam of Calabria) and those who supported Hesychasm (Gregory Palamas). The hesychast faction won a clear victory and Barlaam left Constantinople
1341–1347	Civil war between John V (supported by Serbs) and John VI Kantakouzenos (with Turkish help)
1341–1350	Commune hostile to aristocracy rules Thessalonica
1346	Štefan Dušan crowned emperor of the Serbs and Greeks
1347	Black Death reaches Constantinople; local council at Constantinople confirms decisions of council held in 1341
1351	Synod in Constantinople approves Palamism (hesychasm) in detailed discussion of its theological arguments
1354–5	Civil war between John VI and John V (backed by Genoa). Ottomans employed as allies establish themselves in Gallipoli and Thrace
1355	John VI abdicates and enters a monastery. John V proposes union of churches to pope
1365	Ottomans take Adrianople, which becomes their capital
1366	John V visits Hungary seeking support against Ottoman threat
1371	Ottomans defeat Serbs in battle
1373	John V forced to submit to Ottoman Sultan Murat I; John's son Andronikos IV rebels but is defeated
1376–1379	Civil war in Byzantium: Andronikos IV rebels against John V, who is supported by his younger son Manuel
1379	John V restored with Turkish and Venetian support
1388	Bulgarians defeated by Ottomans
1389	Battle of Kosovo: Serbs forced to withdraw by Ottomans, Serb empire ends. Accession of Bayezit I
1393	Turks capture Thessaly. Battle of Trnovo, Bulgarian empire destroyed
1396	Sigismund of Hungary organises crusade against Ottoman threat, but is utterly defeated at Nicopolis
1397–1402	Bayezit I besieges Constantinople, but army withdrawn when Turks defeated by Timur at battle of Ankyra (1402)
1399–1402	Manuel II tours Europe to elicit military and financial support (December 1400, guest of Henry IV in London)
1422	Murat II lays siege to Constantinople
1423	Governor of Thessalonica (a brother of John VIII) hands the city over to the Venetians
1430	Thessalonica retaken by Ottomans; populace and Venetian garrison massacred
1439	Council of Ferrara moves to Florence; union of churches formally agreed by emperor John VIII, present at Council
1444	Hungarians and western Crusaders, led by Vladislav of Hungary and Poland, defeated at battle of Varna. Vladislav killed in battle
1448	John VIII dies; his brother Constantine, Despot of the Morea, succeeds as Constantine XI, with coronation at Mistra in 1449
1451	Mehmet II becomes Sultan
1452	Union of churches proclaimed at Constantinople
1453	Mehmet II lays siege to Constantinople. May 29: Janissaries break through defences and permit main Ottoman army to enter city. Constantine XI, the last emperor, died in the fighting, and his body was never identified.
1460	Mistra falls to the Turks
1461	Trebizond falls to the Turks

Glossary of Byzantine and Technical Terms

Annona Military rations issued from taxation collected in kind; Gk. *synonê*

Apotheke A state depository for various goods and materials; in the 7th–9th centuries the warehouse, and the district to which it pertained, under the control of a *kommerkiarios*

Archontes Holders of imperial titles or offices; provincial landholding elite dominating towns

Arianism Christian tendency which viewed Christ as man alone. Condemned as heretical at council of Nicaea, 325

Augustus Senior emperor of a group, either a college of rulers (e.g. the tetrarchy) or within a single family

Autokrator Greek equivalent of the Latin *imperator*, emperor, used especially after the 7th century to emphasise the emperor's autonomous and God-granted rule

Basileus Formal title of the Byzantine emperor from the 7th century

Bogomilism Dualist neo-Paulician/neo-Manichaean movement which developed in Bulgaria under a certain priest Bogomil in the mid-10th century, probably deriving from eastern Anatolian roots. By late 11th century had spread across Balkans and into Asia Minor, and was an important influence on later Cathar beliefs in the west.

Caesar During the tetrarchy, a subordinate ruler under the authority of the *Augustus*; thereafter used of a junior emperor, and from the 7th century also the highest court dignity, normally limited to the emperor's sons, but exceptionally granted to another.

Capitatio-iugatio A formula relating land to labour power for the assessment of taxation, 4th–7th centuries

Cenobitic From Greek words *koinos bios*, 'communal/common life', used to describe monastic communities in which prayer and meals are shared

Civitas Gk. *Polis*, 'city', understood as a self-governing unit with its own territory and administration; the basic fiscal administrative district into the 7th century

Codex Justinianus Codification of Roman law produced at the beginning of the reign of Justinian I, and the basis for all later Byzantine law

Comitatenses Soldiers/units of the field armies under their *magistri militum*, 4th–7th centuries (cf. *limitanei*)

Curia/curiales Town council and councillors, governing body of a city

Cursus publicus The public postal, transport and relay system

Despotes High imperial title in the later Byzantine period, generally preserved for members of the ruling dynasty; or designation for the ruler of a semi-independent imperial territory

Diocese Lat. *dioecesa*, Gk. *Dioikêsis*, an administrative unit consisting of several provinces; from the 4th century the episcopal administrative unit of the church

Dioiketes Fiscal administrator responsible for the land-tax, usually in a single diocese, from the 7th century

Diophysitism Belief in two natures (*physeis*) in the person of Christ, the creed adopted and defined at the Council of Chelcedon in 451, and after this time the official doctrine of the orthodox church

Domestikos Senior official in army, state and church, although junior *domestikoi* also existed.

Donatism Rigorist Christian sect chiefly in North Africa, which challenged the validity of sacraments issued by those who compromised with the pre-Christian imperial administration. Condemned on several occasions from the 4th century, it appears to have survived into the 7th century.

Dromos Greek term for *cursus publicus*

Dux/doux In later Roman period, commander of a military unit; commander of a unit of *limitanei*, or garrison troops; in the middle and later Byzantine period the title *doux* reintroduced as a high military rank

Eremetic Solitary lifestyle of the hermit

Exarch The military governors at Ravenna and Carthage

Excubitores Small palace bodyguard recruited from Isaurian mountain people by the emperor Leo I. During the 7th century they became a show troop, but the unit was revived as a larger active elite regiment under Constantine V in the 760s, as the *exkoubita*. It disappears during the later 11th century

Follis Low value copper coin worth 40 *nummi*: there were 288 to the gold *solidus* or *nomisma*

Genikon (sekreton) The general treasury and main fiscal department of government after the 7th century

Hesychasm Late Byzantine mystical approach to prayer and meditation, especially popular in monastic circles

Hexagram Silver coin introduced by Heraclius, lit. 'six grams', twelve to a *nomisma*. Although issued in large quanities under Heraclius and Constans II, its use dwindled until production ceased in the early 8th century

Homoian Modified Arian belief which placed less emphasis on Christ as man alone and stated that, while Father and Son were alike, they were not of the same substance (to be differentiated therefore from the homoiousians who claimed that they were of like substance, and from the homoousians who argued that they were of the same substance)

Hyperpyron The highest value gold coin from the reform of Alexios I Komnenos

Iconoclasm Rejection of the honouring of sacred images, as a form of idolatry. Condemned as a heresy at the council of Nicaea in 787, re-established by Leo

V as imperial policy in 815; condemned again in 843

Kastron — 'Fortress', but after the 7th century also used to mean 'town' or 'city'

Kastrophylax — 'Castle guardian', governor of a fortress

Katepano — Military officer in command of independent unit and/or district (8th–12th centuries); imperial provincial/regional governor (after the 13th century)

Kephale — Provincial military and civil governor, in charge of a *katepanikion*, a military-administrative district, in the 14th–15th centuries

Kleisoura — Small frontier command; district along/behind the frontier (esp. later 8th–10th centuries)

Kommerkiarioi — Fiscal officials responsible for state-supervised commerce and the taxes thereon. During the 7th and 8th centuries had a much expanded role in the fiscal system and the supplying of the armies; from the middle of the 8th century reverted to chiefly commercial functions.

Limitanei — Provincial garrison troops in the later Roman period

Logariastes — Chief fiscal officer following the reforms of Alexius I

Logothetes — Fiscal official, lit. 'accountant'; from the 7th century all the main fiscal bureaux were placed under such officials, who were often very high-ranking

Logothetes ton agelon — 'Logothete of the herds', in charge of imperial stud ranches in the provinces of Asia and Phrygia, and successor of the older *praepositus gregum*

Magister militum — Divisional military commander, replaced by the *strategos* of the period after c. 660

Magister Officiorum — 'Master of offices', leading civil minister and close associate of the emperors in the later Roman period

Metochion — A subordinate or daughter monastery under the authority of a larger or more powerful monastic house

Miliaresion — Lat. *milliarensis*, a silver coin worth one twelfth of a *solidus/nomisma*. Originally struck at 72 to the pound, from the 7th–11th centuries used of the basic silver coin, struck at varying rates from 144 to 108 to the pound, especially of the reformed silver coin introduced under Leo III. Production ceased under Alexius I, but the term continues in use as a money of account.

Monoenergism — A compromise formula developed by the patriarch Sergios, by which the issue of the two natures was made secondary to the notion that they were united in a single divine energy. Rejected by all parties within a few years of its being proposed, and condemned as heretical at the sixth ecumenical council in 681.

Monophysitism — Doctrine of the 'single nature': Christian tendency rejecting the two natures, both human and divine, of Christ, believing instead that the divine subsumed the human nature after the incarnation. Condemned as heretical at council of Chalcedon in 451, but remained the majority creed in large parts of Syria and Egypt, and of the Syrian and Coptic churches today.

Monotheletism — A second attempt at compromise proposed by Sergios and supported by the emperor Heraclius, by which the key issue was acceptance of the notion of a single divine will, within which natures and energy were subsumed. Imposed during the reign of Constans II, but condemned and rejected at the council of 681.

Nestorianism — 5th-century Christian heresy in which the divine and human aspects of Christ were seen not as unified in a single person, but operating in conjunction. Nestorians were accused of teaching two persons in Christ, God and man, and thus two distinct sons, human and divine. Condemned in 431 at the Council of Ephesus, the Nestorians left the empire and established their own church in Persia in 486. Nestorianism established a firm foothold in Persia and spread across northern India and central Asia as far as China. It survives today, especially in northern Iraq, as the Assyrian orthodox church.

Nomisma — Lat. *solidus*, the gold coin introduced by Constantine I which remained the basis for the Byzantine precious metal coinage until the Latin conquest in 1204. Weighing 4.5g, it was reckoned at 24 *keratia*, a unit of account (carat), and its fractions were 12 silver *hexagrams* or *milliaresia* and 288 copper *folleis*. From the middle of the 11th century increasingly depreciated, it was reformed by Alexios I, and more commonly known thereafter as the *nomisma hyperpyron* or simply *hyperpyron*.

Partitio Romaniae — Agreement to partition the Byzantine empire between Venice and Crusaders, agreed before the sack of 1204

Patriarch/ate — The five major sees of the Christian Church and their bishops, at Rome, Constantinople, Antioch, Jerusalem and Alexandria. Constantinople was a 4th-century addition following the establishment there of a new imperial capital under Constantine I.

Paulicians — A dualist sect of the 7th–9th centuries. During the mid-9th century they took over much of eastern Anatolia and fought the empire with the support of the Caliphate. They were crushed by Basil I

Polis — See *Civitas*

Praetorian Prefecture — The largest administrative unit of the empire from the time of Constantine I, under a praetorian prefect (originally a commander of the praetorian guard). Each prefecture was divided into dioceses, then provinces, and had its own fiscal administrative and judicial structure.

Praktikon — Document drawn up by fiscal officials listing obligations of tenants on an estate or estates

Prokathemenos — Town/fortress governor of the Komnenian period

Pronoia — Attribution of fiscal revenues, usually to a soldier in return for military service. Appears first on a limited basis in the 12th century; eventually included lifelong and heritable grants

Protonotarios — Chief fiscal administrator of a theme from c. 820 to mid-11th century

Res privata — Imperial treasury, originating in emperor's private finances. Subsumed during the 7th century into the department of imperial estates

Sacred largesses — Government fiscal department originating in the imperial household, responsible for bullion and coinage until the 7th century

Sakellarios — Senior fiscal officer with oversight over other fiscal departments after the 7th century. Originally in charge of emperor's personal treasury or 'purse' (*sacellum*)

Scholae — In the period from Constantine I until the later 5th century a crack cavalry unit; by the later 5th a show force. The units were reformed and became once more elite regiments under Constantine V, forming until the 11th century the core of the imperial field armies.

Spectabilis — Second-rank senatorial grade

Strategos — A general; in Byzantine times usually the governor of a military district or *thema*, and commander of its soldiers

Stratiotikon Logothesion — Fiscal department which dealt with recruitment, Muster-rolls and military pay from the 7th century

Stylite — An ascetic hermit living on top of a column, such as Sts Symeon and Daniel

Tagmata — (1) Elite field units recruited by Constantine V. They formed the core of imperial field armies until the 11th century; (2) any full-time mercenary unit – used especially of foreign mercenary troops in the 10th–12th centuries

Territory — Lat. *territorium*, the region pertaining to and administered from a city

Tetrarchy — Lit., 'rule of four', the system invented by Diocletian to provide for better administrative and military governance of the empire. It broke down, however, over the period from 305–310

Thema — A 'theme', from the middle of the 7th century the district across which soldiers were quartered, and from which they were recruited; an administrative unit; the army based in such a region

Varangians — Mercenary unit first recruited during the reign of Basil II, consisting of Russian and Scandinavian adventurers and mercenaries

Bibliography

Atlases and Historical Geography

Historical Atlases

C.F. Beckingham, ed., *Atlas of the Arab World and the Middle East* (London/NY 1960)

F.W. Carter, ed., *An Historical Geography of the Balkans* (London-NY-San Francisco 1977)

H.L. Cooke, H.W. Hazard, *Atlas of Islamic History* (Princeton 1954)

H. Kennedy, *An Historical Atlas of Islam*, 2nd revised edn. (Leiden 2002)

A. Mackay et al., *Atlas of Medieval Europe* (London 1997)

C. McEvedy, *The New Penguin Atlas of Medieval History* (Harmondsworth 1992)

N.J.G. Pounds, *An Historical Geography of Europe, 450 BC–AD 1330* (Cambridge 1973)

W.M. Ramsay, *The Historical Geography of Asia Minor*, Royal Geographical Society, Supplementary Papers IV (London 1890/ Amsterdam 1962)

R. Roolvink et al., *Historical Atlas of the Muslim Peoples* (Amsterdam 1957)

British Admiralty, Geographical Handbook Series

Greece, I: *Physical Geography, History, Administration and Peoples*, Naval Intelligence Division, Geographical Handbook Series, B.R. 516 (London 1944)

Greece, II: *Economic Geography, Ports and Communications*, Naval Intelligence Division, Geographical Handbook Series, B.R. 516A (London 1944)

Greece, III: *Regional Geography*, Naval Intelligence Division, Geographical Handbook Series, B.R. 516B (London 1945)

Turkey, I, Naval Intelligence Division, Geographical Handbook Series, B.R. 507 (London 1942)

Turkey, II, Naval Intelligence Division, Geographical Handbook Series, B.R. 507A (London 1943)

Tabula Imperii Byzantini

J. Koder, F. Hild, *Tabula Imperii Byzantini* 1: *Hellas und Thessalia* (Denkschr. d. Österr. Akad. d Wiss., phil.-hist. Kl. 125. Vienna 1976)

F.Hild, M. Restlé, *Tabula Imperii Byzantini* 2: *Kappadokien (Kappadokia, Charsianon, Sebasteia und Lykandos)* (Denkschr. d. Österr. Akad. d Wiss., phil.-hist. Kl. 149. Vienna 1981)

P. Soustal, with J. Koder, *Tabula Imperii Byzantini* 3: *Nikopolis und Kephallenia* (Denkschr. d. Österr. Akad. d Wiss., phil.-hist. Kl. 150. Vienna 1981)

K. Belke (with M. Restlé), *Tabula Imperii Byzantini* 4: *Galatien und Lykaonien* (Denkschr. d. Österr. Akad. d Wiss., phil.-hist. Kl. 172. Vienna 1984)

F. Hild, H. Hellenkamper, *Tabula Imperii Byzantini* 5, 1/2: *Kilikien und Isaurien* (Denkschr. d. Österr. Akad. d Wiss., phil.-hist. Kl. 215. Vienna 1990)

P. Soustal, *Tabula Imperii Byzantini* 6: *Thrakien (Thrakê, Rodopê und Haimimontos)* (Denkschr. d. Österr. Akad. d Wiss., phil.-hist. Kl. 221. Vienna 1991)

K. Belke, N. Mersich, *Tabula Imperii Byzantini* 7: *Phrygien und Pisidien* (Denkschr. d. Österr. Akad. d Wiss., phil.-hist. Kl. 211. Vienna 1990)

H. Hellenkemper, F. Hild, *Tabula Imperii Byzantini* 8: *Lykien und Pamphylien* (Denkschr. d. Österr. Akad. d Wiss., phil.-hist. Kl. 320. Vienna 2004)

K. Belke, *Tabula Imperii Byzantini* 9: *Paphlagonien und Honorias* (Denkschr. d. Österr. Akad. d Wiss., phil.-hist. Kl. 249. Vienna 1996)

Other Literature

M. Cary, *The Geographic Background of Greek and Roman History* (Oxford 1949)

M.F. Hendy, *Studies in the Byzantine Monetary Economy c.300–1450* (Cambridge 1985)

J. Koder, *Der Lebensraum der Byzantiner. Historisch-geographischer Abriß ihres mittelalterlichen Staates im östlichen Mittelmeerraum* (Graz-Wien-Köln 1984)

E. Malamut, *Sur la route des saints byzantins* (Paris 1993)

A. Philippson, *Das byzantinische Reich als geographische Erscheinung* (Leiden 1939)

J.C. Russell, *Late Ancient and Medieval Population* (Philadelphia 1958)

J.L. Teall, 'The grain supply of the Byzantine empire, 330–1025', *DOP* 13 (1959) 87–139

J.M. Wagstaff, *The Evolution of the Middle Eastern Landscapes* (Canterbury 1984)

M. Whittow, 'The strategic geography of the Near east', in *idem*, *The Making of Orthodox Byzantium, 600–1025* (London 1996) 15–37

General Surveys

Norman H. Baynes and H. St.L.B. Moss, eds., *Byzantium: an introduction to east Roman civilization* (Oxford 1969)

G. Bowersock, P. Brown and O. Grabar, *Late Antiquity: a guide to the postclassical world* (Cambridge, Mass.- London 1999)

Robert Browning, *The Byzantine Empire* (Washington D.C. 1980/1992)

Guglielmo Cavallo, ed., *The Byzantines* (Chicago 1997)

Deno J. Geanakoplos, *Byzantium. Church, Society, and Civilization Seen through Contemporary Eyes* (Chicago 1984) (a useful collection of sources in translation)

T. Gregory, *A History of Byzantium* (Malden, MA.-Oxford 2005)

J.F. Haldon, *Byzantium: a history* (Stroud 2000)

J.F. Haldon, *The Byzantine Wars* (Stroud 2001)

Alexander Kazhdan with Giles Constable, *People and Power in Byzantium: an Introduction to Modern Byzantine Studies* (Washington D.C. 1982)

Cyril Mango, *Byzantium: the empire of New Rome* (London 1980/1994)

Cyril Mango, ed., *The Oxford History of Byzantium* (Oxford 2002)

George Ostrogorsky, *A History of the Byzantine State*, trans. J. Hussey (Oxford 1968)

Warren Treadgold, *A History of the Byzantine State and Society* (Stanford 1997)

Political and General History

M. Angold, *A Byzantine Government in Exile: government and society under the Laskarids of Nicaea (1204–1261)* (Oxford 1975)

M. Angold, *The Byzantine Empire 1025–1204. A political history* (London 1984/1997)

P.S. Barnwell, *Emperors, Prefects and Kings. The Roman West 395–565* (London 1992)

P.R.L. Brown, *The World of Late Antiquity* (London 1971)

J.B. Bury, *A History of the Later Roman Empire from Arcadius to Irene (395 A.D. to 800 A.D.)* (London 1889/Amsterdam 1966)

The Cambridge History of Islam, vol. 1: *The Central Islamic Lands*; vol. 2: *The Further Islamic Lands, Islamic Society and Civilisation*, P.M. Holt, A.K.S. Lambton, B. Lewis, eds. (Cambridge 1970)

The Cambridge Medieval History, IV: *The Byzantine Empire*, 2 parts, revised edn. J.M. Hussey (Cambridge 1966)

Averil Cameron, *The Mediterranean World in Late Antiquity A.D. 395–600* (London 1993)

Averil Cameron, Peter Garnsey, eds., *The Cambridge Ancient History*, XIII: *The late empire, A.D. 337–425* (Cambridge 1998)

R. Collins, *Early Medieval Europe 300–1000* (London 1991)

T. Cornell, J. Matthews, *Atlas of the Roman World* (Oxford 1982)

Fred. M. Donner, *The Early Arabic Conquests* (Princeton 1981)

S. Franklin, J. Shepard, *The Emergence of Rus, 750–1200* (London 1996)

J.F. Haldon, *Byzantium in the Seventh Century: the transformation of a culture* (Cambridge 1997)

J.F. Haldon, *Warfare, State and Society in the Byzantine World, 565–1204* (London 1999)

J. Harris, *Byzantium and the Crusades* (London 2003)

A. Harvey, *Economic Expansion in the Byzantine Empire 900–1200* (Cambridge 1989)

J. Herrin, *The Formation of Christendom* (Princeton 1987)

A.H.M. Jones, *The Later Roman Empire: a social, economic and administrative survey*, 3 vols. and maps (Oxford 1964)

W.E. Kaegi, *Byzantium and the Early Islamic Conquests* (Cambridge 1992)

H. Kennedy, *The Early Abbasid Caliphate: a political history* (London 1981)

H. Kennedy, *The Prophet and the Age of the Caliphates: the Islamic Near east from the sixth to the eleventh century* (London 1986)

R.-J. Lilie, *Byzantium and the Crusader States 1096–1204*, trans. J.C. Morris, J.E. Ridings (Oxford 1993)

P. Magdalino, *The Empire of Manuel I Komnenos, 1143–1180* (Cambridge 1993)

C. Mango, *Byzantium: the empire of New Rome* (London 1980/1994)

D.M. Nicol, *The Last Centuries of Byzantium, 1261–1453* (London 1972)

D.M. Nicol, *The Despotate of Epiros, 1267–1479* (Cambridge 1984)

D.M. Nicol, *Byzantium and Venice* (Cambridge 1988)

D. Obolensky, *The Byzantine Commonwealth. Eastern Europe 500–1453* (London 1971)

G. Ostrogorsky, *A History of the Byzantine State*, Eng. trans. J. Hussey (Oxford 1968)

S. Runciman, *A History of the First Bulgarian Empire* (London 1930)

M.A. Shaban, *Islamic History, AD 600–750 (AH 132). A New Interpretation* (Cambridge 1971)

B. Spuler, *The Muslim world, I: the age of the Caliphs; II: the Mongol period*, trans. F.R.C. Bagley (Leiden 1960)

E.A. Thompson, *Romans and Barbarians: the decline of the western empire* (Madison, Wisconsin 1982)

A. Toynbee, *Constantine Porphyrogenitus and his World* (London 1973)

J. Vogt, *The Decline of Rome: the metamorphosis of Ancient Civilization* (London 1967)

Sp. Vryonis, jr., *The Decline of Medieval Hellenism in Asia Minor and the Process of Islamization from the Eleventh through the Fifteenth Century* (Berkeley-Los Angeles-London 1971)

M. Whittow, *The Making of orthodox Byzantium, 600–1025* (London 1996)

C.J. Wickham, *Early Medieval Italy* (London 1981)

State Structures and Administration

M. Angold, *A Byzantine Government in Exile: government and society under the Laskarids of Nicaea (1204–1261)* (Oxford 1975)

M.C. Bartusis, *The Late Byzantine Army. Arms and society, 1204–1453* (Philadelphia 1992)

J.W. Birkenmeier, *The Development of the Komnenian Army: 1081–1180* (Leiden-Boston-Köln 2002)

Averil Cameron, ed., *States, Resources and Armies: papers of the third workshop on late Antiquity and early Islam* (Princeton 1995)

Karen R. Dixon, Pat Southern, *The Late Roman Army* (London 1996)

J.F. Haldon, *Byzantium in the Seventh Century: the transformation of a culture* (Cambridge 1997)

J.F. Haldon, *Warfare, State and Society in the Byzantine World, 565–1204* (London 1999)

A. Harvey, *Economic Expansion in the Byzantine Empire 900–1200* (Cambridge 1989)

M.F. Hendy, *Studies in the Byzantine Monetary Economy, c.300–1450* (Cambridge 1985)

A.H.M. Jones, *The Later Roman empire, 284–602: a social, economic and administrative survey*, 3 vols. and maps (Oxford 1964)

M. Kaplan, *Les hommes et la terre à Byzance du VIe au XIe siècle. Propriété et exploitation du sol* (Paris 1992)

A. Toynbee, *Constantine Porphyrogenitus and his World* (London 1973)

Urban and Rural History, Economy and Society

M. Angold, *The Byzantine Empire 1025–1204: a political history* (London 1984)

M. Angold, 'The Shaping of the Medieval Byzantine "City"', *Byzantinische Forschungen* 10 (1985) 1–37

M. Balard, 'The Genoese in the Aegean', in B. Arbel, B. Hamilton, D. Jacoby, eds., *Latins and Greeks in the eastern Mediterranean after 1204* (London 1989) 158–74

G.P. Brogiolo, B. Ward-Perkins, eds., *The Idea and Ideal of the Town between Late Antiquity and the Early Middle Ages* (Leiden 1999)

T.S. Brown, *Gentlemen and Officers. Imperial administration and aristocratic power in Byzantine Italy, A.D. 554–80* (Rome 1984)

G. Duby, *The Early Growth of the European Economy. Warriors and peasants from the seventh to the twelfth century* (London 1974)

A. Dunn, 'The transformation from *polis* to *kastron* in the Balkans (III–VII cc.): general and regional perspectives', *Byzantine and Modern Greek Studies* 18 (1994) 60–80

C. Foss, *Byzantine and Turkish Sardis* (Cambridge, Mass.-London 1976)

C. Foss, 'Archaeology and the "Twenty Cities" of Byzantine Asia', *American Journal of Archaeology* 81 (1977) 469–86

C. Foss, 'Late Antique and Byzantine Ankara', *Dumbarton Oaks Papers* 31 (1977) 29–87
C. Foss, *Ephesus After Antiquity: a Late Antique, Byzantine and Turkish city* (Cambridge 1979)
C. Foss, D. Winfield, *Byzantine Fortifications. An introduction* (Pretoria 1986)
J.F. Haldon, *Byzantium in the Seventh Century: the transformation of a culture* (Cambridge 1997)
J.F. Haldon, W. Brandes, 'Towns, tax and transformation: state, cities and their hinterlands in the east Roman world, c. 500–800', in N. Gauthier, ed., *Towns and their Hinterlands between Late Antiquity and the Early Middle Ages* (Leiden, 1999)
A. Harvey, *Economic Expansion in the Byzantine Empire 900–1200* (Cambridge 1989)
R. Hohlfelder, ed., *City, Town and Countryside in the Early Byzantine Era* (New York 1982)
A.H.M. Jones, *The Later Roman Empire 284–602: a social economic and administrative survey* (Oxford 1964) 716–19
A.H.M. Jones, *The Greek City from Alexander to Justinian* (Oxford 1967)
A. Laiou, 'The Byzantine aristocracy in the Palaeologan period', *Viator* 4 (1973) 131–151
A. Laiou, 'The Byzantine Economy in the Mediterranean Trade System: Thirteenth-Fifteenth Centuries', *DOP* 34–35 (1980–1981) 177–222
A. Laiou et al., eds., *The Economic History of Byzantium from the Seventh through the Fifteenth Century* (Washington D.C. 2002)
P. Magdalino, *The Empire of Manuel I Komnenos, 1143–1180* (Cambridge 1993)
C. Mango, 'The development of Constantinople as an urban centre', in *Seventeenth International Byzantine Congress. Major Papers* (New York 1986) 118–36
C. Mango, G. Dagron, eds., *Constantinople and its Hinterland* (Aldershot 1995)
D.M. Nicol, *Byzantium and Venice* (Cambridge 1988).
J. Sauvaget, 'Le plan antique de Damas', *Syria* 26 (1949) 314–58.
A. Walmsley, 'Production, exchange and regional trade in the Islamic Near East: old structures, new systems?', in I.L. Hansen, C.J. Wickham, eds., *The Long Eighth Century. Production, distribution and demand* (Leiden 2000) 264–343

Ecclesiastical Structures and Monasticism

A.S. Atiya, *A History of Eastern Christianity* (London 1968)
M. Angold, *Church and Society in Byzantium under the Comneni, 1081–1261* (Cambridge 1995)
P.R.L. Brown, *The World of Late Antiquity* (London 1971)
P.R.L. Brown, *The Cult of the Saints* (London 1981)
Cambridge Medieval History, vol. iv, parts 1 and 2, revised edn. J.M. Hussey (Cambridge 1966)
H. Chadwick, *The Early church* (Harmondsworth 1967)
P. Charanis, 'The monastic properties and the state in the Byzantine empire', *Dumbarton Oaks Papers* 4 (1948) 53–118 (reprinted in his *Social, Economic and Political Life*, essay I)
P. Charanis, 'The monk as an element of Byzantine society', *Dumbarton Oaks Papers* 25 (1971) 61–84
G. Every, *The Byzantine Patriarchate (451–1204)* (London 1947)
W.H.C. Frend, *The Rise of the Monophysite Movement* (Cambridge 1972)
N. Garsoïan, *The Paulician Heresy* (The Hague-Paris 1967)
J. Herrin, *The formation of Christendom* (Princeton 1987)
J.M. Hussey, *The Orthodox Church in the Byzantine Empire* (Oxford 1986)
M. Loos, *Dualist Heresy in the Middle Ages* (Prague 1974)
J. Meyendorff, *Byzantine Theology: historical trends and doctrinal themes* (New York 1974)
R. Morris, ed., *Church and People in Byzantium* (Birmingham 1990)
R. Morris, *Monks and Laymen in Byzantium, 843–1118* (Cambridge 1995)
D. Obolensky, *The Bogomils* (Cambridge 1948)
D. Obolensky, *The Byzantine Commonwealth. Eastern Europe 500–1453* (London 1971)
J. Pelikan, *The Christian Tradition*, 4 vols. (Chicago 1971–83)
S. Runciman, *The Medieval Manichee. a study of the Christian dualist heresy* (Cambridge 1947)
A. Sharf, *Byzantine Jewry from Justinian to the Fourth Crusade* (London 1971)
J. Starr, *The Jews in the Byzantine Empire, 641–1204* (Athens 1939)
Sp. Vryonis, jr., *The Decline of Medieval Hellenism in Asia Minor and the Process of Islamization from the Eleventh through the Fifteenth Century* (Berkeley-Los Angeles-London 1971)
C. Walter, *Art and Ritual in the Byzantine Church* (London 1982)

Appendix 1: Rulers and Princes

Eastern Roman rulers (324–1453)

Constantine I + Licinius	311–324
Constantine I	324–337
Constantine II, Constantius II + Constans	337–340
Constantius II	337–361
Julian	361–363
Jovian	363–364
Valentinian I (+ Valens 367–375)	364–375
Valens + Gratian, Valentinian II	375–378
Theodosius I	378–395
(+ Gratian, Valentinian II	378–383)
(+Valentinian II, Arcadius	383–392)
(+ Arcadius, Honorius	392–395)
Arcadius	395–408
Theodosius II	408–450
Marcian	450–457
Leo I	457–474
Leo II	474
Zeno	474–475
Basiliscus	475–476
Zeno (restored)	476–491
Anastasius I	491–518
Justin I	518–527
Justinian I	527–565
Tiberius II Constantine	578–582
Maurice	582–602
Phokas	602–610
Heraclius	610–641
Constantine III and Heraclonas	641
Constans II	641–668
Constantine IV	668–685
Justinian II	685–695
Leontios	695–698
Tiberios III	698–705
Justinian II (restored)	705–711
Philippikos Bardanes	711–713
Anastasios II	713–715
Theodosios III	715–717
Leo III	717–741
Constantine V	741–775
Artabasdos	741–742
Leo IV	775–780
Constantine VI	780–797
Eirene	797–802
Nikephoros I	802–811
Staurakios	811
Michael I	811–813
Leo V	813–820
Michael II	820–829
Theophilos	829–842
Michael III	842–867
Basil I	867–886
Leo VI	886–912
Alexander	912–913
Constantine VII	913–959
Romanos I Lakapenos	920–944
Romanos II	959–963
Nikephoros II Phokas	963–969
John I Tzimiskes	969–976
Basil II (+ Constantine VIII)	976–1025
Constantine VIII	1025–1028
Romanos III Argyros	1028–1034
Michael IV the Paphlagonian	1034–1041
Michael V Kalaphates	1041–1042
Zoe and Theodora	1042
Constantine IX Monomachos	1042–1055
Theodora (again)	1055–1056
Michael VI Stratiotikos	1056–1057
Isaac I Komnenos	1057–1059
Constantine X Doukas	1059–1067
Eudokia	1067
Romanos IV Diogenes	1068–1071
Eudokia (again)	1071
Michael VII Doukas	1071–1078
Nikephoros III Botaneiates	1078–1081
Alexios I Komnenos	1081–1118
John II Komnenos	1118–1143
Manuel I Komnenos	1143–1180
Alexios II Komnenos	1180–1183
Andronikos I Komnenos	1183–1185
Isaac II Angelos	1185–1195
Alexios III Angelos	1195–1203
Isaac II (restored) + Alexios IV Angelos	1203–1204
Alexios V Mourtzouphlos	1204
Constantine (XI) Laskaris	1204 (Nicaea)
Theodore I Laskaris	1204–1222 (Nicaea)
John III Doukas Vatatzes	1222–1254 (Nicaea)
Theodore II Laskaris	1254–1258 (Nicaea)
John IV Laskaris	1258–1261 (Nicaea)
Michael VIII Palaiologos	1259–1282
Andronikos II Palaiologos	1282–1328
Michael IX Palaiologos	1294–1320
Andronikus III Palaiologos	1328–1341
John V Palaiologos	1341–1391
John VI Kantakouzenos	1341–1354
Andronikos IV Palaiologos	1376–1379
John VII Palaiologos	1390
Manuel II Palaiologos	1391–1425
John VIII Palaiologos	1425–1448
Constantine XI (XII) Palaiologos	1448–1453

Empire of Nicaea

Constantine (XI) Laskaris	1204
Theodore I Laskaris	1204–1222
John III Doukas Vatatzes	1222–1254
Theodore II Laskaris	1254–1258
John IV Laskaris	1258–1261
Michael VIII Palaiologos	1259–1282 (from 1261 at Constantinople)

Principality (Despotate) of Epiros

Michael I	1204–1215
Theodore	1215–1230 (emperor from 1224 in Thessalonica)

Thessalonica

Manuel	1230–1237
John	1237–1244
Demetrios	1244–1246
(defeated by John Vatatzes in 1246)	

Thessaly

John I	1271–1296
Constantine	1296–1303
John II	1303–1318

Epiros

Michael II	c. 1231–1271
Nikephoros I	1271–1296
Thomas	1296–1318
Nicholas Orsini	1318–1325
John Orsini	1325–1335
Nikephoros II	1335–1340

Grand Komnenoi of Trebizond

Alexios I	1204–1222
Andronikos I	1222–1235
John I	1235–1238
Manuel I	1238–1263
Andronikos II	1263–1266
George	1266–1280
John II	1280–1297
Alexios II	1297–1330
Andronikos III	1330–1332
Manuel II	1332
Basil	1332–1340
Eirene	1340–1341
Anna	1341
Michael	1341
Anna (again)	1341–1342
John III	1342–1344
Michael (again)	1344–1349
Alexios III	1349–1390
Manuel III	1390–1416
Alexios IV	1416–1429
John IV	1429–1459
David	1459–1461

Despotate of the Morea

Manuel Kantakouzenos	1348–1380
Matthew Kantakouzenos	1380–1383
Demetrios Kantakouzenos	1383
Theodore I Palaiologos	1383–1407
Theodore II Palaiologos	1407–1443
Constantine and Thomas Palaiologos	1443–1449
Thomas and Demetrios Palaiologos	1449–1460

Latin emperors at Constantinople

1205-1205	Baldwin I of Flanders
1216-1216	Henry of Flanders
1217	Peter of Courtenay
1219-1219	Yolande
1228-1228	Robert of Courtenay
1228–1261	Baldwin II
(1231–1237 John of Brienne)	

The Bulgars

First Bulgarian empire 681–971

Asparuch	681–702
Tervel	702–718
Anonymous	718–725
Sevar	725–739
Kormisoš	739–756
Vinech	756–762
Teletz	762–765
Sabin	765–767
Umar	767
Toktu	767–772
Pagan	772
Telerig	772 (c.)–777
Kardam	777–c. 803
Krum	c. 803–814
Dukum, Dicevg	814
Omurtag	814–831
Malamir	831–836
Presiam	836–852
Boris I Michael	852–889
Vladimir	889–893
Symeon	893–927
Peter	927–969
Boris II	969–971

The Bulgars' 'Macedonian' empire 976–1018

Samuel	976–1014
Gabriel Radomir	1014–1015
John Vladislav	1015–1018

The second Bulgarian empire 1186–1396

Asen I	1186–1196
Peter	1196–1197
Kaloyan	1197–1207
Boril	1207–1218
Ivan Asen II	1218–1241
Kaloman Asen	1241–1246
Michael Asen	1246–1256
Constantine Tikh	1257–1277
Ivailo	1278–1279
Ivan Asen III	1279–1280
George I Terter	1280–1292
Smiletz	1292–1298
Čaka	1299
Theodore Svetoslav	1300–1322
George II Terter	1322–1323
Michael Šišman	1323–1330
Ivan Stephen	1330–1331
Ivan Alexander	1331–1371
Ivan Šišman	1371–1393 (at Trnovo)
Ivan Stracimir	1360–1396 (at Vidin)

Grand Župans/Kings of Serbia (from 1168)

Štefan Nemanja	c. 1168–1196
Štefan I	1196–1217

Štefan Radoslav	1217–1227/28
Štefan Vladislav	1227/28–1234
Štefan Uroš I	1234–1276 (emperor from 1345)
Štefan Dragutin	1276–1282
Štefan Uroš II	1282–1321
Štefan Uroš III	1321–1331
Štefan Uroš IV Dušan	1331–1355
Štefan Uroš V	1355–1371

Islamic rulers

Caliphs

The four 'rightly-guided' Caliphs, direct descendants of the Prophet

Abu Bakr	632–634
'Umār I	634–644
'Uthmān	644–656
'Alī	656–661

Umayyad dynasty

Mu 'āwiyya I	661–680
Yazīd I	680–683
Mu 'āwiyya II	683–684
Marwān I	684–685
'Abd al-Malik	685–705
Walīd I	705–715
Suleimān	715–717
'Umar II	715–720
Yazīd II	720–724
Hishām	724–743
Walīd II	743–744
Yazīd III	744
Marwān II	744–750
Ibrāhīm	744

Abbasid dynasty

as-Saffāh	750–754
al-Mansūr	754–775
al-Mahdī	775–785
al-Hādī	785–786
Hārūn ar-Rashīd	786–809
al-Amīn	809–813
al-Ma'mūn	813–833
al-Mu'tasim	833–842
al-Wāthiq	842–847
al-Mutawwakil	847–861
al-Muntasir	861–862
al-Musta'īn	862
al-Mu'tazz	862–866
al-Muhtadī	866–869
al-Mu'tamid	869–892
al-Mu'tadid	892–902
al-Muqtafi	902–908
al-Muqtadir	908–932
al-Qāhir	932–934
al-Rādī	934–940
al-Muttaqī	940–943
al-Mustakfī	943–946
al-Mutī'	946–974
at-Tā'i'	974–991
al-Qādir	991–1031
al-Qā'im	1031–1075
al-Muqtadī	1075–1094
al-Mustazhir	1094–1118
al-Mustarshid	1118–1135
ar-Rāshid	1135–1136
al-Muqtafiī	1136–1160
al-Mustanjid	1160–1170
al-Mustadī	1170–1180
an-Nāsir	1180–1225
az-Zāhir	1225–1226
al-Mustansīr	1226–1258
al-Musta'sim	1258

Seljuk Sultans of Rum

Suleiman I	1077–1086
Kilij Arslan I	1092–1107
Malik Shah	1107–1116
Masud I	1116–1156
Kilij Arslan II	1156–1192
Kaikhusraw I	1192–1996
Suleiman II	1196–1204
Kilij Arslan III	1204
Kaikhusraw I (again)	1204–1210
Kaikawus I	1210–1220
Kaikubad I	1220–1237
Kaikhusraw II	1237–1245
Kaikawus II	1246–1257
Kilij Arslan IV	1248–1265
Kaikubad II	1249–1257
Kaikhusraw III	1265–1282
Masud II	1282–1304
Kaikubad III	1284–1307
Masud III	1307–1308

Ottoman Sultans (to 1453)

Osman	1288–1326
Orhan	1326–1362
Murad I	1362–1389
Bayezit I	1389–1402
Mehmet I	1402–1421 (sole ruler from 1413)
Suleiman	1402–1410
Musa	1411–1413
Murad II	1421–1451
Mehmet II Fatih 'the Conqueror'	1451–1481

Armenia from 885 (until its incorporation into the Byzantine empire in 1042–1045)

Ašot I the Great	885–890
Smbat I the Martyr	890–914
Ašot II the Iron	914–928
Abas I	928–952
Ašot III the Merciful	952–977
Smbat II the Conqueror	977–989
Gagik I	989–1020
John-Smbat III	1020–1040
Ašot IV the Valiant	1021–1039
Gagik II	1042–1045

NB: The Armenian princes who ruled or governed Armenia or parts thereof nominally for the eastern Roman emperors and for the Persian kings in the 5th to 7th centuries, or for the Byzantine emperors and Caliphs from the 7th to 9th centuries, are not included.

Georgia

Iberia

Adarnase IV	888–923
David II	923–937
Smbat I	937–958
Bagrat II	958–994
Gurgen I	994–1008
Bagrat III	1008–1014

Abasgia

Leo II	767–812
Theodosios II	812–838
Demetrios II	838–873
George I	872–879
John Šavliani	878–880
Adarnase Šavliani	880–888
Bagrat I	888–899
Constantine III	899–917
George II	916–961
Leo III	961–970
Demetrios III	970–977
Theodosios III	977–979
Bagrat III of Iberia	979–1014

Georgia (Abasgia and Iberia together)

Bagrat III	1008–1014
George I	1014–1027
Bagrat IV	1027–1072
George II	1072–1089
David III	1089–1125
Demetrios I	1125–1156
David IV	1155
George III	1156–1184
Thamar 'the Great'	1184–1212
George IV	1212–1223
Rusudan	1223–1245
Interregnum	*1245–1250*
David V	1250–1258 (secedes in Imeretia/Abasgia)
David VI	1250–1269
Interregnum	*1269–1273*
Demetrios II	1273–1289
Vakhtang II of Imeretia	1289–1292
David VII	1292–1301
Vakhtang III	1301–1307
George V	1307–1314
George VI	1314–1346
David VIII	1346–1360
Bagrat V 'the Great'	1360–1395
George VII	1395–1405
Constantine I	1405–1412
Alexander I 'the Great'	1412–1442
Vakhtang IV	1442–1446
Demetrios III	1446–1453
George VIII	1446–1465
Bagrat VI	1465–1478
Constantine II	1478–1505

NB: The kings of Georgia generally ruled in association with a junior co-ruler who was often one of their immediate successors.

Appendix 2: Patriarchs and Popes

Patriarchs and Popes

Archbishops of Constantinople (324–381)

Alexander	324–337
Paul I	337–339
Eusebius	339–341
Paul I (again)	341–342
Macedonius I	342–346
Paul I (again)	346–351
Macedonius I (again)	351–360
Eudoxius	360–370
Demophilus	370–379
Gregory I (of Nazianzos)	379–381

Patriarchs (381–1456)

Nectarius	381–397
John I Chrysostomos	398–404
Arsacius	404–405
Atticus	406–425
Sisiinius I	426–427
Nestorius	428–431
Maximianus	431–434
Proclus	434–446
Flavianus	446–449
Anatolius	449–458
Gennadius I	458–471
Acacius	472–489
Fravitas	489–490
Euphemius	490–496
Macedonius II	496–511
Timothy I	511–518
John II the Cappadocian	518–520
Epiphanios	520–535
Anthimos I	535–536
Menas	536–552
Eutychios	552–565
John III Scholastikos	565–577
Eutychios (again)	577–582
John IV the Faster	582–595
Kyriakos	595–606
Thomas I	607–610
Sergios I	610–638
Pyrrhos	638–641
Paul II	641–653
Pyrrhos (again)	654
Peter	654–666
Thomas II	667–669
John V	669–675
Constantine I	675–677
Theodore I	677–679
George I	679–686
Theodore I (again)	686–687
Paul III	688–694
Kallinikos I	694–706
Kyros	706–712
John VI	712–715
Germanos I	715–730
Anastasios	730–754
Constantine II	754–766
Niketas I	766–780
Paul IV	780–784
Tarasios	784–806
Nikephoros I	806–815
Theodotos	815–821
Anthony I	821–837
John VII Grammatikos	837–843
Methodios I	843–847
Ignatios	847–858
Photios	858–867
Ignatios (again)	867–877
Photios (again)	877–886
Stephen I	886–893
Anthony II	893–901
Nicholas I Mystikos	901–907
Euthymios I	907–912
Nicholas I (again)	912–925
Stephen II	925–927
Tryphon	927–931
Theophylaktos	933–956
Polyeuktos	956–970
Basil I	970–974
Anthony III	974–979
Nicholas II	979–991
Interregnum	*991–996*
Sisinnios II	996–998
Sergios II	1001–1019
Eustathios	1019–1025
Alexios	1025–1043
Michael I Keroularios	1043–1058
Constantine III	1059–1063
John VIII Xiphilinos	1064–1075
Kosmas I	1075–1081
Eustratios	1081–1084
Nicholas III	1084–1111
John IX	1111–1134
Leo	1134–1143
Michael II	1143–1146
Kosmas II	1146–1147
Nicholas IV Mouzalon	1147–1151
Theodotos II	1151–1154
Nephytos I	1153–1154
Constantine IV	1154–1157
Loukas	1157–1170
Michael III	1170–1178
Chariton	1178–1179
Theodosios	1179–1183
Basil II	1183–1186
Niketas II	1186–1189
Dositheos	1189
Leontios	1189
Dositheos (again)	1189–1191
George II	1191–1198
John X	1198–1206
Michael IV	1208–1214
Theodore II	1214–1216
Maximos II	1216

Manuel I	1217–1222
Germanos II	1222–1240
Methodios II	1240
Manuel II	1244–1254
Arsenios	1255–1259
Nikephoros II	1260
Arsenios (again)	1261–1264
Germanos III	1265–1266
Joseph I	1266–1275
John XI Bekkos	1275–1282
Joseph I (again)	1282–1283
Gregory III	1283–1289
Athanasios I	1289–1293
John XII	1294–1303
Athanasios I (again)	1303–1309
Niphon I	1310–1314
John XIII Glykys	1315–1319
Gerasimos I	1320–1321
Isaias	1323–1332
John XIV Kalekas	1334–1347
Isidoros I	1347–1350
Kallistos I	1350–1353
Philotheos Kokkinos	1353–1354
Kallistos I (again)	1355–1363
Philotheos (again)	1364–1376
Makarios	1376–1379
Neilos	1379–1388
Anthony IV	1389–1390
Makarios (again)	1390–1391
Anthony IV (again)	1391–1397
Kallistos II Xanthopoulos	1397
Matthew I	1397–1410
Euthymios II	1410–1416
Joseph II	1416–1439
Metrophanes II	1440–1443
Gregory III	1443–1450
Gennadios II Scholarios	1454–1456

Popes (314–1455)

Sylvester I	314–335
Mark	336
Julius	337–352
Liberius	352–366
(Felix II	355–365)
Damasus I	366–384
(Ursinus	366–367)
Siricius	384–399
Anastasius I	399–401
Innocent I	401–417
Zosimus	417–418
Boniface I	418–422
(Eulalius	418–419)
Celestine I	422–432
Sixtus III	432–440
Leo I the Great	440–461
Hilarius	461–468
Simplicius	468–483
Felix III	483–492
Gelasius I	492–496
Anastasius II	496–498
Symmachus	498–514
(Laurentius	498, 501–505)
Hormisdas	514–523
John I	523–526
Felix IV	526–530
Boniface II	530–532
(Dioscorus	530)
John II	533–535
Agapetus I	535–536
Silverius	536–537
Vigilius	537–555
Pelagius I	556–561
John III	561–574
Benedict I	575–579
Pelagius II	579–590
Gregory I the Great	590–604
Sabinianus	604–606
Boniface III	607
Boniface IV	608–615
Deusdedit I	615–618
Boniface V	619–625
Honorius I	625–638
Severinus	640
John IV	640–642
Theodore I	642–649
Martin I	649–655
Eugenius I	654–657
Vitalianus	657–672
Deusdedit II	672–676
Domnus	676–678
Agatho	678–681
Leo II	682–683
Benedict II	684–685
John V	685–686
Conon	686–687
(Theodore	687)
(Pascal	687)
Sergius I	687–701
John VI	701–705
John VII	705–707
Sisinnius	708
Constantine I	708–715
Gregory II	715–731
Gregory III	731–741
Zacharias	741–752
(Stephen II	752)
Stephen III	752–757
Paul I	757–767
(Constantine	767–769)
(Philip	768)
Stephen IV	768–772
Hadrian I	772–795
Leo III	795–816
Stephen V	816–817
Pascal I	817–824
Eugenius II	824–827
Valentinus	827
Gregory IV	827–844
(John	844)
Sergius II	844–847
Leo IV	847–855
Benedict III	855–858
(Anastasius	855)
Nicholas I	858–867
Hadrian II	867–872
John VIII	872–882

Marinus I	882–884
Hadrian III	884–885
Stephen VI	885–891
Formosus	891–896
Boniface VI	896
Stephen VII	896–897
Romanus	897
Theodore II	897
John IX	898–900
Benedict IV	900–903
Leo V	903
(Christopher	903–904)
Sergius III	904–911
Anastasius III	911–913
Lando	913–914
John X	914–928
Leo VI	928
Stephen VIII	928–931
John XI	931–935
Leo VII	936–939
Stephen IX	939–942
Marinus II	942–946
Agapetus II	946–955
John XII	955–964
Leo VIII	964–965
Benedict V	964–966
John XIII	966–972
Benedict VI	973–974
Benedict VII	974–983
John XIV	983–984
John XV	985–996
Gregory V	996–999
Sylvester II	999–1003
John XVII	1003
John XVIII	1004–1009
Sergius IV	1009–1012
Benedict VIII	1012–1024
John XIX	1024–1032
Benedict IX	1032–1044
Sylvester III	1045
Benedict IX (again)	1045
Gregory VI	1045–1046
Clement II	1046–1047
Benedict IX (again)	1047–1048
Damasus II	1048
Leo IX	1049–1054
Victor II	1055–1057
Stephen X	1057–1058
Nicholas II	1059–1061
Alexander II	1061–1073
Gregory VII	1073–1085
Victor III	1086–1087
Urban II	1088–1099
Pascal II	1099–1118
Gelasius II	1118–1119
Calixtus II	1119–1124
Honorius II	1124–1130
Innocent II	1130–1143
Celestine II	1143–1144
Lucius II	1144–1145
Eugenius III	1145–1153
Anastasius IV	1153–1154
Hadrian IV	1154–1159
Alexander III	1159–1181
Lucius III	1181–1185
Urban III	1185–1187
Gregory VIII	1187
Clement III	1187–1191
Celestine III	1191–1198
Innocent III	1198–1216
Honorius III	1216–1227
Gregory IX	1227–1241
Celestine IV	1241
Innocent IV	1243–1254
Alexander IV	1254–1261
Urban IV	1261–1264
Clement IV	1265–1268
Gregory X	1271–1276
Innocent V	1276
Hadrian V	1276
John XXI	1276–1277
Nicholas III	1277–1280
Martin IV	1281–1285
Honorius IV	1285–1287
Nicholas IV	1288–1292
Celestine V	1294
Boniface VIII	1294–1303
Benedict XI	1303–1304
Clement V	1305–1314
John XXII	1316–1334
Benedict XII	1334–1342
Clement VI	1342–1352
Innocent VI	1352–1362
Urban V	1362–1370
Gregory XI	1370–1378
Urban VI	1378–1389
Boniface IX	1389–1404
Innocent VII	1404–1406
Gregory XII	1406–1415
Martin V	1417–1431
Eugenius IV	1431–1447
Nicholas V	1447–1455

Index

Compiled by Sue Carlton